One Man Show

Henry Miller In the San Joaquin

Edited by Charles Sawyer

 The Ralph Milliken Museum Society would also like to acknowledge the following for use of materials in which they hold a copyright interest.

The Henry Madden Library at Fresno State University, Fresno, California

The Bancroft Library at University of California, Berkeley, California

Ralph Milliken Museum Society
P.O. Box 2294, Los Banos, California 93635
ISBN: 0944707-07-6

Cover Design by Susan D. Klassen
Production Services:
Word Dancer Press, Sanger, CA. 93657 (800) 497-4909

Published in association with and distributed by:

Loose Change Publications, 936 6th, Los Banos, CA. 93635
(209) 826-3797 Fax (209) 826-1514
Email: nco4242@mercednet.com

Table of Contents

Photographs iv
Introduction v

Part I

Chapter One — Heinrich Kreiser to Henry Miller 3
Chapter Two — In the San Joaquin 13
Chapter Three — Miller the Stockman 27
Chapter Four — Irrigation and Litigation 37
Chapter Five — One Man Show 53
Chapter Six — Bancroft Interview 69
Chapter Seven — Family Matters 91
Henry Miller's Family 112
Sheldon Family of Rhode Island 113

Part II

Chapter Eight — Letters–1860 to 1890 117
Chapter Nine — Clippings–1860 to 1890 153
Chapter Ten — Oral and Written Interviews–1860 to 1890 161

Part III

Chapter Eleven — Santa Rita County 203
Chapter Twelve — Milliken's Notes–Odds and Ends 207

Part IV

Chapter Thirteen — Letters 1890–1916 229
Chapter Fourteen — Newspaper Articles & Clippings - 1890-1916 . 235
Chapter Fifteen — Interviews Covering 1890–1956 245
Chapter Sixteen — End of an Era 305

Bibliography 311
Index 313
RALPH LEROY MILLIKEN 329
THE MILLIKEN MUSEUM SOCIETY 330

Photographs

Photos have come from various sources. The line drawings are from the Annals of San Francisco, Appelton Co., New York 1855. The photo of early San Francisco, a view from Kearny Street is from the Bancroft Library, UC Berkeley, the earthquake photos are from the archives of the San Francisco Public Library. The balance of the photos are from the collection of the Ralph Milliken Musuem.

A special thanks is in order for the great work done on the Museum photos by Don Hughes of Don Hughes Photography.

Cover Photo: Henry Miller seated on the porch at Bloomfield Farm

Introduction

As editor of these works it is not my intent to give an in depth biography of Henry Miller; others have done that. What I hoped to do is to relate a portrait of the man and the times and events that shaped him. The Ralph Milliken Society has wanted to publish some of its many documents relating to Henry Miller, Miller & Lux, and his adventures in the San Joaquin.

With this book the Society hopes to accomplish its goal of making these documents and information accessible to the reader and to give a brief background of events that shaped the San Joaquin of today. I have tried to put these events, letters and interviews in a sequence of time dating from the early events of the 1860's until Miller's death in 1916.

Letters, text and oral interviews were gathered from friends, foes, and employees of Henry Miller as well as of the firm Miller & Lux. Miller gave one interview during his lifetime. He was interviewed by Mr. Fowler of the History Company. This was in connection with Bancroft's epic work "The History of California."

Ralph Milliken had worked for Miller & Lux as a boy and young man. He later interviewed many of these people between 1916 and the late 1960's. Many stories are told several different ways; dates in some events vary from a few days to a few years. The facts of an event occurring are true, the details are sometimes from the teller's perspective. Chapters one through five are a condensed collective work of Ralph Milliken and David Lawrence. They corresponded for several years with each section going back and forth, being written over and marked out. They also had others look the work over and add their thoughts to the mix. Sixty years later it is hard to tell whose work was whose. Names and places are sometimes spelled in a rather creative manner. I have chosen to leave them as written as most are easily identifiable to the reader.

I would like to acknowledge the generous help of staff members of the Milliken Museum, Gilroy Museum, Bancroft Library at the University of California, Henry Madden Library at Fresno State, Pat McNally, Bob Edminster, Charlie Alamo, Joe Cox, Don Hughes and Joanne Hoefer whose help was invaluable to this project.

Charles Sawyer
Editor

Part I

Chapter One

Heinrich Kreiser to Henry Miller

Henry Miller as he is known today started life as Henry or Heinrich Kreiser in the small German town of Brackenheim. Henry came from a family that was long established in the province of Wurtenberg. His father Christian was from a long line of stock raisers and a master butcher. His wife Christine was from a family of vintners. They married and produced several children, three daughters, Frederike, Elizabeth and Karoline, and Henry or Heinrich Alfred Kreiser was born to them on July 21, 1827.

From his early boyhood Henry knew more of work than play. He early on became familiar with the work required of the butchers trade. At five he was watching calves and sheep graze, but the age of eight he was trusted with the responsibility of sharing the family labor. His responsibilities were varied. He was often to be seen on the roads around Brackenheim, chasing after his fathers calves on the way to and from the pasture. By the time Henry was 12 he was sometimes making a purchase of a likely calf or goat and driving it back to his fathers shop.

He was also at the age of 12 already an apprentice and was on his way to becoming a master butcher.

In those days as an apprentice he was required to serve seven years before he could be considered a master butcher and had to know everything about the slaughter and use of the animal parts even down to the making of fiddle strings. His first year as an apprentice he made the sum of ten Prussian dollars from which a sum was taken for his food and clothing made of homespun. To Henry it seemed the years of being his father's apprentice would never end. He was restless and chafed under the apprenticeship.

He did attend school from the age of 7 through 14 and that is the only formal education he ever had. His school reports noted his singular aptitude for figures, his excellent memory and for ready answers to any questions put to him by his teacher or minister. Henry was not interested in school and when asked in later life his comments made it clear it held no fond recollections for him.

In 1842, when Henry was 14, his mother passed away. His grandmother and sisters took over running the household. This did not sit well with Henry as he often fought with his sisters and chafed under parental training and by now the severe labor load his father exacted from him. The next two years were ones of great unhappiness for Henry. He spent his time going from hamlet to hamlet buying stock for his father's business and while his apprenticeship was completed by this time (he had started it at eight years old) he was still under his father's thumb.

It has been told that the reason he left home was his father's insistence that he herd a flock of geese. This so infuriated and humiliated him that he came home and told his sisters he was through.

In reality there were many reasons for his leaving. He was sick of working under his father's harsh rule and was ready to make his own way; his father was often in trouble and in jail, forcing Henry to take responsibility for the family affairs; and like most young men the time had come to leave home.

He left in short order and went to Holland where he worked at the butchers trade. Not much is known of his time spent there and he later went on to England where he worked again in the butchers trade, and learned the invaluable skill of understanding English, though he did not speak it that well. In 1847 at the age of 19 he sailed for New York. He arrived with a bundle of clothes and $50 in American dollars. Once in New York he set out to find work. But his speaking skills were poor and he could hardly make himself understood and workers were plentiful.

The New York that Henry had arrived in was a town filled with gangs. The Butchers had long held sway over the lower part of the city and for the most part worked shops out of the Washington Market. Unable to break into this closed group, Henry at first took a job as a gardener at $4.00 per month because it came with a place to sleep and his board. After hours he continued to look for a job in the butchering trade.

After a time he found a stallholder who was willing to give him a chance. His job consisted of working sixteen hours a day and doing everything from the killing of the hogs, dressing them to attending to the stall at the Washington Market in the mornings. For this he received the princely sum of $8.00 per month and as a prerequisite he got the intestines of the slaughter animals, which he sold as sausage casings for about $1.00 per day. He slept in the back of the shop. He later left the Washington Market and went to work for an old butcher who owned a poor shop and had an equally poor business. Because the man did not want to "take on help" at the time, Henry agreed to work for half price, and again sleep in the back of the shop.

With his characteristic industry, Henry cleaned, painted and rearranged the inside of the shop, which in turn brought in more business. He remained on this job for more than a year and it was from this shop that he left for California. When he discovered the delivery boy was not doing his job Henry took over the deliveries after hours and pocketed his wages as his own. As the delivery boy he had the opportunity to become acquainted with a large number of people and to improve his spoken English. His acquaintance usually resulted in the gaining of new trade and continued improvement of the once failing business.

It is supposedly on one of these delivery tours that he accidentally met a girl he knew in Brackenheim who had come to America before he had left home. "Linda" was surprised to see Henry and as a result the old friendship was revived and she agreed to help him with his English pronunciation and correctness of speech. While Henry never fully mastered English, her help was invaluable.

He had also met a young man by the name of Henry Miller while making deliveries. This Henry was a shoe salesman, about Henry's age but an American. Henry Kreiser cultivated his acquaintance because Henry had small feet and had difficulties in getting boots. He was proud of his small foot and often wore boots too tight causing foot problems due to his being on his feet all day. The friendship grew and soon these two young men were about town on their off hours, Henry the butcher would often invite Henry the shoe salesman for beer and pretzels and practice his English on him. The major topic of the day was California, for gold had been discovered there. It was all anyone could talk about.

Both young men eagerly sought every bit of news about the distant California. Soon they were gripped with "gold fever" both making endless plans to go to California and become rich. Henry the shoe salesman was able to save for his passage and soon had enough for the ticket. Henry the butcher saved but he could scarcely save from his meager wages. Eventually Henry the shoe salesman purchased his ticket from New York to San Francisco via Panama. Henry the butcher promised he would meet his friend as soon as he had passage money. It is at this point the story varies depending on who was telling it. One version has it that on the Monday following the purchase of the ticket Henry the shoe salesman had a change of mind and suggested that Henry the butcher take his ticket as it was non refundable. It has been written that Henry the shoe salesman missed the boat and tried to sell the ticket to Henry the butcher. Whatever the truth of the matter, it is clear that Henry the butcher, after some shrewd bargaining, bought the ticket of Henry the shoe salesman, quit his job, drew his wages, gathered his belongings and headed for the docks. He did not even say goodbye to his childhood friend "Linda."

When it came time to board the ship he looked at the ticket and saw it was in the name of "Henry Miller" and the words "Not transferable." He was anxious to go to California, unsure and feared if he told the authorities he was not Henry Miller but had purchased his ticket from Henry Miller that he would not be allowed to use it. On the spot he made up his mind to go as "Henry Miller" and so he remained the rest of his life. He sailed for California on April 13, 1850 on the steamer *Georgia*.

The trip to the Isthmus of Panama was slow and tedious. Henry found himself in the company of total strangers, mostly men of all different classes with the one goal in common to get to the gold fields of California. Some substantial businessmen, who hoped to take advantage of the opportunity the gold rush provided, others had their minds strictly on the mines. It can be assumed in such close quarters that Henry naturally turned to the first as he knew little of mining and cared less about making quick easy money. He knew the butcher business from top to bottom and it is on the *Georgia* that he heard of the vast herds of cattle in California and that most of them were slaughtered merely for the hides and tallow, the remainder of the animals left to the wild beasts that infested California.

On arrival another journey had to be made across the Isthmus to Panama City where a ship bound for California had to be found. The crossing was a nightmare. In order to cross the Isthmus one had to hire a man or a mule to pack one's goods. The problem was that the greenhorns would be robbed along the way, losing both their goods and the mule. Miller had heard these stories. Some of the men along with Miller hired a man and a pack mule. Miller when later discussing the trip stated, "We had a long sack with clothes in it, at night the natives took their mules back and with them took a great deal of the stuff belonging to the other men. I took special precaution to watch the mule that carried my goods and I sat up and watched him all night." By the time they reached Panama there were no ships, the reason being that as soon as the word of the strike in California reached Panama, ships weighed anchor and sailed to California.

Due to the over crowding of the town and the thousands waiting for ships, the accommodations were poor at best. Panama fever was raging. Miller on arrival learned that the town's only butcher had been stricken with the fever. Never one to miss an opportunity, with the assistance of his fellow passengers, he bought the butcher's tools and went into business providing meat for the stranded population of Panama City. Miller later estimated that there were four to five thousand in a village that not a year before had contained a few hundred. All went well for a short time and Henry as well as his partners made money hand over fist.

Suddenly the fever got Henry Miller and he lay in a private hospital from May to August of that year. His inexperienced partner carried on the business but when Miller was sufficiently recovered to hobble down to his establishment, he discovered his partner had swamped the business beyond all hope of any recovery. He decided to sell out. When the bills had been paid there was scarcely enough for his passage to San Francisco. He later told people, "If I hadn't got the fever when I was in Panama, I would have stayed there because the opportunities were very great to make money."

Ships began to arrive at long last and Miller along with 1,500 other passengers boarded the steamer *Republic*. Having no idea of how long it would take to get to San Francisco Miller was relieved when the steamer stopped at Acapulco for two days. He walked out into the countryside where he washed and dried his clothes. Mexico held little charm and he was happy to board and head north. After another 24 days the *Republic* steamed into San Francisco Bay on September 24, 1850. Miller landed with his gold watch, his clothes, walking stick and six American dollars in his pockets. He had need of a walking stick, as he was still weak from the fever.

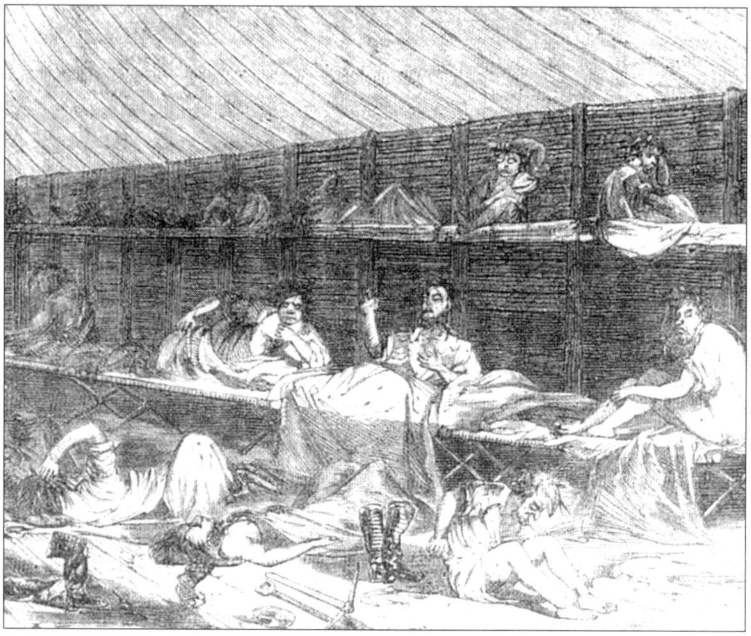

Sleeping conditions in San Francisco 1850—*Annals of San Francisco*

The San Francisco at which he landed was a rough and ready place. Less than 15 years before there had not even been a tent in sight; now the hills were covered with them. There were three what you could call streets but for the most part the roads were mere rutted pathways. There was no government, no law and certainly no order. The citizenry consisted of a horde of gold seekers, gamblers, harlots, politicians and other parasites of mankind.

San Francisco had been founded in 1776 when a mission and later a garrison known as the Presidio was established. After the Bear Flag revolt, Captain Montgomery with his sloop of war came into the bay and took it over. Prior to that it had been known as Yerba Buena. On January 30, 1847 the Alcalde ordered the name changed to San Francisco. The school board took a census in 1848 finding a population of 700 whites. Half were Americans, 150 were Indians, Negroes or Sandwich Islanders, the balance Mexicans. The census stated that there were 200 buildings, which included tents, outhouses and sheds.

There is a famous quotation about San Francisco's streets of that time. "This street is impassable; not even jackassable." Due to the influx of people and stock, the streets, which had never been graded, were seas of mud sometimes, one side being 10 to 15 feet higher than the other. Many times during the winter of 1849-1850 animals actually drowned in the mud. Many people nearly lost their lives as well. Tree limbs were put down to improve the road surface but it only made matters worse. There were no dwellings in the City as there was no wood close at hand. Tents were few and far between and many lived under the open sky. To sleep in a bunk or under a tent cost as high as $15 per night and many charged $8 per hour to sleep on a plank bench or table. The cost of just about any commodity was on par with sleeping accommodations. Wash-ing of clothes cost $20 per dozen pieces; porters refused to work for less than $30 a day in wages; food and vegetables in early days of the City were luxuries that only the very rich could afford. Yet there were farms nearby that grew the best vegetables in the world. Large supplies of carrots, beets, cabbage, apples, and eggs were to be had but at a price. Apples sold for $5 each, eggs were a steal at $10 a dozen but more often brought $50. Tea and coffee was $400 a barrel and as much as $10 a pound in pound lots. For wheat flour and salt pork, the going rate was $40 a barrel. Bread, which in New York sold for four cents a loaf, sold for 75 cents in San Francisco. The tools of the gold miner ran $15-$25 for a shovel, a tin pan to wash and pan for gold with ran $20, a pair of boots could not be gotten for less than $40 and any kind of medicine was at best $10 a dose while laudanum sold for $1 a drop.

One of San Francisco's many fires 1850-51—
Annals of San Francisco

The housing consisted of wood frames covered with Osnaburg partitions and with luck the outside covered with sails taken from the abandoned ships left in the harbor as the crews deserted them. This material had to be tacked up, as there were no nails to be found. At one time a pound of tacks cost a pound of gold. Owners of these buildings made fortunes in renting them. A 15 foot building on Portsmith Square rented for $3,000 a month, another similar in size a block away brought its owner $40,000 for the year. The Parker house which was a two story frame built stretcher which had cost $30,000 to build rented for $120,000 a year. At the El Dorado, a saloon and gambling hall next door, which was nothing more than a canvas tent, 15x20 feet rented for $40,000 a year. Gambling paid for most of this both in the case of the Parker House and the El Dorado. The United States Government paid $7,000 month for a 4x4 board shanty used as a customs office.

In October of 1850, less than a month after Henry Miller had set foot on San Francisco's shores, a cholera epidemic broke out. California had been admitted to the Union on September 9, but word did not reach San Francisco until October 29, when the schooner *Oregon* arrived, flags flying. This was the San Francisco to which Henry Miller had come. While the beginning was less than auspicious, Miller was a young man who could and would take advantage of what was before him. His story up to this point is no more or less than that of many an immigrant to the golden shores of California. What he made out of that opportunity made him one of the biggest cattle barons and landowners in America.

The question is sometimes asked why did Miller not go to the gold fields. In New York he and his friend had been gripped by gold fever. It was the reason for going to California. There are perhaps many reasons known and unknown. Miller was not dumb. Seeing and experiencing the hardships of getting to California and on landing seeing the price of land and food, he realized that he could make his way without digging for gold. He had little money and could not finance himself to the gold fields even if he wanted to go. He also realized that he needed to live today and people had to eat and they had to eat daily. The answer is most likely it was his circumstances at the time that dictated what followed.

He left the ship in the morning and immediately set out to find a job. "I asks a man where is a butcher shop," he later stated. The man had pointed to a shack on a hill near where California and Kearney streets are today. "I finds a man sitting on a box in front of a lean to. Flies were buzzing around him. He was reading a newspaper, which had come on the boat. He took no notice of me, so I coughed. When he looked over the top of the papers I says I just come on the boat, I am a butcher and wants a job." He refused Miller, and Miller moved on, later that day taking a job as a dishwasher. He had his start.

Miller did not wash dishes very long. In a few days he knew the town and went to a Frenchman named Edward Barron who had a butcher shop at the head of Dupont Street (Grant Avenue today) and was immediately employed to butcher sheep for Barron. He did not endure the problems of getting work that he had in New York. San Francisco was not a city of gangs and any man who wanted to work could find work. Miller again worked for a small wages for both Barron and other butchers and

Fire that burned out Menches Bros. May 1851—*Annals of San Francisco*

to save the price of a room slept behind the slaughterhouse. He did his own cooking from the scraps. On occasion he would splurge and eat at a Chinese restaurant. The charge of a dollar bought him a steak, potato and a cup of coffee.

Fire was common in San Francisco. Tents daily went up in smoke. Between 1850 and 1853 there were six total burn downs of the city. The fire of June 1851 set the stage for Miller to branch out on his own. Miller found and leased a lot on Jackson Street between Dupont and Kearny. He erected a one story building, obtained a few hooks, got a block and set up shop. He had saved nearly every penny he had made since arrival and had $150 left on which to start his business. He started by purchasing some calves, slaughtered these and packed them on his back to the other butchers from North Beach to Clarks Point. As such he started his wholesale business on a small scale from day one.

A break came when John Center asked Miller to kill some hogs. Center had gotten a shipload of hogs and some steers from the Sandwich Islands and had hired Henry Lux to kill them, as he was out of feed for them. Lux was only able to kill 6 or 7 a day and Center wanted them killed as fast as possible. Center was involved in many aspects of the city, had many men working for him, he also had a restaurant. He hired Miller and provided him with a mule to ride out to the Mission area where the hogs were being kept. Miller arrived early in the morning and set to work. In three days Miller killed and dressed over 100 head of hogs. Center was so impressed he offered to set Miller up in a bigger business. Miller declined because he didn't know the man well enough and wanted to be his own boss within his own business.

Miller did not drink and seldom smoked. For two years he worked and increased his business gradually and carefully. The business was built step by step. By 1853 he had a small slaughterhouse and butcher shop on the corner of Fifth and Howard streets. At this time he decided to take in a partner the man was named Zimmerman. Miller was no longer able to do it all on his own and needed someone in the shop. Miller furnished the capital and Zimmerman took care of the shop and office and ran the business for which he received a percentage of the profit. This lasted about a year at which time he took on another younger man, Westir, again on a similar labor for capital percentage of profits deal. Westir remained for about a year, the time it seemed to take one to get his start, and he left. At that point Miller acquired a full partner named Joseph Lawler. That summer they purchased a large lot of Spanish cattle. Most of the cattle in the City were driven from the south, sometimes as far as from San Diego. They were pastured at the Guadalupe Valley and the cattle would then be driven in small lots to the City. In the meantime Miller purchased from other butchers the brains, kidneys and other parts, which he dressed and furnished to his French customers as a specialty thus gaining the bulk of their trade.

By his careful building of his business and with his quick astute mind and the close economies he practiced, Miller was able to increase his capital of $150 into thousands in two short years and to firmly establish himself as a wholesaler and master butcher. In later 1853 he purchased from Livingston & Kincaid a herd of 300 prime American oxen for $33,000. That sum was at the time reputed to have been more than any other butcher in the city could command. This herd was said to be the first band of American cattle driven into San Francisco and this herd became the start, which would crowd out the Mexican cattle, which had been the main supply of meat in California.

In 1857 Henry Miller and Charles Lux purchased a band of cattle. Lux knew Miller through his brother Henry and made the proposal to Miller as a fellow German. This herd was purchased from a man named Erskins who had driven them from

Texas and now was in a hurry to return due to other business. There were approximately 2,000 head that he had turned loose in the Pacheco Pass area while he arranged for their sale. At the time Lux proposed to Miller they go in and buy the herd, there were about 1,500-1,600 left. Miller & Lux agreed to join and buy the herd. The price was $67.50 per head; each partner paying a one half cash share with profits of the sales split the same.

It has been reported that this was the start of the partnership of Miller & Lux. This is not true. This was a one shot deal. Lux not long after decided to dispose of his butcher and cattle business. Lux's wife was from Rhode Island. Her first husband, Potter, had been killed in a boat explosion on the bay and she and her son Jessie had moved in with friends who lived south of the city. Lux had met her going back and forth to his ranch. Early in 1858 he had completed his deal with Miller and the Luxes had decided to go east. By the end of the year Lux had returned to San Francisco and contacted Miller.

Lux returned in late September and by October 5 they had agreed to purchase nearly 2,000 head of cattle from Thomas Dias at Milpitas and pastured them on rented land. It was this business deal that developed the co-partnership of Miller & Lux. This partnership continued until Lux's death in 1887, nearly 30 years. The partnership was suggested by Lux and for a short while the firm was called Lux & Miller. It was also the start of another long term relationship, that of Thomas Dias and the firm of Miller & Lux. Charles Lux had similar experiences to Miller's. He had gone from his home in the Alsace region of Germany to New York where he too had worked as a butcher. He had come to California several months before Miller. However, it soon became clear that Miller was the moving force in the firm and it soon became known as Miller & Lux.

Because of the nature of the business and the volume of it and in order to avoid what they had both seen happen to other firms, they drew up an agreement that should cover either of them should things go wrong. They had worked together for years without any problems but the business was getting so large they both felt they needed this. On June 24, 1875 and again on May 22, 1884 they drew an agreement that stated:

Witnessed: "Whereas, said Miller & Lux have been co-partners for many years last past, to wit: from the 6th day of October 1858 and as such co-partners have been engaged in the business of buying land and rearing, grazing, and buying and selling cattle and horses; and their co-partnership business has extended into other kinds, but generally pertained to the business above named, and they are now continuing the same and propose to go on as heretofore:

"And whereas their said business and property is of such character and in such condition. That whenever they may conclude to close, settle and terminate the same, several years will be necessary to do so without injury, loss and sacrifice."

"Now in view of the foregoing and in consideration of the matters and things pertaining to said business and for their mutual benefit and advantage that thereby agree to and with each other as follow:

"That in the event of the death of either of the said copartners, then the survivor shall continue to carry on the said business for the joint use of himself and the estate heirs and legatee of the deceased for such period of time after the decease as he many deem necessary in order to close and settle up the business without detriment or sacrifice; provided that the said time may be extended to but not beyond the period of seven years after such decease.

"During the time that the surviving party shall carry on and the said business, all of the co-partnership property, real, personal and mixed, and the control, management, sale and disposition thereof shall be under the exclusive charge and dominion of the said survivor and the executors, heirs and legatees or trustees of the deceased or the trustees of any of the legatees of the deceased shall have no power or control over the same except such as may be given by the will and consent of the survivor.

"——The surviving partner shall have the power to hire, lease, sell, and convey all the personal property of the partnership without let or hindrance at such process and on such terms as he may see fit, and the real property of the partnership may be leased, sold, and conveyed by the survivor as he may deem proper, and the executors of the deceased shall join in the conveyance.

"The said surviving partner shall control and manage all said real estate during all of said seven years except such as shall be sold by him during that time. ——Both parties hereto agree to make proper provision in their wills to carry out this agreement."

The above agreement proved most fortunate to Henry Miller, and was his key to the success of his business empire, because as it turned out, Charles Lux preceded Miller to the grave by nearly thirty years. Armed with this agreement Henry Miller was able to stave off partition of the estate for at least seven years after Mr. Lux's death in 1887. There were many Lux heirs, both in America and Germany. Henry Miller was constantly harassed by their lawyers and agents, who were anxious to represent them in any capacity. Whatever Miller did with the properties and the management of the business, in their eyes, was prejudicial to their interests. Miller was in lawsuits constantly from 1887 to 1897, defending himself against the Lux heirs, individually and severally. If he declared a stock dividend from the surplus, they complained and said he should put money back into the business. If he did just that, they complained that he should have declared a stock dividend. Miller was benefiting alone to their great disadvantage and at their expense.

Eventually, an agreement was made whereby a company called the Miller & Lux Divided Lands Inc. was formed, with a capital stock of twelve million dollars and to be operative for fifty years; with the object and purpose of "acquiring all the property and assets of the late firm of Miller & Lux. Also to purchase, lease or otherwise acquire, hold, improve, sell or otherwise dispose of every kind of property." The articles of incorporation were dated May 5, 1897 and were duly certified by the Secretary of State. By this agreement Henry Miller automatically held fifty percent of the company's stock while the Lux heirs held their half. As the years went by Henry Miller bought the Lux heirs' half, so that by 1916, the year of his death, he virtually became sole owner of Miller & Lux Divided Lands Inc.

It was fortunate for Henry Miller and Charles Lux that they became associated. Both men spent all of their active business lives in the West and knew it thoroughly. Each was a totally different type, filling out the requirements of the other. Henry Miller was aggressive and combative whenever the interests of the firm were served by such a course. Charles Lux was invariably genial and urbane. But he was also well known as a sharp trader, land speculator and a behind the scene deal maker not adverse to skirting the law. Miller once said, "There is no similarity between Lux and myself—not in sympathy nor in view nor nothing." In the division of the business the outside work fell naturally on the powerful shoulders of the younger man, Miller, while the administration and city work received the attention of the older man, Lux. Henry Miller must have soon been impressed with the easy manner in which Charles Lux seemed to be able to obtain money and credit in order to expand the business. On the other hand, Miller disliked city life and enjoyed the rough life of the country. It is not surprising, under these conditions, that the partnership continued for nearly thirty years without any serious differences between the two men.

To Miller was always conceded the general control of affairs. He apparently never recognized Lux's share of contributions to the success of the business, for once he was so ill mannered to say, "I have made three fortunes. One for myself, one for my partner, and one for my lawyers." Apart from their business relations they had little in common, though related by marriage on two occasions.

With the consummation of the partnership between Miller & Lux the elements requisite to the success and expansion of a foundational industry were brought together. The combination of the capital and business ability of these two men made success easier.

Miller & Lux stock horses and cattle, 1890's

Chapter Two

In the San Joaquin

When Henry Miller first set foot in the Great San Joaquin Valley is unknown. He had visited and seen the Pacheco Pass area in 1857 when he and Lux bought Erskins' herd. He may have seen the north end of the valley when he purchased the herd of oxen that had been driven from Salt Lake. Prior to that time the Spanish and Mexican governments were most generous with California land, and made extensive grants. Frequently a single grant comprised thirty to forty thousand acres. The land was not surveyed and the boundaries of the grants were extremely general, vague and uncertain. They were usually measured by leagues rather than by feet, yards, rods, or miles, and bounded by natural objects, such as the ocean, rivers and mountains. The owners were monarchs of all they surveyed. They lived in comparative luxury and maintained haciendas of considerable wealth. Fond of horses, riding and racing, they spent hours at these sports. They engaged principally in raising horses and cattle, which ranged on the natural grasses of the extensive ranchos. The agricultural and mineral wealth of the country was unexploited by these easygoing rancheros.

Before the passing of the 1850's, however, California had been infected by trappers, traders, and settlers of different languages, races and institutions. The "gringoes," a term applied by the earlier Spanish colonists to all subsequent settlers not belonging to the Latin race, came in increasingly larger numbers, in spite of the Mexican protestations and impotent threats. The "gold fever" brought them to California by the thousands and before long they had literally flooded the country. It was into this California, vast, undeveloped, agriculturally undiscovered, swamped in parts, and semi-arid in others, that Henry Miller had set foot in 1850, and it was into its principal city, San Francisco, teeming with excitement, growing by leaps and bounds, thinking gold, talking gold, and seeing gold, and needing to be fed, that Miller settled.

Miller's early years had been spent mainly in San Francisco, traveling occasionally as far south as San Jose. By the late 1850's San Francisco extended southward as far as San Francisquito creek, which flows just north of present day Palo Alto. As the demand for meat grew, the distance from which the cattle came became greater. This increase in the distance cattle came from put Henry Miller in touch with rancheros from the central valley. He learned that many of the cattle he purchased came from there, yet the bulk of the supply was being driven to the mines.

Miller would often see cattle with a HH brand and as a group they were often superior; they were bigger, fatter and dressed out at higher weights than other stock. He resolved to ride into the Great Central Valley and see it for himself and see if there was the possibility of securing cattle from there for his trade.

Miller made his journey into the valley via the El Camino Real south through the Buri Buri Rancho, which was then in San Francisco County. As he rode though this rancho, which was part of the vast private estate of Charles Lux, his thoughts were on cattle and holding them for sale. The Buri Buri had water access through its creek, which had a landing on the bay. Lux had purchased it in 1854. He went south to the sleepy pueblo of Santa Clara, past San Jose, going south to the Las Animas Rancho, which one day Miller would own and which today is a part of Gilroy and Mt. Madonna State Park.

His next stop was San Ysidro or Old Gilroy. Following the road around Soap Lake, he turned east to cross the Pacheco Pass. The pass and its surrounding lands were part of a Mexican land grant and known as the San Luis Gonzaga Rancho. Francisco Pacheco was its Don. Pacheco had a vast grant and numerous cattle. The rancho had bear, wild boar, deer, antelope and other wildlife. It was also a way stop for travelers crossing the pass on the way to the southern gold fields.

At first glance the valley seems filled with promise. Between the hills of the Coast Range and the towering ramparts of the Sierra Nevada Mountains lies the Great Valley of California. In the main it is a vast plain, stretching over five hundred miles from Mount Shasta to Tehachapi Pass, and averaging nearly fifty miles in breadth. Studied from the summit of Mount Diablo, it presents an unequalled panorama. On a clear day practically the entire valley appears spread out over a vast and seemingly level table. It readily appears to consist of two great drainage basins, a northern and a southern. The rivers, flowing from the north and south, are separated near their mouths by an imperceptible divide, which in itself breaks out and is broken by a network of sloughs and cross currents so that the two streams appear to mingle their waters before finally emptying them through the Strait of Carquinez into the Bay of San Francisco. As Miller topped the Pacheco Pass and looked down on this vast treeless plain, its grass tall and high, one can only wonder at what his thoughts were. Miller was soon to learn that while the San Joaquin Valley was ideal for cattle it was less so for man. The valley was hot and dusty during the summer months, with temperatures over 100 degrees for days at a time. There was little if any rain from April to November. When it was not hot it was cold, damp, windblown, and had a tule fog that prevented any movement of man.

The plains extended as far as he could see and there were few trees to break the monotony of the dry, dusty valley. Heading for a clump of willows that could be seen in the distance, he came to a creek, which he later learned had been named Los Banos. Refreshing himself and his horse from the waters of this stream he continued his southeastward course toward the San Joaquin River. After Miller had descended the ridge of the mountains into the valley he realized he had not seen any signs of human habitation. Drawing nearer the river he soon came upon a large herd of cattle that bore the "double H brand." Shortly, he rode up to the ranch house of the Rancho Sanjon de Santa Rita, which was then owned by Thomas Hildreth and William Dumphy.

Thomas Hildreth and Henry Miller directly became interested in each other and for several days they exchanged views and explored the country around. Henry Miller learned that Hildreth was the proud possessor of a quarter of a Spanish grant of over 48,000 acres along the west side of the San Joaquin River, known as Rancho Sanjon de Santa Rita. It appears that Thomas Hildreth was not very enthusiastic about the country; he wanted to get away to the mines. Miller as a lad had once tended some cattle that belonged to his father. Being a young boy he was daydreaming and fell asleep. In his dream he saw a vast plain and a herd of red cattle with white faces, with the mark HH on them. Here, years later, were that vast plain and 7,500 head of those cattle with the HH brand. Miller knew he had to have them. The meeting ended by Miller securing an option to purchase two square leagues of land (some 8,835 acres) of the rancho, for a consideration of $10,000, and seven thousand five hundred head of cattle at $5.00 per head, together with the "double HH brand."

It is interesting to know that Thomas Hildreth informed Miller that it was too far to drive cattle from the San Joaquin to the San Francisco market. Miller apparently thought otherwise, for he not only secured an option on Hildreth's fat cattle, but on his return journey to San Francisco he visited other ranchos, in the San Joaquin, Santa Clara and Salinas valleys, and usually left with an option to buy all their fat cattle and in some cases the land as well. Miller said nothing about his options to other butchers of San Francisco but they soon realized that he had practically cornered the market. Armed with his options, it was easy for Miller to deal with other less fortunate butchers who needed meat for their business. The following year the firm of Miller & Lux was formed. Mr. Lux probably was influenced in his proposal of a co-partnership because of this episode.

It soon became clear to one and all that Miller & Lux, as well as being butchers and meat wholesalers, were land proprietors on a large scale. In 1854, Charles Lux bought a tract of 1,700 acres near Baden Station in partnership with Alfred Edmondson, an Englishman. A short time afterward Lux bought Edmondson's interest when the latter decided to terminate his business and return to England. In 1858 Henry Miller bought a tract of land near Gilroy of about similar extent as that owned by Mr. Lux. This was all the land, not including some city property, owned by Miller & Lux when they became partners. In 1858 the combined land holding of the two partners was less than three thousand five hundred acres.

Henry Miller's reason for purchasing the Gilroy property was for a stopping place for the bands of cattle from the south. Stockmen called it the center place. Not only this, the place was well watered, had

Thomas Hildreth sold the Santa Rita to Henry Miller in 1863.

abundant grass, and possessed excellent facilities for camping large bands of cattle. Stockmen have a peculiar faculty for picking out nice camping places, and Miller's discerning eye must have immediately seen the merits of the place.

It was not until 1863 that Henry Miller plunged into a veritable orgy of cattle raising and land buying that continued unabated for fully a half century. What were the reasons that prompted this action and made it one of Miller's life policies? Before 1863 the firm had been buying numerous small lots of cattle, some of which were not in condition for slaughter. Land, in many instances, had to be leased on which to graze these cattle in order to fatten them for the market. The cost of renting per acre often amounted to more than what the land could be bought for outright. Miller said that he had been paying more for rent for the Lomas Muertas pasturage than he paid for the land when he bought the property. Guided by economy and common sense, Miller formulated his land policy of buying and holding his property.

The first purchase of land by the firm of Miller & Lux was made in 1863 from Thomas Hildreth and William Dumphy. Miller had, as previously noted, secured an option to purchase two leagues of the Santa Rita Rancho. At this time the firm saw fit to exercise the option and on May 22, 1863, secured title to the land. This initial purchase represented the extreme two southeastern leagues of the rancho. The Sanjon de Santa Rita grant had originally been made to Francisco Sobranes on September 7, 1841, by Juan B. Alvarado, then Governor of the Department of both Californias, and was for "eleven Spanish square leagues." According to the records, Sobranes first sold the nine northern leagues in 1853 and reserved the two southeastern leagues for himself. In 1858, Sobranes sold these two leagues to Salisbury Haly. Haly in turn sold them to William Dumphy for a thousand dollars in cash and a mortgage of $5,000. Dumphy deeded a half interest to Thomas Hildreth and they in turn later deeded the two leagues, as we have noted, to Miller & Lux in 1863. The nine northern leagues had any number of owners. Eventually, a division deed was made out for the various claimants and each received a definite part of the ranch for their share. From time to time, and in rather close order, Henry Miller purchased these nine adjoining leagues, until the whole of the eleven league Mexican grant fell under the control of Miller & Lux. The whole rancho (some 48,878 acres)

Keyes Ferry, 1870's

Santa Rita in the late 1870's

cost on an average about four dollars an acre. The Santa Rita Rancho was located along the west side of the San Joaquin River. Over forty-six thousand acres are within the county of Merced. The remainder, slightly over two thousand acres, are in Fresno County adjoining Merced County.

Of the Santa Rita, Miller had the following to say, "When I saw this Santa Rita Ranch, I thought that is just the place I am looking for. The grass is green nine months of the year there and it has a great many natural advantages—handy to all part of the state, nice level country—and I thought that it was the most favorable place that I had struck yet." The Santa Rita became Miller's favorite ranch and was commonly referred to as the "Pride" of Henry Miller.

With the purchase of large land holdings in the San Joaquin Valley cattle were constantly moving from the Santa Rita through the Pacheco Pass to Gilroy, and along "El Camino Real" to San Francisco. With characteristic foresight, Miller either rented or bought properties at convenient intervals to rest, feed, water, and fatten the cattle on the way. Thomas Hildreth had informed Miller that it was too far to drive cattle from the San Joaquin Valley to the San Francisco market. Henry Miller's properly spaced watering, resting and feeding grounds proved that not only was it possible but it was decidedly profitable.

In 1854 Miller purchased property in Monterey County (later San Benito County) from George W. Crane. Crane had married a widow lady who had inherited two Spanish grants. In this same year and in 1866 they purchased what was later called Tequesquite and the Lomerias Muertas Rancho. The latter purchase consisted of about 24,000 acres and joined the south of the Las Animas Rancho, which Miller began buying in the year 1858. To the properties already mentioned were gradually added all or part of the following Mexican grants of ranchos Santa Rita, Buri Buri, La Laguna, Bolsa de San Felipe, Juristac, Salivpuedos, San Justo, Las Lomas Muertas,

Aromitas y Agua Caliente, San Antonio, San Lorenzo, Orestimba, Las Animas and Tequesquite.

In getting together such a large amount of land one might well ask if Henry Miller did not have a method or possibly a technique in bringing these extensive holdings under the dominion of the firm of Miller & Lux. Miller frankly admitted, "It required no small amount of engineering to get all that property together." Not only one method was used by various simple and demonstrably effective practices were followed.

One of Miller's favorite methods was to buy out one or more of the heirs of the Spaniard who formerly owned the grant. This gave him as tenant in common the right to range his cattle on the land. Often the other heirs had no cattle on the land, so he would soon be practically in sole possession, but he would be careful to in no way exclude the others so as to make it necessary to account to them for the profits or use of the land. Gradually he would buy them all out, or if any of them became troublesome, he would compel a partition. Another method was the outright loan. If one of the Hidalgos got into trouble with a Hebrew storekeeper or small money lender, Miller appeared on the scene, offering, with a gruff but friendly word, a loan, for which the firm charged the then prevailing heavy rate of interest. The Mexican or Spaniard, fond of spectacular display, seldom repaid the loan, for his way of doing business was generally lax.

Miller & Lux advanced a good round sum, at one percent a month or more, compounded monthly, and took a chattel mortgage on the stock. Time of payment came; nothing had been paid, for "tomorrow" (manana) was always the answer to a demand for a settlement. The Hidalgo would then ask for an additional loan, would receive it and pledge as security the livestock on the rancho. After six months a demand for payment was made; again it was not met. The chattel mortgage was foreclosed and the property was denuded of the livestock, thereby destroying its income-bearing quality, for land had little rental value in those days. This land was then leased by the firm at its own figure, and a stiff advance on the rental was made to cover the outstanding store bills, debts or other indebtness.

After an elapse of a short period of time a mortgage was taken on the property. For a few years no demands were made for payment. Then the request for payment was presented, could not be met, and the family lost its all. Such tragedies took place all over the state. The morals and business ethics of the times did not condemn these practices and many of the most respectable and influential families of the state got "their start" by making such advantageous loans. Thus he laid the foundations of his land empire.

During periods of financial difficulty, the United States Government issued land scrip in order to pay many of its obligations. In this manner it paid claims for expeditions against the Indians as well as soldiers' claims for services during the Civil War. This land scrip passed as currency in California, one hundred and sixty acres of scrip being worth about one hundred and eighty dollars. In exchange for lands owned by the half-breed Sioux Indians, this transaction took place in 1833. Many years afterwards, towns were built and white people settled on the west side of Lake Pepin. This white settlement on Sioux lands was in the nature of a trespass. Agitation followed for a number of years, and finally, Congress passed a bill authorizing the purchase by the Government of the tract, and the issue as payment to the Indians "of as many acres of scrip as there was land in the tract." After an enumeration of the natives was taken, the scrip was equally divided among the Indians. This amounted to four hundred and eighty acres to each individual. The four hundred and eight acres were issued in five pieces—two pieces of forty acres each, one piece of eight acres, and two pieces of one hundred and sixty acres, to every Indian who had an interest in the Lake Pepin tract. One very advantageous feature of this scrip was that it was usable on any Government land. In the transfer of the scrip to a second party the Indian had to issue a power of attorney to the second party in order that the purchase might locate the land. For this reason the scrip was always accompanied by the power of attorney; otherwise it was valueless expect to the original scripees, who were all half-breed Sioux Indians. Quantities of this scrip were located in California, and Henry Miller bought up any scrip that gave him clear title.

As Untied States Government lands were opened up to settlers, much of the surveying was done by contract, and even before the plat was filed, favored ones were "tipped off" so that they might pick out the choice lands to file upon. In the wake of the surveyors came the land speculators. The men who held the contracts to do the surveying also later became land speculators. The careers of Hyde, Chapman, Friedlander, Renson, and others date back to this time. They procured people to enter government land and to buy land from the state. They dealt in land scrip, and Miller was ever ready to buy the land, at his price, after clear title had been acquired.

Temple Warehouse, twin of the Salt Slough Warehouse built in 1875

Henry Miller was industrious in buying scrip, and for a dollar and a quarter an acre, and usually less, he obtained land ultimately worth hundreds of dollars. Some surviving employees of Miller & Lux deny that Henry Miller ever had his employees homestead one hundred and sixty acres and then deed the land to Miller & Lux after securing clear title. In the early days of California under American rule, the federal land laws were notoriously loose. It is a well-known fact that there has been more or less land of all kinds taken up by means of "dummy" entries. Even the best families indulged in this practice, at least to the extent of having members of their families make nominal residences on a piece of land. Frequently a family would have a house set on wheels and it would be moved from location to location. A member of the family, or some trusted employees, would make a nominal residence on the piece of land for the necessary length of time, and then the home would be moved to another piece, and this process would be repeated.

The "dummy system" was commonly practiced and the county records of Merced, Fresno, and Kern possess abundant evidence of the system being in vogue in the San Joaquin Valley. As an example of what took place during the early settlement period, the finding for one county is presented. The records of any of counties mentioned are equally convincing. The deed records of Merced County from April 27, 1863, to January 29, 1887, show that Henry Miller and Charles Lux were grantees in two hundred and eighty-seven different instances. The conveyances were often of one hundred and sixty acres, although instances of more and instances of fewer acres transferred appear. Fully half the number of deeds during this period were for one hundred and sixty acres. The names of Miller & Lux's employees appear as grantors in many instances. In many other instances, however, it is impossible to trace the names to the company's payroll. There are still some men living who privately admit being parties to this type of transaction. The entries were made in the Deed Index Book according to the year, month, and day the deeds were executed. In some instances whole pages of the Grantee's Deed Index are devoted to Miller & Lux exclusively.

The following statement appeared in the Bakersfield *Californian*. "Miller & Lux needed lands to graze their thousands of heads of cattle. A simple idea originated in the head of the cattle king. His cowboys entered homesteads from the vast domains and when they had perfected title the land was turned over to Henry Miller." Indeed Miller & Lux did not get all of their land in this manner; only a very small portion was so acquired. Miller spent money for land, vast tracts of it, and it was immediately incorporated into his enormous ranches.

Mr. William J. Stockton, one of the oldest surviving pioneers in the Los Banos area, had the following to say in this connection: "Henry Miller got hold of so much land around here by buying up Government land, by putting vaqueros on quarter sections and having them prove up, and by buying everybody out that wanted to sell."

There were still other effective methods by which Henry Miller acquired additional lands, very cheaply. In order to aid the reclamation of swamp and overflowed land, Congress on September 28, 1850, passed the Swamp and Overflow Land Act. This act provided that in order "to enable the several states.... to construct the necessary levees and drains, to reclaim the swamp and overflow land therein" there should be granted to various states "the whole of the swamp and overflowed land, made unfit thereby for cultivation...The proceeds of said lands, whether by sale or direct appropriation in kind, shall be applied exclusively, as far as necessary, to the reclamation of said lands, by means of levees and drains."

Section three of the act further provided that in making out a list and plats of the land aforesaid, all legal subdivisions, the greater part of which is "wet and unfit for cultivation" shall be included in said lists and plat, but when the greater part of the subdivision is not of that character, the whole of it shall be excluded therefrom. Thus it's clearly evident that land that was not overflowed or swamp, in some instances, would be included under the category of swamp and overflowed land.

By this act and a subsequent act passed on July

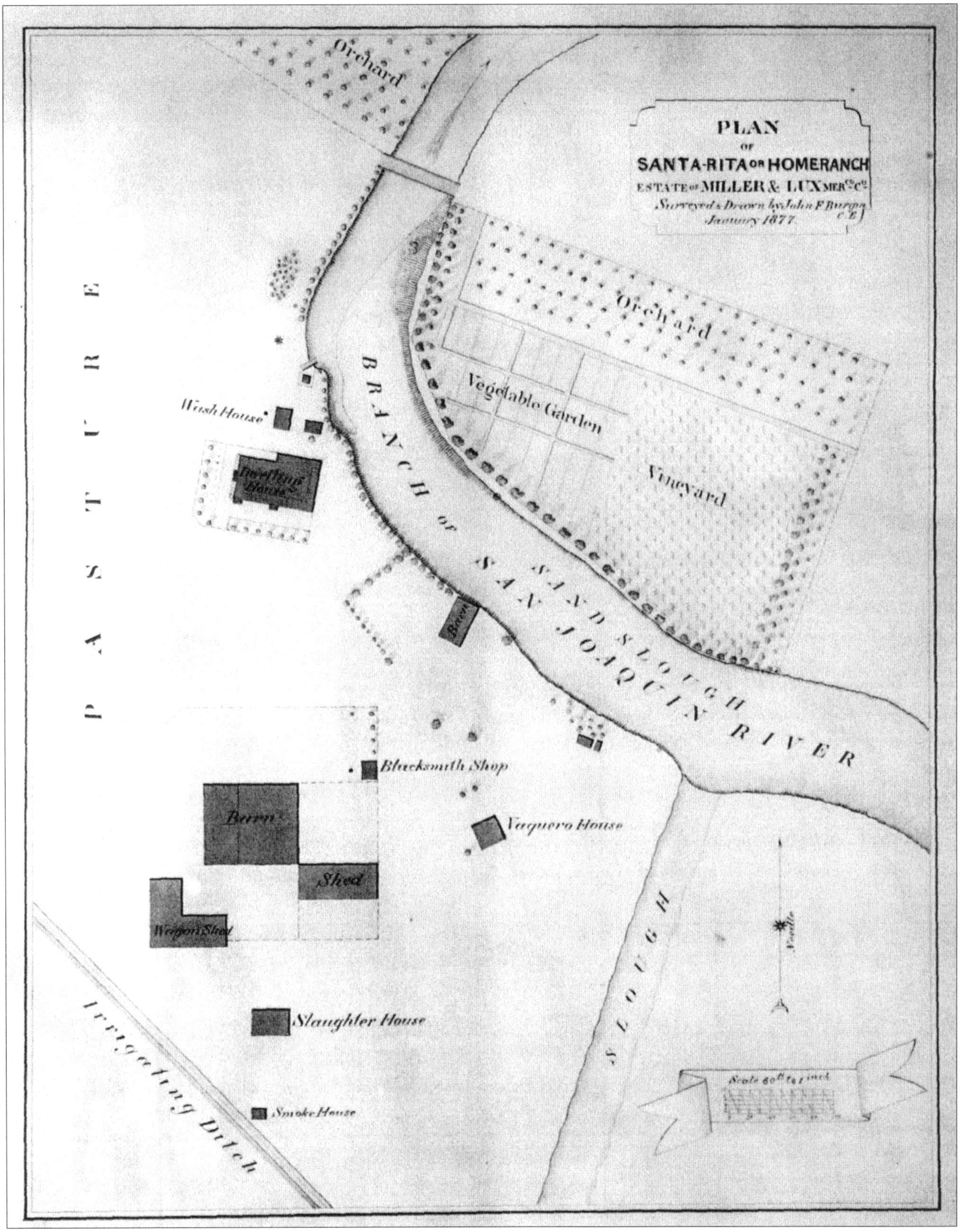

Plat plan of the Santa Rita Ranch, 1877

23, 1866, California eventually received from the Federal Government upward of two million acres of land. By January 17, 1874, there had already been listed to the State of California one million five hundred thousand acres of this land. The State first provided for the sale of the land at one dollar and twenty-five cents an acre, and later provided that the owner of a body of swamp land make proof before the Board of Supervisors of the county of the expenditure of a dollar and twenty-five cents an acre in the reclamation of the land and obtain repayment of the purchase price. This presented a grand opportunity to acquire large tracts of land. Henry Miller availed himself generously of the golden opportunity and acquired land that was just ideal for cattle. The swamp land bordered the rivers, and along the San Joaquin he acquired a continuous strip of nearly one hundred miles in extent.

Corrals at Santa Rita made of river willows

It is also true that Miller was not always first there with the money or means to obtain the land and when he was beaten to the land, he either made an offer or stood aside. A story has often been told of one such land deal. Henry Miller was acquiring land around Los Banos. David Wood was grazing sheep on the land that Miller had earmarked for himself. Wood too had intentions of eventually becoming owner of the land. From his sons, David Wood learned that while he had been away to Visalia on business, Miller and his surveyors had been over the land locating section corners, etc., preparatory to making purchase. David Wood wasted no time but immediately made a hasty trip to the land office at Stockton. The arrival of Wood and Miller at the Government Land Office was very close, but Wood beat Miller and had just made his purchase when Miller came up. Miller fumed and stormed when he learned that Wood had beaten him to the purchase. He had his heart set on the seven thousand two hundred eighty acres that Wood had just purchased.

In 1875, ten years afterwards, Miller had the personal satisfaction of buying from David Wood the land he had wanted so much, but failed to get, because he was a few minutes late in arriving at the Government Land Office in 1865. The purchase price, however, was somewhere in the neighborhood of fifty thousand dollars, and not the one dollar and twenty-five cents an acre that he would have had to pay had he secured the land from the government directly.

Thus, Henry Miller, through the purchase of Mexican grants, school lands, swamp land, homesteads, and preemptions, acquired extensive land holdings. Scrip was laid upon large areas along the rivers and creeks, across the plains, and even far into the mountains. Immediately as new land was acquired it was stocked, and Miller & Lux's "double H brand" was to be seen in every valley, plain and ravine of the San Joaquin Valley.

The firm of Miller & Lux purchased lands in enormous quantities. In 1869, in Merced County alone, their land holdings jumped from 62,818 acres to 160,118 acres, an increase just short of one hundred thousand acres. In the neighboring county of Fresno the increase for the same year was about half as great as that in Merced. In 1872, in Fresno County over twenty thousand acres were added to their holdings. From 1871 to 1877 their land possessions increased, in Fresno County, from 102,495 acres to 186,150 acres. In Merced County from 1870 to 1879 the increase was from 164,551 acres to 201,307 acres. The seventies were bad business years throughout the San Joaquin Valley generally, because of recurring droughts and floods; these phenomenal land additions of Henry Miller became proportionally more impressive.

In Kern County from 1881 to 1887 the increase was from 62,969 to 80,073 acres. Henry Miller's interests in the extreme southern reaches of the San Joaquin developed slowly and came nearly a decade after his purchase of the Santa Rita. From 1887, the year of his partner's death, to 1890, his holdings in Kern County jumped from 80,073 acres to 120,587 acres.

Workhorse and wagon barn at Santa Rita, 1880's

Seldom do the records of Merced, Fresno, Kern and Madera counties show the same assessment in acreage for two successive years. Generally, with the exception of an occasional big spurt, the increases are gradual and consistent from year to year. Paul Vandor, author of a history of Fresno County, says, "In Visalia once, that Miller made on one day entry upon six townships of land." It does not require a great deal of calculating to discover that this represents something just short of one hundred forty thousand acres of land. A study of the assessment rolls of Merced and Fresno counties for the year 1869 conclusively confirms this statement.

As early as 1875 the unabated land activities of Henry Miller gave rise to some very interesting stories. The following are typical. One day Henry Miller was observed by two men, in a small valley town, leaning against a watering trough, apparently in deep meditation. These men, who stood on the sidewalk a short distance away, knew Henry Miller and his great ability. The first one said, "There is Henry Miller. I wonder what he is so absorbed about?" The second man replied, "He probably is figuring out a little scheme to buy the rest of the State."

Often the statement has been made that Henry Miller testified in court that he "rode in a boat all over the land," thereby making good his claim to certain grasslands under the swamp and overflow act. Only too late was it ascertained that indeed he had, but that the boat had been placed in a wagon and pulled over the dry soil by of his sturdy mules.

The newspapers in 1916, after Henry Miller's death, made the astounding and astonishing statement that Henry Miller had left an estate of over fourteen million five hundred thousand acres of land. In a measure, this is probably exaggeration in order to impress readers. The reason, however, why great inaccuracy resulted in calculating Miller's land holdings can be attributed to the fact that his lands were strategically located. Miller possessed an uncanny ability to locate and buy lands that were riparian to important streams—lands possessing springs he took up with calculated foresight. Furthermore, Miller had an eye for discovering the most vital spots. When people saw that the lands held by Miller controlled the important water supplies, as well as held the key position to the surrounding land, they unthinkingly assumed that Henry Miller owned the intervening lands between the vital locations. In many instances this was not true; consequently, the wild guesses to his holdings.

The actual truth was that at the time of Henry Miller's death no one knew just how much land he really possessed. Even Henry Miller, years before his death, had frankly admitted his inability to state the exact amount of his land possessions. In so far as he owned 119,781.25 shares of the total of 120,000 shares issued, Henry Miller by 1916 had virtually become sole owner of Miller & Lux Divided Lands Inc. At this time the land holding of the company in the San Joaquin were 920,274 acres.

The San Mateo, Santa Clara, Monterey, San Benito, and a few other scattered holding in other California counties (Miller had land in nineteen California counties) would bring the grand total for California holdings to near the million and a quarter mark. The possessions outside of California, primarily in Oregon and Nevada, were variously estimated at anywhere from two million acres to six and one half million acres.

Henry Miller was a land merchant and would have become the greatest real estate man in the world if had lived long enough. He was always acquiring more than he disposed of. When he sold one acre he bought two. His plan was for a gradual transformation of his vast land holdings into business enterprises, such as banks, stores, etc. As he gradually sold off his best lands to small settlers he would build up business enterprises to get the trade of the settlers, in the meantime buying up cheap lands to continue his stock business. In this way his firm would be a constantly growing and expanding institution.

Primarily, however, Henry Miller was a cattleman and the land to him was simply a part of the whole of his business in the beginning. Later, when he saw the days of the cattleman were numbered, it became the means to keep his business expanding. No single individual on the Pacific Coast ever held in name clear of all encumbrances a greater tract of land.

Miller & Lux buckaroos at work, 1890's

Due to the swampy nature of the San Joaquin, pulling a cow from the mud was an everyday occurrence. Many were lost during times of heavy rains and floods, as they could not be gotten to higher ground. In dry times stock would crowd the water holes and become mired in the mud.

Chapter Three

Miller the Stockman

Cattle had first come to California with Portola in 1769. Both Mexico and Spain had been free with land grants and many large ranchos dominated the state. Until the gold rush cattle were raised and killed for their hides and were the main source of income of the rancheros. The Spanish cattle were small, thin deer-like animals with long horns, and by nature quite wild, a major reason the Spanish and Mexicans were such excellent horsemen, as these cattle were unsafe to work with from the ground. The rancheros made a living with these cattle, as they did not use them for milk and sold the hides and tallow to the ships that put into California ports at the time.

Some blooded stock had been imported into California as early as 1852 but progress was slow in changing the makeup of the herds. It was during the decade of the 1850's that the newcomers demanded a better grade of beef. Prior to that time cattle sold for $2.00 a head for hide and tallow, and the meat was stringy and tough. A horse at the time sold for $5-$10. The inrush of miners and others who followed changed all this. Prices skyrocketed. By 1851 the killing of cattle for hide and tallow had stopped totally. Cattle that the year before had brought $2 a head now sold for up to $150 a head. The demand was constantly increasing. The cattle in the San Joaquin were mostly Spanish but in the mid 1850's herds were being driven in from Texas and points east. Some of these herds contained fine bulls and were a far better quality of cattle overall.

Miller had had experience with the Spanish cattle as his early trade had been made out of them. He also had an eye for cattle and knew that these new herds coming into the state would be the foundation of his business. He had in 1853 purchased the herd from Kincaid and Livingston and knew his profits were greater from this higher class of cattle and the satisfaction of his customers was high. He resolved at that point that the best meat was none too good for his trade and let the others take the rest. His next purchase was in partnership with Charles Lux of 1,500 to 1,600 head of cattle that had been driven from Texas.

In the early 1860's Miller & Lux decided that they would undertake the business of raising their own stock. Prior to that they had mainly purchased others' herds, kept them on pasture and killed them when ready. This new policy was fraught with danger and uncertainly because neither Miller nor Lux knew that California was subject to periods of flooding and drought. Both had arrived in 1850; in 1851 there were heavy rains somewhat normal in the City but not in the Valley. That year 35 inches of rain fell in a valley whose normal rain is 7 to 9 inches. The valley from Sacramento to Visalia had water across its entire width. A decade later in 1861 the same occurred. It started to rain in December of that year and by 1862 over 50 inches of rain had fallen.

The valley lay under two feet of water its entire width. Damage was estimated at 50 million dollars. Miller & Lux took the loss and planned to be more calculating in the future.

Just two years later, in 1863, the great drought started. There was scarcely a drop of rain in the winter of '63-'64. When spring came the land was dry and little hay had been made as most were sky farming. Hay that had been made was going for $60 a ton; wheat was scarce at five dollars a bushel. Horses, cattle and sheep perished by the thousands. The plains of the San Joaquin were strewn with their carcasses. In the low marsh areas, the last to dry up, travelers years later would report still seeing stacks of bones. It has been estimated that more than a quarter of the state's livestock perished. There was nothing that could be done to save the starving animals, as there was not a railroad at the time and to attempt to drive them to markets was out of the question. Many cattlemen resorted to skinning the dead stock and shipping the hides to market, realizing some money from them.

Miller & Lux were quick to grasp the situation and salvaged what they could from the disaster. But the market was soon flooded and the price fell out of the hide market. A newspaper report of the day states that in Santa Barbara in one week 18,000 head were killed for the hides. In 1864 Miller & Lux had on pasture in the valley seven or eight thousand head of cattle on the Santa Rita Rancho. There was no feed left anywhere for them but tules. Miller & Lux drove what cattle they could to Antioch and Stockton but they lost most of the herd. Miller later stated, "1864 was very disastrous for us on account of the drought because we did not know how to take advantage of it and we lost two-thirds of our cattle because of this."

These were trying times for the firm because while there was a general depression the cattle market was high. Cattle was selling at 25 to 35 cents a pound. However, there was little good beef to be found. The meat and cattle markets soon became depressed as well. Charles Lux became discouraged and wanted to get out of the business. Miller would not hear of it. "We have found it here and we have lost it here. We are not in condition to sell out." To encourage Lux to hang on, he said that they would sell out when they had recovered the money and had made some money and then they would discuss quitting the cattle business.

Miller was thinking ahead. In 1864 everyone was getting out. Cattle fell to $5 a head. Steers could be bought from $8 to $10 a head. Miller perceived his opportunity and bought a large number of cattle at $5 a head. From Able Sterns, one of the state's largest stockmen, he purchased 8,000 head. He knew now what California could do to a stockman and he did not doubt the future of the state and its growth. During the panic caused by the terrible drought ranges were allowed to go along with the cattle for a song and the firm of Miller & Lux took due advantage of the opportunity offered.

To do this the firm needed to increase their obligations to their faithful friend, John Parrott, to between $50,000 and $60,000. Parrott had such faith in Miller & Lux as businessmen as well as their integrity that they were welcome to whatever funds they needed from him without security. Miller & Lux did give Parrott personal notes but they never mortgaged their stock or properties.

The move proved to be a smart one. The value of the stock doubled in less than a year. In 1867 some of the cattle they had bought for $5 a head was now worth $20 a head and in 1868 they doubled again, bringing $50 a head. The property that had come with some herds was doing the same. The crisis and calamity of 1864 that had destroyed others had been turned to Miller & Lux's advantage.

In 1850 the state had passed a brand law and in 1851 it had been modified to the extent that in addition to registration of the brand, ear and flap cuts made in the cattle that a piece of leather with the same marks had to be delivered to the county recorder in which the herd was based. Miller & Lux, in order to conform to this law, had a problem. They had acquired several brands along with the many herds they had purchased. In order to make their herds conform with the law the bulk of the California cattle were branded or rebranded with the HH brand that had been purchased with the Santa Rita Rancho. The Oregon cattle were branded with an "S wrench with the S reversed," Nevada with a "7" and "L.F." when Miller & Lux acquired the Overfelt Company in Oregon.

Spanish Longhorn cattle, 1860's

These brands were used in courts of law to determine who actually owned the cattle. Since in the San Joaquin cattle grazed over miles of land, identification was important and roundups or rodeos sorted cattle. In 1851 the Legislature passed a law regulating rodeos. The rodeo was a Spanish institution in operation when the Americans arrived in California. It consisted of a roundup of all the cattle on a given range at more or less fixed periods and then the cattle would be sorted, each going to its owner. Calves were branded and a "Judge of the Plain" would sort out the unbranded ones and decided who got what if two parties claimed a calf or steer. Every owner must hold a general rodeo on his land between the first of May and the first of August. The act also provided that any unbranded calves without mothers would go to the holder on whose land the rodeo was held.

Miller & Lux held rodeos twice a year, one in the early spring and one in the fall. The spring rodeos would start in the south near Tehachapi and move northward. Miller & Lux had a number of rodeo corrals made of logs sunk into the ground with lighter poles intertwined to make stout round enclosures. The poles were laced together with green leather strips, which tightened as they dried. The cattle would be run into the corral in small groups and one by one they would be detached from the group by the vaqueros. If they were on the open plain the vaqueros would make a circle with the riatas and hold the stock in a circle until the entire group had been worked.

Cattle were then branded on the hip to show ownership. If a cow were sold she would be reframed on the shoulder to show she had been sold. Some old cows had numerous brands containing a perfect record of her ownership. This method of marking cattle left something to be desired, as the open range and unbranded cattle made some honest men cattle rustlers. A small owner unable to attend all the rodeos held by his larger neighbors saw his cattle disappear into their herds. He could square the account in only one way. He became a cattle thief.

Rustling was common on Miller & Lux's range. It was assumed to be all right to rustle the rich man's property. If a rustler was caught it was practically impossible to get a jury to convict. Miller & Lux employees stated that rustling occurred in broad daylight. Even herds of horses were taken. At times Miller was powerless to do anything about it. South of Los Banos was a great rodeo grounds. It was here that the bulk of the herds were sorted out twice each year. A bad crowd of Spaniards, Mexicans and Americans would come down from Livermore to work the rodeo. They were so bad that no one would interfere with them. Miller was forced to sit back out of sight watching them "fat brand" his cattle. A fat brand cow was one that was fat and ready for market. This group would boldly sort out the cattle they wanted and drive them off to Oakland to sell to the highest bidder. Will Wood recalls at one such event in 1868 Miller sat on his horse and watched them ride off with several hundred head of the 10,000 gathered there that day. The Livermore crowd had threatened to kill Miller, so he let them take what they wanted. Miller bided his time and when he had the land he built his fence. By that time settlers had come into the valley and the local county officers had become a stronger force and the threat was eliminated. There is a story told of Miller and a rancher. Once when Miller was taking the train to the City he ran into this rancher on the train. Miller had long known that the rancher was not above picking up a few head here and there of Miller & Lux stock and changing the brand. "Well" said Miller good-naturedly, "I've got a brand that you can't change now. It is an S wrench." The rancher replied, "I have a new brand as well, its an S wrench with a nut on it."

Rustling was a thorn in the side of the cattleman but the settler was a far worse danger and a serious menace to the open range. The nester or, as he was called in the San Joaquin Valley, the sandlapper came to exercise his rights to 160 acres under the Homestead Act which granted anyone who improved the land title and ownership of the land. Most were sky or dry farmers. This type of operation displaced the cattleman, as soon fences were everywhere, leaving the cattleman with the less desirable and for the most part remote lands.

New settlers were shocked. Many had come from the Midwest and the tales of the cattlemen were not the encouragement they had expected. The land speculators had hyped California to them. The truth was that the thermometer could and did reach 115 in the shade, if there was shade, birds would drop dead in flight from the heat. Bugs, did you say bugs—the valley had its share and more. Fruit? Oh you could grow it but it either baked on the trees or was destroyed by the spring wind and rains. Vegetables, well if you could find water but more than likely the jackrabbits that overran the plains would take care of them for you. Water, there was water but for the most part salty and undrinkable, and you could not catch enough to save from the rainfall. Potato crops failed three out of four years; few ever were raised. Vines, well there were the rabbits. Butter and cream, only in winter;

the spring and summer were far too hot to keep them from going bad. But the settlers came and many stuck it out. The early settlers on Miller's lands had brought with them the love of fun and a grotesque sense of humor peculiar to the Midwest. This they displayed in giving names to the places they settled. "Cradle of Innocence," "Hells Half Acre," "The Refuge," and "Tranquility," to name a few.

One of Miller's policies was to deal with everyone fair and square. When the old Dos Palos Colony that was located on his land west of the delta region on the Fresno-Merced County line was in trouble, he had it moved. Miller had become involved with Bernhard Marks, who in turn sold the land to the settlers as the ideal spot to raise stock and alfalfa. In reality the land was swamp land that was good for little but grazing. The settlers soon discovered that nothing would do well there. They sent a group to meet with Miller who told them he had nothing to do with the deal, that Marks had their money, not he. Yet he sent a man to see just what the problem was. On reporting back to Miller, Miller decided that he would move them about five miles north east on the present site of Dos Palos. This was a major operation. Miller gave the settlers acre for acre new land and moved their stock and houses to the new location. His men moved 30 houses in 31 days to the new site. The settlement succeeded and it was a means of success for Miller to sell additional stock and have a source of quality employees.

Again Miller was looking ahead. He knew that there was no way he could keep settlers out and so he early on faced the situation and was prepared. He quickly saw it was better to work with than against the settler. He gave the settlers valuable advice as to the land and what the weather could do to them, as well as extending them credit at his stores. He soon found it was profitable to sell some land to the best of these farmers. But it was not all smooth going. There were outbreaks of trouble and

Miller & Lux buckaroos on the range, 1890's

the "No Fences" law went down to defeat. Starting in 1866 in Sacramento and ending in the San Joaquin in 1874.

Miller & Lux had fenced a large section of the west side of the San Joaquin in 1860's in order to gain control of the vast herds that grazed on the west side as well as to discourage rustling. They had a string of fences 5 boards high that ran for more than 70 miles with many cross fences inside the main fence. This board fence was in constant need of repair. Bulls would charge it, breaking boards; travelers and teamsters on the way to and from Stockton would take out boards to pass through. Miller stopped some of this by placing gates every so many feet along the fence so that one could clearly see where to go through but many still just took down boards and went on their way. When Miller & Lux changed to barbed wire fence is not known. Barbed wire fence is dangerous to man and beast. Cattle in a panic or horses can be lacerated by the sharp pines and even be killed. It is thought that it was around 1882 when they first switched to barbed wire. This meant the hiring of fence riders whose only job was to ride the fence line and keep it in repair. Miller again installed gates so that it was easier to open a gate than cut through the fence. He also installed windmills and a tank, along with a small corral at the gate so the travelers could water themselves and their stock and camp overnight. Soon it became easier to stop and go through the gates; thus did Miller save his fences.

Settlers brought other problems. When Miller's vaqueros drove a herd of cattle through the country they often were careless about sorting out stock that would sometimes get mixed in with the herd as they went. These were not herds of a few dozen cattle but herds sometimes of one or two thousand head moving north from southern ranges. Often a farmer would discover some of his stock gone and would hurry north to catch the herd and get his stock back.

Remnant of Miller's board fence, 1870's

Remnant of early Miller board and wire fence, 1880's

Miller had instructed his men to pay fully for what the cattle were worth as it was far more costly to stop the herd and try to sort out a few head. Usually the owner would take the money because whatever price he set he would be paid. Some would even put out poor cattle near the road in the hopes that they would mix in with Miller's herds, become good fat cattle and bring a good price. But this worked both ways and many animals Miller had to pay for.

The first purebred bulls had come to California in 1852. By the 1870's Miller & Lux were improving their stock to a great degree. By the 1880's the stock they ran consisted of a mix that was three-quarters Devon stock and in some herds seven-eighths Shorthorn and Devon mix. Miller also dealt in hogs and sheep as well and went about improving them with the same will that he had the cattle herds. Things did not always go well. Hogs in California for the most part were wild boar with very few domestic hogs in the state. Miller at first rounded up thousands of these wild hogs and tried breeding from what he considered the best of them. He was unsatisfied and decided to bring in some meat hogs from the east. No expense was spared on the pens and quarters for these wonders. Miller purchased two prime breed sows in the east for the sum of $300 each, and had them shipped west. With these hogs he would be ready to breed the improved Miller hog. The hogs arrived and all seemed well for a time. Miller was called east again and left his nephew, Albert Long, who was a trusted cattle buyer, but not a hog man, in charge of his sows. While Miller was gone both sows sickened, most likely from some form of hog fever. Long had taken good care of the hogs but Miller was gone several weeks longer than expected and it was in that time they became ill and then died. Both Long and L. Nance, who was Miller's driver, were afraid to tell him. They talked the bookkeeper, an Englishman named Payne into telling Miller. Miller was known for his temper, often taking off his hat and stomping on it in a fit of temper. Payne, who was devoid of any sense of humor, walked into Miller's office. Miller had been gone weeks longer than expected and knew that something had to have gone wrong. He asked, "How are things?" Payne replied "Fine, Mr. Miller." Miller, knowing Payne, asked, "What is the first troubles?" "One of the sows died," replied Payne. Miller exploded at the revelation of this news, but when it was over demanded, "Now, what is the second troubles?" Payne's blunt reply came without the slightest hesitation: "The other sow died." One can well imagine Miller's consternation at this piece of news—his plans for an improved hog killed at one stroke.

Miller had started from next to nothing when he arrived in California. His thrift and hard work had set him on the road to his fortune. His rule of watching every penny and every detail of the business as closely as possible was the cornerstone of the Miller & Lux millions. Nothing was wasted nor was a string ever allowed to be tossed away. If a place was suitable for feeding hogs, then hogs must be kept there. After the cattle had pastured a field they were to be followed by sheep. If the men taking care of the herds had surplus time they must give a hand at farming. If any department or section of the operations failed to make a profit an examination would be made to find whether the land, the system or the manger was to blame. Ranches were bought after careful consideration as to their capabilities. Speculation was avoided.

As soon as grain was harvested cattle would be turned on the stubble, after them sheep and then hogs. In this manner the gleanings were utilized to the upmost. The straw after thrashing was put in huge stacks to be used in dry years. Some of these stacks contained upwards of 300 tons. The numbers of livestock owned by Miller & Lux at any one time is hard to determine. Among their vast herds a few thousand more or less did not matter. In 1881 a party attempted to count the stock. He later stated, "I am unable to give the exact number of stock but let me say that at the rodeo 25,000 calves were branded and that this year the firm supplied 83,332 animals to the San Francisco market alone."

In 1883, Charles Warfield, on the Santa Rita contained under his control 50,000 head cattle, 80,000 head sheep, 8,000 horses, 5,000 to 8,000 head hogs. In 1888 the cattle had risen to 80,000, sheep to 100,000 and hogs were estimated at 10,000 head. For several years Miller's sales of meat averaged around one and half million dollars a year.

Miller began to have problems with the East Coast meat dealers around 1890. The cattle business had become industrialized. The big five of Swift, Armour, Cudahy, Morris and Wilson controlled the bulk of the meat markets. Miller had zealously guarded the San Francisco meat trade, as it was the source of income for all his operations. Many firms were forced into mergers or went out of business. The arrival of the big five in San Francisco in the form of the American Cattle Trust challenged Miller's hold on the market.

At Charles Lux's death his widow had put the Buri Buri on the market. This had been the last holding stop for Miller & Lux cattle prior to going on to the city. The Swift interest, through the American Cattle Trust, purchased the land in 1887. Why Miller did not purchase it is unknown. Miller had by his sharp economy kept each phase of his operations at its highest profit. This was to be a chink in his armor in that the American Cattle Trust had huge sums at its disposal and went after Miller in such a way that he could not compete effectively.

In 1891 interests of the American Cattle Trust formed the South San Francisco Land & Improvement Company and later they established the Western Meat Company, which was strictly a meat packing operation. They next built a modern, large slaughterhouse with modern machines, labor saving devices and that new fangled invention, refrigeration. They were then able to undercut and kill any competition and could take losses for some time if need be.

Miller was not the only one hurt by this move. The butchers with the help of Miller & Lux organized the Butchers Board of Trade in 1893. They started a boycott and a propaganda campaign against the American Cattle Trust and Western Meat Company. Charges and countercharges were tossed back and forth as to the quality of the meat, it preparation and use of refrigeration (unheard of in the west) prior to marketing. They called on loyalty to the local butcher as well. San Franciscans responded; the tactic was effective but Miller, shrewd as ever, knew the capital of the Trust was tremendously large. He covered himself by buying into the Land & Improvement Company and was elected to its first board. This put him in a position to gauge the competition. Miller had placed himself in such a position as to be able to be a broker between the Butchers Board of Trade and the interests of the trust.

In the meantime, starting in early 1900 and waxing and waning until 1908, San Francisco was cursed with the plague. The first event occurred in early 1900 and since most of the deaths were in the Chinese community not much action was called for. In August of that year a white drayman died. Dr. Joseph Kinyoun had been trying for months to make the city fathers and residents see the danger of the plague. It had gotten to the point where calls for Kinyoun to leave town were common. By 1901 there was still doubt and denial that the plague existed in San Francisco. The result was that Kinyoun was forced out of his job as Federal Public Heath Officer. The plague seemed to wax and wane with the seasons. By 1903 California and the City of San Francisco were still in denial. A public call was made to acknowledge the plague. Politics played a major role in the situation.

At this point Rupert Blue arrived on the scene. An experienced health officer, he was quick to grasp the situation. He was quick to get things moving and to get community action. One place focused in his site was Butchertown. For years the slaughterhouses had been built over the bay, the excess and waste left to fall into the bay. Rats abounded. Blue was shocked by the conditions. At this time the boycott and propaganda campaigns of both the Butchers Board of Trade and the Western Meat Company were in full swing. Butchers did not take kindly to Blue's statements that if the butchers did not join in the clean up he would print the details of the methods of handling the meat in the evening papers. Miller's operations were run the same as any of the others. Though his firm was never mentioned, it can be assumed that his firm as well as the others in the end complied with Blue's mandates.

While the plague continued, another disaster was waiting in the wings—April 18, 1906, a day San Franciscans would not forget. An earthquake with magnitude estimated at 7.7 struck the city in the early morning hours. Houses and buildings all tumbled to the ground. Soon uncontrollable fire broke out. People were not the only ones fleeing the buildings; the rats came out by the tens of thousands. While the quake and fire killed many of the rats, the plague continued to be a problem until 1908. By that time businesses, residents and politicians had come to agreement. A strict program of rebuilding, cleanliness, inspection and poison effected control of the rat population and the eradication of the plague.

Miller survived the battle with the Western Meat Company by his shrewd purchase into the South San Francisco Land and Investment Company as well as by modernization of his slaughterhouses after the quake. He also bought Western's meat at depressed prices and resold it at still lower prices until they gave up.

This ownership of stock and land also made Miller & Lux one of the largest taxpayers in

California. If one checks the assessment rolls of some of the counties in the San Joaquin Valley they are seen to be many counties' major support. The firm was the largest taxpayer in Merced, Madera, Fresno and Kern counties. At Miller's death in 1916, in Merced County the list of personal property was so long that the tax collector gave up and valued the total at $309,745.00, with the tax paid $6,194.10. The land was valued at $2,404,825.00, with tax of $48,095.25 paid. Today this does not sound like much but one needs to remember that at that time today's dollar was worth ten dollars.

After Miller's death the rise in land values became noticeable. These advances were caused by and automatically reflected in the assessed valuation of Miller's possessions and in the amount of tax collected from his estate.

Miller's improved slaughter operations after the 1906 earthquake

CHAPTER FOUR

Irrigation and Litigation

Life in California was and is based on water. The early rancheros had settled along the coast and had avoided the San Joaquin Valley. The early grants had everything one needed: abundant land, vegetation and natural springs and other water sources. Drought was not unknown to these rancheros. It was thought that droughts occurred at seven to ten year intervals and could be gotten through. Many thought that irrigation was unnecessary and in fact was considered a serious disadvantage to farming and ranching. As late as 1871 only 90,344 acres of the state were under any sort of irrigation and most of that was in the south in Los Angeles County. Less than one in one thousand acres was under irrigation.

Henry Miller realized early on after the drought years of the 1860's that water was going to be key to any success he would have in the San Joaquin Valley. As he acquired land he made sure to acquire as much as he could along the San Joaquin River as well as the Merced, Kings and other smaller tributaries. Dams were put in the sloughs, and levees were built so the water spread out across the large areas of pasture land. The economy and effectiveness of this method of flooding the land was remarkable. A few levees across low places would often spread the water over miles. To prevent continuous flooding or inundation and yet have fresh water but not making it too deep, the water was allowed to flood the land behind one levee and then flow into the next check. This work could be done with a minimum amount of labor because at that time water could be used in almost unlimited amounts. A water tender with this method could cover exceedingly large acreage under this system.

By 1897 Miller & Lux had over 300 men employed leveling large areas for this type of irrigation in Merced County alone. This method is still in use in Merced County. The grass was abundant and stood well up on the side of feeding animals. Miller was not a land buyer merely on speculation but as an end to the means of feeding his stock and trading cattle, sheep and hogs. In order to control water, the lifeblood of his lands, Miller would often acquire land on both sides of a river for a distance of nearly one hundred miles, and he would actually construct levees along both sides of the river, so as to entirely control the flow of the water, sometimes for a distance of 50 miles. His men or farmers did this work primarily with his own horses, mules and scrapers during non harvest periods.

It was not until 1871 that the largest and most far-reaching project of irrigation in California was started in the San Joaquin Valley. Many of these same canals are in use today. The enterprise was started by John Bensley and Orrin Simmons. Bensley has suggested the scheme to Simmons. Before long they had posted a notice near the confluence of the San Joaquin River and the Kings

River stating that they had appropriated a certain amount of water to be used through a canal to be constructed for irrigation, transportation and mechanical purposes. Directly they proceeded to interest San Francisco capital in the scheme. So everyone wanted a piece of the action. The San Joaquin and Kings River Canal and Irrigation Company was organized with a capital of one million dollars; 100,000 shares of stock were issued and public calls for subscriptions to the stock were made. Among the original stockholders were: Bensley, Simmons, Miller & Lux, Haggin & Tevis, William Hamilton, and W. C. Ralston. The point that not only would land be irrigated but also that transportation facilities would be afforded was stressed. The growing hostility of the farmers towards the railroad was played upon and the project was pictured as a great benefit to all.

In a pamphlet issued at the time, the scheme was stated as being "the construction of a system of canals in the San Joaquin and Sacramento Valleys, in the Great State of California. These would lead from the San Joaquin River, the Kings River and their tributaries and also draw from Tulare Lake and Buena Vista Lake, making it possible for passengers and freight to move with ease as well as provide irrigation and water power as well as supply the cities with fresh water for domestic purposes."

The company charter was to run for 50 years and the stated place of business was San Francisco. At about this stage of the proceedings Henry Miller let it be known that the enterprise could not go on until some understanding was had with him as to his riparian rights. As a result the company took a lease from Miller & Lux for a term of five years on a tract of remarkably smooth land, lying immediately below the grade of the canal and containing about six thousand acres, on which it was undertaken to lay out a system of ditches for the purpose of exhibiting the value of irrigation. The company obtained the tract without payment of rent for the five years and Miller & Lux were also required to fence the property leased. The company was to have the proceeds of the tract, but was required to put in one-fourth of the land under alfalfa; the same was to be ready for grazing purposes by November 1874, when Miller & Lux were to have grazing thereon for their cattle, for three months; the same for three months of each successive year during the remainder of the lease. The company had to have the whole seeded down to alfalfa by the end of the lease.

The company also held a contract with Miller & Lux, dated February 7, 1872, which gave the company the right of way for the canal and works over Miller & Lux's lands; also the use of several sloughs, and the willow brush on their lands along the river. The company, on the other hand, under this agreement was required to give Miller & Lux water from the canal sufficient to irrigate 16,667 acres in 1872, 33,334 acres in 1873, and 50,000 acres in 1874, at the price of one dollar and twenty-five cents per crop per acre. The company also had to furnish Miller & Lux and their employees, tenants, and cattle, with water for drinking purposes free of charge, during the years of 1872 to 1874.

Miller & Lux, under this agreement, had to give the company as a subsidy to encourage the construction of the canal, the sum of $6,667 in 1872, and an additional sum of $6,667 in 1873, and a further sum of $6,666 in 1874, provided that the canal or a branch was extended to the fifty-eighth mile by August 1874. On the other hand, should the company fail to supply Miller & Lux with water as required by the contract, it would receive a proportionate less amount of subsidy to the less amount of water furnished in each of the above mentioned years.

Work was started on the excavation of the canal and as there was no railroad reaching the territory, the supplies were brought up the San Joaquin River by barge, a distance of over a hundred and fifty miles. This, owing to the condition of the channel, could only be done during the short period of high water. At other periods in the year, supplies were carried over the Pacheco Pass, a distance of fifty miles from the railroad, which then ran from San Francisco to Gilroy.

Henry Miller soon learned that the enterprise was controlled by the promoters rather than by the financiers. Over one hundred thousand dollars was paid to John Bensley for alleged rights, which consisted of a notice posted upon a tree appropriating the waters of the rivers. The prompters quarreled among themselves and spent large sums of money in attempting to secure a subsidy from the national government and from the state. Miller did not try to control the matter, but merely guided things so that the canal in which he was interested was dug. He held before them the prospects of immediate income from his land under his contract. He showed them how to build the canal and where to build it and it was his horses, men and tools that actually constructed most of it. For implements, all they had were plows and the old style slip scraper, and horses and mules.

As a result of Henry Miller's aid and encouragement, the canal was dug during 1871 to 1873, a distance of seventy-eight miles, and a dam, three hundred and fifty feet long, made of brush and sand, was put across the river. The river often carried an enormous volume of water and many scoffed and laughed at the idea that such a river could be

Miller's 'Best Mule Skinner," James Hall, working on the canal, 1890's

controlled in this manner. Miller insisted that it could and for a quarter of a century the dam was used to force the water into the canal.

While the work of excavating the canal was going forward, Miller & Lux were getting their own land into shape to receive the water. Miles of lateral canals were constructed; thousands of acres were plowed; levees were thrown up and everything was made ready for throwing the water upon the land just as soon as it was available. A continuous line of freight wagons traveled the road from Gilroy to Los Banos with machinery, foodstuffs and lumber.

As the work proceeded, another obstacle did appear. The legislature had declared the San Joaquin River navigable to a point above the location of the dam, so that they had to construct it in such a way that if perchance a boat did come up to the dam, it could get through and the dam would not act as an obstruction to navigation. This obstacle was overcome by constructing a collapsible wooded weir on one end of the dam.

On the early work of the canal a writer said, "The work on its construction is going on. Times would be very dull were it not for that work. Many who would otherwise be idle, find a profitable employment on the canal. Men are paid thirty dollars per month and their board and a man with two horses, received fifty dollars and upkeep for himself and team." The canal was described as beginning at a point six miles south of Firebaugh's Ferry and extending in a northwesterly direction from Fresno slough, almost parallel with the western bank of the San Joaquin River. For the first thirty-eight miles it was sixty-eight feet wide at the top and forty-five feet at the bottom. And for its remaining length, fifty-three feet and thirty-five feet respectively. The canal water in the early days was used for raising grain and not for alfalfa. Consequently, there was need of a great deal of water from the middle of March until the middle of April. The grain had to be irrigated then or not at all. After that there was little demand for water. In order to carry as much water as possible, Miller had the canal doubled for the distance of about twenty-eight miles. At the same time, Miller also spent over seventy-five thousand dollars cleaning out the old canal. After Miller acquired the canal it was eventually extended to a distance of slightly over one hundred miles. Miller had no lands north of Gustine, but he extended the canal northward in order to give irrigation to the dry farmers in Stanislaus County.

Under the control of the speculators, the charge of two dollars and fifty cents per acre for water was called an outrageous extraction. The company was called a monopoly. Everyone connected with the project was abused. R. R. Brereton was appointed superintendent of the work, at one thousand dollars a month. He was sent to England and commissioned to sell the stock of the company at seven and eight dollars a share, while it could be readily bought in San Francisco at half that price. However, the proposed speculation in a foreign market did not succeed. Ralston, Friedlander, Luning, and the other heavy stockholders who had the control managed affairs badly. John Bensley hypothecated his 33,125 shares, lost them, and went bankrupt. For six or seven years Miller & Lux had practically no voice in the concern, but in 1877 they picked up the stock that Bensley was compelled to let go and a few other lots. It was a hard fight that they had to make against a coterie of very wealthy men and shrewd manipulators, but they won. At the annual election in the year 1877, they turned up with a few more shares than a majority of all the stock, but quite enough to place them in control.

Henry Miller made a trip to Germany in 1877, after he had suffered a breakdown following the death of his daughter, and the canal was far from his mind. He also was suffering from a problem with his throat that he wanted solved, and felt that he could get the proper treatment there.

When he returned he found things in a very bad state. In addition to the drought, when he arrived in Los Banos he found that the canal superintendent was refusing to allow his men to water the stock in the canal and that the superintendent had shut off the water from his lands. Miller immediately recognized that this was bad faith on the part of the canal company as well as a violation of his agreement with the company. Henry Miller was incensed. Without hesitation he took an ax and with a laborer proceeded up the canal bank, chopping open every canal gate that they came upon, letting the water pour out upon the land. The canal superintendent promptly swore out three warrants for Miller's arrest, each warrant charging Miller with breaking open a canal gate. Miller, however, had gone back to San Francisco and bought enough shares to give him controlling interest in the canal. Sheriff Meany went to San Francisco and brought Miller to Merced for trial. When the sheriff appeared before the judge, there was no one to prosecute Miller because the first thing Miller did after securing control of the canal company was to fire the superintendent. The case was dropped for want of anyone to prosecute.

Upon gaining control of the canal company, Miller ousted his scheming and spendthrift predecessors. When Miller got control of the canal, the superintendent was being paid four hundred dollars a month and the two engineers were each getting two hundred a month. Miller fired them all and put in J. Q. Drummond at one hundred dollars a month as superintendent. A surveyor at forty dollars a month was hired in the place of the two engineers. For many years Drummond was general superintendent of the canal. He was excellently fitted for the job because he had been superintendent of the construction when it was built. Under the control of the promoters the assessment had been constant and regular. After Miller gained control of the canal, dividends were paid to the stockholders instead of assessments being levied on them. The canal, as it existed in 1881, cost over one million three hundred thousand dollars, which Henry Miller said was "twice as much as it should have cost."

The canal was of immense value to Miller, as well as to many others whose land depended for its life upon this source of irrigation. Without it a vast area from which regular crops of grain and grasses, and later alfalfa and fruits, are produced, would lie, as formerly, a barren waste, so uncertain and unreliable are the rains in this semi-arid portion of the state. By means of the canal, however, through an excellent system of irrigation, every one of the company's customers got all the water he needed, just at the time he required it, with unfailing regularity, at one dollar and twenty five cents per acre of grain land. For alfalfa, of which three, four, or five crops are grown per year, the charge was so much per acre for each crop.

During the lively years of contest between Henry Miller and the canal speculators, a peculiar situation developed. In his contract with the canal company, Miller was entitled to spread over his grassland any water which was wasted by the canal company because its customers did not use it. As settlers were few, and as many of them would not irrigate at night, and as others would not irrigate on Sunday because of religious reasons, water had to be wasted. The result was that the water had to be turned out of the canal and run down the channels to be used to grow grass for Miller without compensation. Years later this resulted in an amusing incident in a case in court regarding waste from this canal. It was contended that the non-use of water on the Sabbath was a necessary and legitimate loss. The opposing attorney, who possessed a peculiar ability to talk through his nose, turned with withering scorn and said, "I can visualize this German squatting on the top of Camp Thirteen waste, seeing this water flowing out on

Sunday to enrich his grasses, fatten his steers and increase his bank account, holding his hands over his expansive stomach, turning his eyes devoutly toward Heaven, and singing: "Day of all the days the best, Emblem of eternal rest."

Henry Miller had unbound faith in irrigation and what it could do to semi-arid and parched lands. Other canals were constructed, primarily to irrigate Miller's own land. The Chowchilla canal derives it water from the right bank of the San Joaquin River, at a point two miles above the mouth of the Fresno Slough and the head of the Great Canal. This canal flows in a generally northerly direction for thirty miles and terminates at the Chowchilla Slough. It was constructed in 1873 by Miller & Lux and William S. Chapman, whose lands it irrigated exclusively. The great Columbia ranch was irrigated by the water from this canal. The cost of construction was approximately one hundred thousand dollars.

The Dos Palos and Temple Slough canals were opened by Henry Miller and drew water from the San Joaquin River on the west side. The channels were originally natural sloughs breaking out from the river and traversing the Santa Rita Rancho, and running parallel to the river. The sloughs were simply deepened and improved for the purpose of affording water for the irrigation of wild grasslands on the rancho. The total cost of these two canals, including the structures and the excavations, was slightly upward of seven thousand dollars.

Sometime in the 1890's it dawned upon Henry Miller that the Great Canal had too much fall, and that if it had been built with less grade it would have covered more land if it had not been built higher. One day he came to his engineer Mr. Frank McCray and said "Mr. McGray (Miller pronounced his c's like g's), I wants a canal out of the main canal for the land above the main canal." McCray stated that a canal above the main canal would run uphill. Miller said, "You builds me a canal," and forthwith marched out of the room. Independent surveys were made, on the basis of a fall of four inches to the mile, but for a while uncertainty prevailed, because there was disagreement among the findings of the surveyors. James Huston, a resident of the district, says that the outside canal was surveyed first by a Mr. Wanzer, but Miller was not satisfied with his survey, and had Mr. McCray resurvey the line of the proposed canal. Miller wanted every acre possible taken in by the new canal. McCray's survey took in thousand of acres more than Wanzer's. Miller's surmise was fully justified. In 1897, during a depression period, when wages were very low and when men were tramping the country looking for work, the outside canal was constructed for a distance of about thirty-five miles. This new or outside canal runs parallel to the Great Canal and from one to two miles above it. The cheapness with which the canal was dug bothered the company in fixing its value for the rate fixing purposes. When Henry Miller saw how cheaply it was dug, he determined to build a new river weir and get the advantage of cheap prices on labor and material.

Thus it was that Henry Miller came into possession of irrigating canals representing a system capable of irrigating 340,000 acres of land. In 1911 it was reported that about one third of this area was being served by Miller's canals, located in Fresno, Merced and Stanislaus counties. With control of the water system, Miller & Lux acquired a public utility. That Henry Miller discharged this public service honestly and fairly is evidenced by the fact that no attempt was ever made to overthrow private control. Dry districts asked for the extension of the system in order to share the benefits of irrigation.

The value of irrigation to the San Joaquin has been clearly recognized. No other element (water) outside of land is of such general and primary necessity or is destined to exert an equal influence on the growth of the state in wealth and population. Before the first irrigation of grain was attempted near Fresno, the land could scarcely be sold for two dollars and fifty cents an acre, but as soon as the results of irrigation became known, land sales increased twenty-five to thirty dollars per acre for the raw land. When in deciduous trees and vines, the same land is worth two hundred and fifty dollars an acre. The citrus land of the foothill, that sells for fifteen hundred dollars or over an acre when in full bearing groves would be valueless without irrigation. The unirrigated fifteen-dollar wheat land becomes the one hundred dollar alfalfa land.

In order to encourage tenants and settler to take up his irrigate lands, Henry Miller offered liberal and advantageous terms. Where the land was prepared for irrigation before the tenant occupied it, the tenant paid one-fifth to one-half of the crop as rent, according to what the owners furnished. If Miller & Lux furnished teams, groceries, feed and seed, they received one-half of the crops as rental. If the tenant furnished everything, one-fifth of the crop delivered to Miller & Lux satisfied the rent. The tenant paid one-half the water bill, receiving the benefit of the special contract made between Miller & Lux and the canal

J. Q. Drummond keeps an eye on workforce building the canal, 1889

company before the canal was built, by which Miller & Lux paid but one dollar and twenty-five cents per acre per annum for all the water required. Where the land was not prepared for irrigation, the owners were furnished lumber at the nearest landing on the river for necessary building and for fencing ten acres of ground to be devoted to alfalfa. The tenant erected his own house and made his ditches and check levees. The tenant paid no rent the first year. Grain growing eventually gave way to alfalfa and one of the finest dairying districts in the state was developed on Henry Miller's cattle ranges on the west side of the San Joaquin River.

It is in the southern reaches of the San Joaquin Valley where Henry Miller played an equally important and successful role. Although he had acquired permanent interests south of the great Tulare Lake before 1871, it was not until after this date that the aggressive Miller and his agent, James Crocker, began to quietly gather up hundreds and thousands of acres on the Kern River. Miller, however, was not alone in the work of reclaiming and acquiring the fertile lands adjacent to the river. J. B. Haggin and W. B. Carr, aided and abetted by the financial support of Lloyd Tevis of San Francisco, proved they were strong, capable and worthy opponents of Miller & Lux. The Haggin-Carr land interests were spread along the south bank of the Kern River, whereas Henry Miller took up the land farther north and west.

The history of the water disputes in Kern County would in itself cover several volumes to relate. This focus is on what particularly concerned Henry Miller, and trying to point out the successes or failures that attended his activities. In order to explain certain events and make others clear and understandable, short accounts of some of the other characters who played important roles will necessarily be given.

While the Haggin-Carr and Livermore-Chester fracas was under way near Bakersfield, the Miller & Lux lands holdings increased on the west side of Kern County. With the dissolution of the older company, Miller & Lux acquired the Kern Valley Water Company, which had built an extensive canal system over one hundred thousand acres on the west edge of the swamp lands. While acquiring these new interests, Miller was also gathering about him able leaders and captains, of whom J. C. Crocker, S. W. Wieble, and Captain John Baker were key men.

It has been already noted how Miller acquired great ranches, from Stanislaus County in the north, clear to Kern and Buena Vista lakes in Kern County in the south. Also noted is his interest and finally, absorption of the largest canal in the state and its development into a great irrigation system. Henry Miller's activities in Kern County were an extension of his operations along the San Joaquin. It is not unlikely that Miller had pleasant visions of a great cattle and sheep ranch extending from one end of the valley to the other. During the fight with Haggin and Carr, Miller was commonly reported as assuring them that he would make them "pack their blankets out of Kern County," and there were not lacking admirers of the doughty and vigorous Miller who fully expected to see him make his threat good.

In the early days of irrigation in Kern County, it was the custom to build wing dams of sand or of sand and brush in times when the Kern River was low, to force the water into the canals. These wing dams would start just below the head of the canal and extend at an angle upward and across the river nearly to the farther bank. A freshet sufficient to raise the water above the top of these dams would speedily melt them away, scattering the brush to form impeding islands in the riverbed, and the work would have to be repeated as soon as the river fell again. The intake from the Kern Valley Canal was moved higher up the river two different times. Still later, the head of the Kern Island Canal was moved still farther up the river, and an artificial canal was substituted for the old natural channel south from some distance.

All these changes were made the excuse for a number of lawsuits over water rights. The questions involved turning on use, priority and the right of riparian owners to have a natural watercourse maintained. The dry year of 1877 gave cause for the first great fight over water rights. The contests between Haggin-Carr and Livermore-Chester were not so fierce as those that came later between Haggin and Carr and Miller & Lux. During these early preliminaries Miller & Lux lingered at the back of the stage. Their lands lay mostly to the north of Buena Vista Lake, twenty miles or more west of Bakersfield, and about the same distance from the center of the contests between Haggin and Livermore. Miller's interest in the disposal of the water of Kern River, however, was quite as great as was that of Haggin, and his position on the river bore the same relation to that of Haggin as the position of Haggin bore to that of Livermore and Chester. When Livermore and Chester put a dam across the river to force the water into the Kern Island Canal, it left dry the canals in which Haggin and Carr had acquired the controlling interest. Later, when Haggin and Carr build the Calloway weir to force the water into the Calloway canal, the result was to dry up Miller's newly planted alfalfa fields, and the tule swamps where his herds gathered rough forage.

Teams excavating the bottom of the canal, 1889

"The sloughs and natural water courses though which remnants of Kern River had meandered leisurely thought the broad, flat trough of the valley to Tulare Lake changed from clear, though limpid and leisurely streams, to green and slimy sinks of stagnate water. Then they became nothing but streaks of mud in which the feet of the weakened cattle were held fast until the vaqueros came to drag the poor beast out with riatas about their horns. A little later all the sloughs and swamps were parched as dry as the naked, gray expanses of alkali desert that bordered them. Where the water had been, great cracks opened in the earth down which a walking stick could be thrust its entire length. Only in deep holes, puddled by the feet of many starving cattle and fouled by the carcasses of dead brutes, was any water left in all the fifty miles of swamp land between Buena Vista and Tulare Lakes."

Of course such a state of affairs could lead but to vigorous defensive action on the part of Henry Miller. Consequently, the famous suit of Miller & Lux versus Haggin was filed and brought to trial on April 15, 1881, before Benjamin Brundage, judge of the superior court of Kern County. This suit not only marks an epoch in the history of Kern County but marks an epoch, also, in the history of irrigation in the State of California. It began with little more notice from the public than any of the other hundred or more suits that had been filed by rival claimants to waters of Kern River. Before it had gone far, people realized that it was the battle royal, and before it was over, every wire was pulled, every politician was approached, and every local election in Kern County was bitterly contested. It ranged practically every California newspaper of general circulation on one side or the other, resulted in calling of two state irrigation conventions and a special session of the legislature, and started a movement to amend the state constitution so that the Supreme Court, which rendered an unpopular decision in connection with the suit, might be reorganized. The latter movement did not succeed.

In brief, the contention of the plaintiffs, Miller & Lux, was that they were the owners of riparian lands along the lower reaches of Kern River, that Kern River was a natural stream flowing in an established and continuous channel throughout their lands, and that under the common law of England they were entitled to have the waters of the river flow over, through and upon their lands undiminished in quantity and unimpaired in quality. The defendants, Haggin and Carr, et al., claimed that they were entitled by right of appropriation to divert the waters from the river for purposes of irrigation, to develop water power, and for domestic and other purposes. It was a contest in short between riparian rights and the right of appropriation. In addition to setting forth the rights of the plaintiffs, the complaint alleged that the defendants, by diverting the water in their canals, had rendered the lands of the plaintiffs dry and barren to such an extent that their cattle had neither grass to eat nor water to drink.

On the morning of April 15, 1881, the trial began with Judge Brundage on the bench and a formidable array of counsel for both parties before the bar. Louis Haggin was in charge of the case for the defendants and was assisted by John Garber and Geo. Flournoy, Sr. Hall McAllister was nominally the chief counsel for Miller & Lux, but R. R. Houghton, then a young attorney, was the active man and really the one who outlined and carried on the campaign.

Several of Henry Miller's men figured prominently in the suit. One of these was S. W. Wieble, who had built a number of the early canals for the Kern River. When Livermore, Redington and others formed the Kern Valley Water Company, to undertake the reclamation of swamp land district number 121, Wieble was placed in charge as engineer. Wieble built the great canal which extended north from Buena Vista Lake for a distance of some twenty-six miles, when first constructed, but which since has been carried to Tulare Lake. The canal follows the western edge of the swamp and overflowed district, and was one hundred twenty-five feet wide on the bottom and calculated to carry a stream seven feet in depth. It was designed to carry all the waters of Kern River that might flow so far, and also was to serve as the feeder for irrigation ditches that would cover 100,000 or more acres of land.

When Miller & Lux acquired the Kern Valley Water Company's interests, Wieble went to the new management. He was later made general superintendent for Miller & Lux in Kern County. Wieble was noted as one of the few men who stood in no awe of Henry Miller when the latter flew into his celebrated fits of passion. It is related that on one occasion when Miller had made the discovery that one of his warehouses had leaked and wet a quantity of wool and was dividing his time between furiously jumping on his hat when he found new evidence of destruction, Wieble followed his employer along the warehouse wall and jumped on the hat while Miller chopped the holes until the ludicrousness of the performance finally appealed to the cattle king and appeased his wrath.

Another of Miller's employees who figured most prominently in the suit was James (Jim) Crocker. Crocker had been in Miller's employ in the San Joaquin and was sent to Kern County to lay the foundation of the Miller occupancy there. Crocker was the sort of man Miller would be expected to choose for the job—a quiet, self-contained man, but a good mixer in spite of his reserve and a man of native force and personality that made him a natural leader. He was bred to get up in the morning at four o'clock and go out on hard jaunts with the vaqueros. Chasing down and breaking up organized bands of horse and cattle thieves appears to have been his favorite pastime. If a field or fellow stockman was in trouble, financially or otherwise, Crocker was ready to go to his bail to the extent of his own possessions. Men rallied to the standard of Crocker because of their friendship and confidence.

Captain John Baker got into the Miller-Haggin fight partly because he was a riparian owner, although his lands were higher upon the river than the intake of any of the irrigation canals, and "partly because, like an old war horse, he could not remain inactive when his nostrils caught the scent of battle." Captain Baker was described as a bluff, outspoken man, a vitriolic writer when his righteous wrath was stirred, and an offhand orator, the sarcasm of whose phrases was dulled only by the sledge hammer methods of their delivery. It was said that Captain Baker would roast his victim alive, pour carbolic acid over his withered remains and end by quoting a few pages of Shakespeare, Byron or Bobby Burns to give a classic flavor to his philippic. He entered no less fervently into his friendships, and between his battles and his benefactions was a leader in his county.

For the opponents, W. B. Carr, the general of the Haggin forces, was described as fat, aggressive, determined, and absolutely unabashed, with bulldog courage and endurance. He was a typical political boss of the large and more perfect type. Frequently and fervently cursed and hated, he could walk into a saloon in a hostile ward and in ten minutes have enough sworn allies to insure the victory of his candidate. If a delegation of angry farmers in the early days of the bitter water troubles came after Carr with the intention of puncturing him with bullets or stringing him up to a high branching cottonwood, he met them with outstretched hand and slaps on their back and sent them away wreathed in smiles of hope and assurance.

Walter James, chief engineer for the Haggin interests, figured in the water disputes, in court and out, mainly as an expert witness. His long and intimate association with everything that had to do with the appropriation and use of Kern River's waters, aided by a retentive memory and a logical, consecutive manner of stating the salient facts, made him invaluable as an authority. Frank McCray, Henry Miller's chief engineer, if not as intimately associated with diversion of the waters of Kern River, was a worthy match for James when it came to presenting accurate surveys of streams, channels, and the general topography of the county.

The witnesses included everybody in Kern County who was supposed to know anything about the history and habits of Kern River, the location of its various courses and the dates when these courses were changed, or anything concerning the appropriation of water from the river. In addition to these, there were expert witnesses who had read in books what happened in Calcutta or about the river Nile in the days of the Pharaohs. This kind of testimony, of course, was duly objected to because they had not been present at the times and places mentioned nor seen with their own eyes the things they pretended to describe. It is said that John Garber carried a potato in his pocket for luck, and that he developed a habit of taking it out and shaking it at the witness when he asked a question of especial moment. R. R. Houghton, on a like occasion, would stand up, reach across the table and dip his pen in the ink as though he intended forthwith to write the answer down in plain black and white so that it could never be denied, altered, or evaded evermore.

Teams shaping the canal, 1890's

The witnesses were equally eccentric and picturesque. An old man by the name of Stevens, who came from the head of the South Fork Valley, made a speech in response to every question that was put to him, and finally, as he was leaving the stand, he swept his long arm out over the big assemblage of pioneers who crowded the space behind the attorneys and remarked, "I'm getting to be an old man, and I don't know if I'll ever see you all here together again; and I want to say to you now, while I've got you all together, that I'm oldest settler in Kern County." Of course, one of the attorneys took and exception to the statement and asked that it be stricken from the records.

In the meantime, if the local papers were not doing much in the way of reporting the trial, they were, however, sparing no effort to prove what the judgment should be. After forty-nine days the last witness was heard on June 2, 1881, and all the testimony, when it was written, made a stack of paper four feet high. For the convenience of the lawyers, the court consented to hear the arguments in San Francisco. The speech-making began on June 20, 1881, and on November 3, 1881, Judge Brundage rendered his decision in favor of Haggin and Carr, which was to the effect that the appropriators were entitled to the water of the river, as against the riparian owners, represented by Miller & Lux. Promptly, Miller & Lux appealed to the Supreme Court.

Somehow Kern River seemed possessed of a certain titanic sense of humor, and no one will be surprised to read that while the Supreme Court took its time in considering the mass of evidence, the gist of which was that neither party to the suit was willing to let the other have any water, the river began to increase its flow, and in the early part of 1884 the two chief parties to the suit were engaged in a fiercer fight than ever to keep the swollen river from flooding their lands, even though it involved the turning of the excess water on the other.

Previous to the suit just reviewed, Haggin and Carr had reclaimed the beds of Kern and Buena Vista lakes and had built the Goose Lake Canal to carry off any excess water that the Calloway and other irrigation canals could not handle. The Goose Lake Canal led off to the north, and on the south side of the river, Haggin had built the Cole Levee farther to prevent the river from breaking over and flooding his reclaimed lake bottoms. By far the greater part of Haggin's reclaimed lands lay to the south of the river, and by far the greater part of Miller & Lux's reclaimed lands lay to the north. Miller & Lux had built levees along the north bank to protect their lands, and had constructed the great Kern Valley Water Company's canal to carry any excess waters to the north of their cultivated fields.

As the snows melted in the mountains and the river rose higher and higher against the levees, it became a most absorbing question as to whether the waters would break on Miller's side or on Haggin's. On May 17, 1884, it broke on Haggin's side and in a short time there was a hole forty feet wide in the Cole Levee, through which a steam of muddy water, nearly twenty feet deep, was rushing to cover all the lands that Haggin had reclaimed with so great expense. In short order a great force of men from the Haggin ranches were rushed to the break with teams, scrapers, shovels and sand bags to repair the gap with bags of sand. The broken ends of the levee were riprapped to prevent further washing and a row of piling was driven across the break.

Early in these proceedings Henry Miller arrived with R. E. Houghton. Having a suit in the Supreme Court in which their contention was that they were entitled to the full flow of the river, over, through and upon their land at all times, Miller was hardly in a position to object to Haggin's men repairing a break in their levee that would then throw the full force of the stream over on Miller. Houghton, however, was fully equal to the emergency. It happened that Miller owned forty acres of land in the bed of the Buena Vista Lake (surrounded by Haggin sections), and Miller set up the claim that he was entitled to the full flow of the river unhindered and unimpaired through and upon his land. Miller strode up to the break in the levee where Walter James was superintending the driving of the piles. "What are you doing here? What are you doing here?" he demanded. "I'm just carrying out my instructions," drawled James. "We thought we'd put a few piles in here, because we may want to build a bridge across, or something." "Well, I don't want you to stop my water. I don't want you to stop my water. Do you understand? I don't want you to stop my water" Miller repeated.

As soon as the train got him back to San Francisco, Houghton went to Judge Hunt of the superior court, and on a petition setting forth that Miller was the owner of a piece of land, to wit, forty acres, etc., and that whereas when the waters of Kern River were allowed to flow over it unhindered, etc., large quantities of tules and other plants and grasses valuable for feed grew thereon, and whereas one Haggin had a force of men at work with piles, a pile drive, brush, etc., endeavoring to restrain the said water from flowing over Miller's said land, etc., and whereas Miller would be greatly damaged etc., etc., and injunction was duly secured. By the time the injunction was served, the ends of the levee were pretty well protected with sand bags, and most of the piling had been driven, but the water was flowing through the break almost as rapidly as ever.

It was no small task to get the Miller injunction set aside for the reason that after he had issued it Judge Hunt had gone on a fishing trip back into the mountains, leaving orders for no one to interfere with any matter in his court during his absence. Louis Haggin, however, prevailed on another judge to set aside Judge Hunt's order. Meanwhile, Walter James got all the Haggin forces together, about five hundred men, and immediately proceeded to fill bags with sand to gap the break. The force was so organized and was of sufficient size to permit day and night shifts. The hoisting engine used in driving piles was rigged to haul wagons loaded with sand along the levee. Heavy cables were laced back and forth among the piles, and the work of building in a wall of sand bags proceeded with system and dispatch.

The Miller forces were not inactive. While Haggin was getting rid of Judge Hunt's injunction in San Francisco, Houghton was getting another injunction out of the superior court in Napa County. This was issued at the request of George Cornwell, who owned a small piece of land on the south side of the river and many thousands on the north side and who made the same representations as Miller had made before Judge Hunt. James and his great crew of men were swarming over the levee, working in a frenzy of haste to place the last sand bags that would stop the torrent of water when the sheriff arrived with the injunction issued in Napa. The sheriff waved the injunction and ordered the work stopped, but everyone was too busy to hear. It was an intense moment, for many months of work, tens of thousands of dollars, and victory or defeat depended upon a few more minutes of time. The sheriff handed the injunction to Carr and explained its purport, but Carr had to read the document, and his glasses were over in the tent. He went to the tent, got his glasses, sat down and read the injunction and the complaint, which accompanied it. All the while S. W. Wieble was demanding haste. When Carr finished studying the order, he desired James to read it. James read it quite slowly and carefully as Carr had done. Wieble stormed over to where Dave Coffee was rushing in the sand bags with redoubled haste and energy, and commanded him to desist in the name of the law. Coffee knew nothing of law or injunctions and he kept right on shoving the sand bags down to the men who were building them, now, just above the surface of the water. Finally Carr sauntered back from the tent, saw that the gap in the levee was closed and the bags of sand rose clear and dry above the surface, he held up his hand as a signal of submission to the court's decree.

One thing had been neglected, however. James had buried logs or "dead men" on the upper side of the levee and had attached to them loops of cable ready to slip over the tops of the piling to help them carry the great weight of the water pressing on the narrow dam. These loops of cable had not been adjusted, and the upper ends of the piling were without support. For a little while the piles and the wall of sand bags stood, and then, as the water lowered on the outer side, they leaned and swayed, the sandbag wall splashed out of sight, the broken piles bobbed to the surface, and the flood leaped through the breach once more to spread over section after section of Haggin's reclaimed swamp land, and "undiminished in quantity and unimpaired in quality, flowed over, through and upon Miller's forty acres of Buena Vista Lake bottom until it covered twelve or fifteen feet in depth." On July fifth, more than a month after the wall of sand bags washed out, the water was still pouring through the Cole Levee upon Haggin's land. Later, Miller's attorney, Houghton, had Carr and James hauled before Judge Hunt on the ground that no court had any authority to set aside Judge Hunt's injunction, and that all the five hundred men were in contempt when rushing in the sand bags. "The defendants are fined one thousand dollars each," said Judge Hunt.

Another victory was coming to the Miller forces. The Supreme Court on October 27, 1884, reached a final decision, by a vote of four to three, in favor of the riparians. Of course, many irrigators in the state sat up and howled. It was not long before Carr had them organized and holding big irrigation meetings, drafting laws for submission to the state legislature. He was also responsible for arousing the members of the legislature to petition the governor to call a special session.

The extra session of the legislature convened in August 1886, and with the din of the statewide battle in their ears, the members of the Assembly passed the irrigation bills as per schedule. The Senate balked. It would not defeat the bills nor would it pass them. On September 11, 1886, the legislature adjourned with the question of water legislation immersed and fathoms deep in "status quo."

There was only one possible result for a death struggle between two such determined fighters as these great companies; made up of aggressive individuals never willing to acknowledge that they had had enough. To prevent the extermination of both, a compromise agreement was achieved on July 28, 1888. This agreement is the present basis for the division of the water of the Kern River. The agreement provides that the water of Kern River is to be measured at two points, the first one about five miles above Bakersfield (sec. 2, T. 29S., R. 28E.), and the other about twenty-three miles farther down the river, (sec. 23, T. 30S., R. 25E.). From March to August, inclusive, if the water at the first point exceeded three hundred cubic feet per second, Henry Miller, et al., were to receive one-third of the excess, measured at the first point, delivered undiminished at the second point.

During the remaining six months of the year, J. B. Haggin, et al., were entitled to all the water above the second point except three hundred cubic feet per second, which belongs to Kern Island. Any water passing the second point from September to April belonged to Miller, et al. The agreement also permitted the use of Buena Vista Lake as a reservoir, defined its boundaries, and prohibited storage in Kern Lake. The reclaimed dry bed of the latter lake is now cultivated.

In this agreement Henry Miller showed his generosity. The law gave him and his associates all of the water of the Kern River, yet he immediately said, "There is more water than we can use, and it does not come at the right time of the year, but comes in a great flood early in the spring, and in the hot months of summer the river is dry."

"You builds me a reservoir, and I gives you two-thirds of the water," he said, and as we have noted, the difficulty was solved.

By being a good general, as well as his astute use of statesmanship, Miller had laid a solid foundation for his great possessions in the southern reaches of the San Joaquin Valley. Miller & Lux canal operations would give impetus to the irrigation district movement in the decade after Miller & Lux took possession of the canal. The presence of this canal caused the failure of California's first serious proposal for public ownership of an irrigation system.

Today the battle rages on between the politicians, farmers, environmentalists and homeowners.

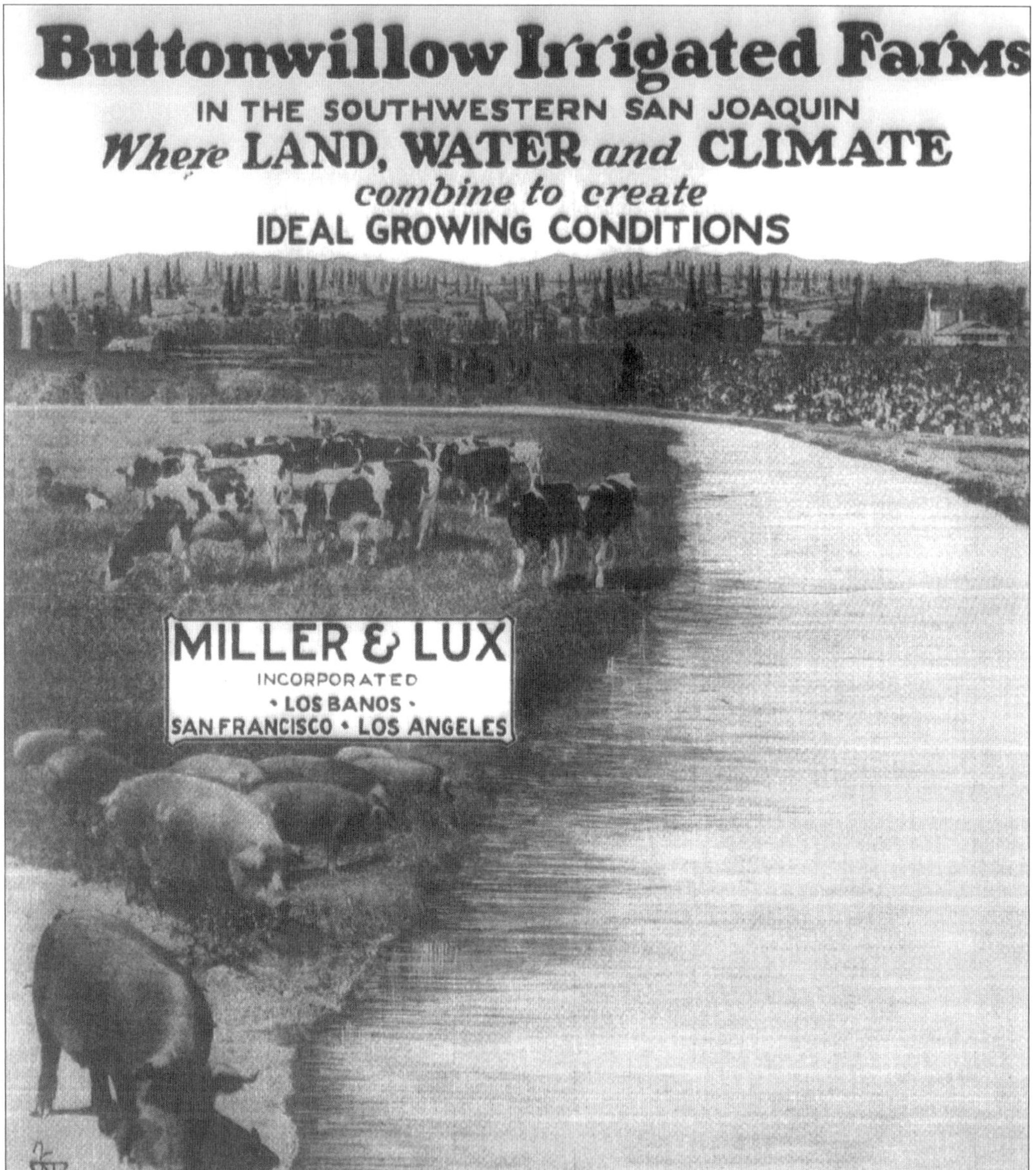

CHAPTER FIVE

One Man Show

Who was Henry Miller the man? Cattle and land baron or robber baron? Penniless immigrant butcher boy who made good? Empire builder? He was all of these things and much more. The firm of Miller & Lux controlled slaughterhouses, banks, stores, and hotels in addition to ranches, dairies and farms. He employed scores of managers, clerks, and foreman, vaqueros by the hundreds. His firm accounted for a full 90 percent of the farm equipment purchased in the San Joaquin Valley. Machinery, farm implements, barbed wire, seed and engines he bought not by the five and ten's but by the railroad car loads. He had the youth's dream of wealth and he achieved all he had dreamed of and more. He had been in the right place at the right time, saw his opportunity and took it with both hands.

Henry Miller was about five feet eight inches tall and exceptionally neat in appearance. He hated display, always wore black clothes cut after rather an old style, a spotless white shirt with string tie, and below it, a small stud rather a clerical look in his appearance not that of a cowman. He was quick as a flash, sure and rapid in his judgments of land, beasts and men; courteous in manner, quick tempered and inclined to be fiery when opposed; altogether a man of caliber who could succeed in any business or profession.

Wallace Smith, the well known historian and writer, knew Miller personally for a long period of time as an employee, described him as "possessing a masterful mind, indicated by a high, broad forehead; an iron will revealed by keen unwavering eyes, and endowed by a tremendous vitality. Nature bestowed upon him the insight to see situations from all sides and left him unburdened with a ballast of too finely spun ethical feelings. Still stocky and powerfully built at seventy years of age, his carriage was erect and he walked with an athletic stride."

Henry Miller was never a snob and he remained democratic and open in his way of acting and thinking. He rode at the head of the master and journeymen butchers' parade through the streets of San Francisco, admired by his fellow craftsmen, admired by the rank and file of the people, and only ridiculed as "the butcher boy" by those aristocrats who were removed one generation from their father's and mother's workbenches or washtubs.

He kept a low profile and let Charles Lux be the face of the firm. Not for him was dealing with the City snobs; he hated dealing with the money men and considered them all crooks of one kind or another, and felt the less he had to do with them the better.

Generous, liberal, charitable in manner, quiet and unassuming, and in spite of the pressure of his vast interests, kindly and accessible. He was a shrewd businessman, honest and upright. In his business dealings Miller was decisive and you knew where you stood. His word was his bond. He expected nothing less from those he dealt with as well.

Although stern in makeup Miller was generous. He left orders with the foreman of every ranch that no man should leave the property hungry. A man who needed meat could kill a sheep from Miller's herd provided he used the meat himself and did not sell it. Miller's keen eye never overlooked so much as an empty barley sack out of place. Any loose end of unfinished business was like a challenge to him. Stories are told of him sending a man a mile to pick up some trifle. Charles W. Smith relates that when he was in Miller's employment before 1878, Miller once sent him pretty much the length of the Santa Rita at night with two dollars for someone to whom it was due.

S. C. Cornett tells how Miller came into the company store at Firebaugh, trailed by several workless men; ordered the storekeeper to give one a shirt, another a pair of pants, another shoes, and so on; and then, when the job was done, asked, "How much is it?" and paid for it out of his pocket.

In speaking of the bad years in the 1890's, W. J. Stockton said, "There wasn't a man in Los Banos or Dos Palos who could have paid his debts to save his life, and there were no exceptions. The only thing left to any of them was a credit account in the Miller and Lux store. For five solid years they didn't even send out store bills. If a man was of any account they helped him out, if he wasn't they helped him anyway. If it hadn't been for Henry Miller the people would have suffered for the necessities of life. I met Miller one day and he said to me, 'I'm an old man and I don't want any of my old friends to be in trouble after I pass away. Now, Mr. Stockton you owe me $9,000. How much can you pay?' Mr. Miller you know my business as well as I know it myself. 'Well,' he said 'Give me two pieces of property that you hold which are mortgaged to me and $1,000 in cash, and I'll call it square.'"

When the Western Meat Company started to exploit San Francisco as they had the eastern cities, the price of beef went down to almost nothing. Miller had at that time forty thousand head of beef cattle. He turned them out on the range, and without the meat company knowing what he was doing he started buying beef from them, and reselling it. At times he took a whole shipload. He had been buying about six months before they realized what he was doing. Then they undertook to raise the price. Miller told them that the people of San Francisco were his friends, and that they should not be robbed. The end result was that beef sold from three to five cents cheaper in San Francisco than in Kansas City, Omaha, Saint Louis, Chicago, or any of the eastern cities.

When the first settlers came into the Badger Flat country, Miller was very generous with beef. Several times during the years of 1869, 1870 and 1871 the camp boss at the San Luis Camp would come to the ranch of some settler with his vaqueros and bring a big fat Miller & Lux steer. Word would be sent to all of the settlers to be at the particular ranch at a certain time. A vaquero would drive up the steer and butcher it and then the meat would be divided free of cost among the settlers. Any settler in the early days could have all the cows he wanted to milk provided he would raise the calves for Miller until they were old enough to wean.

At Christmas employees of the company were given gifts. William T. White, sheriff of Merced County for years, had the responsibility of drawing up the list of names and scouring all pertinent information about each employee. After the list was complete Miller would go over the names with White and comment on the various employees. An employee who had a good, dependable, clean housewife was always regaled with additional gifts. Often it was the good wife who saved the erratic employee when the Christmas list was made.

For Thanksgiving friends of the firm were gifted with at turkey raised on the company's

Santa Rita buildings thought to be in 1880's

ranches. Public officials were not forgotten when the turkey lists were prepared. This, no doubt, was good policy. Miller did not expect better treatment from government officials but he expected equal or similar consideration as that given his less fortunate neighbors. Miller felt the turkey would put the official in the proper frame of mind to give him a square deal.

The poor in the community were not forgotten. One did not have to be an employee of Miller to receive his bounty. A list of the poor and needy was kept and on Christmas Eve necessities of various kinds were sent to these people. Sidney Shannon, a United States District Marshal at Fresno, was on the inside of many of Miller's business affairs. During the bad years Shannon stated that he scratched $350,000 off the books, in mortgages and notes. Miller got his pleasure and recreation in life from handling his vast possessions. With him it was a game and his great interest in life was in playing it. Without attempting the impossible he set out to see what he could accomplish. He did not worry over temporary losses but was always thinking ahead at what the next and future moves would be.

Miller had the vision to see that the county was going to pass into the hands of little farmers. He often spoke of this in his latter days. He didn't exactly like the thought, but he foresaw what was coming and his idea was to have things well in hand and divide his vast domain in an orderly way. He expected to control the transaction.

Miller showed himself a genius in controlling his great, complicated, widely flung industrial-agricultural organization. One of the reasons behind Miller's success was his phenomenal ability to handle detail matters. He never carried even a pocket memorandum, but depended entirely on his marvelous memory, and knew to the exact number his many herds of cattle, even when the numbers had grown to thousands and tens of thousand. On trips from ranch to ranch extending over thirty days he would borrow from one to pay up another keeping no memorandum and no accounts, carrying the transactions in mind, giving accurate account to the bookkeeper upon his return to the City and the monthly statements to foremen were always correct. He had all the details of his business at his fingertips. He built up a rural administrative machine, which was entirely personal, having acquired his vast knowledge of matters agricultural gradually as the properties grew; and his great success being based on practical experience and close observation. He promoted almost exclusively from those men who had started to work for him as cowboys, teamsters and laborers. He seldom made a mistake in his selections and commanded the loyalty of his men, who relied entirely on him.

In judging the man and conditions one should not apply a present-day gauge to those years. In the early development of the West, most great enterprises were dominated by the genius of one, or at the best, of a few men, who had to be ruthless autocrats. The new ideas of public control, participation of workers in profits, if not in the capital of a corporation, let alone in its management, were still unborn.

During the many years that Miller conducted his business he had occasion to write many letters. A compilation of these letters actually constitute Miller's philosophy on how to conduct and manage agriculture, raise stock, carry on a dairy, attend an orchard, combat pests, irrigate land, get on with the neighbors, secure and keep good men, etc., etc. In fact, these letters touch upon every conceivable activity in which Henry Miller was engaged.

It was not uncommon for Miller to arise at four or four-thirty in the morning, according to the season, and before sun-up, write a dozen or more letters giving specific orders and directions as to what was to be done on a particular ranch. He often complained that he wrote many letters but felt that no letters were so little heeded as his own. Miller wrote in a small cramped style. Often grammatical errors appeared, yet every phrase, clause and sentence was loaded with meaning. Miller was direct and outspoken; his letters clearly reflected this characteristic in the man. He went directly to the gist of any matter without wasting words or paper.

Miller often reminded his superintendents of the facts concerning the conduct of the business. His intimate contact with the various properties, coupled with an exceptional memory, made these reminders doubly significant.

The following reminders commonly appeared during the month of January. "This is a good month to get rid of old cows, bulls and stags. All that are available should be in the market before the grass cattle start to come in March. Put plenty of stock on the land at Merced Falls in order to eat the grass until it is partly green. We cannot use the feed after June on account of the hot weather, scarcity of water and hard ground. Cattle get very foot sore there after the ground hardens, and they fall away badly. If the grass is not eaten early, when it dries it is liable to catch fire and cause considerable damage. Cattle will soon commence to work back into the Hills, catch them carefully. Lambing will soon commence, so get all the tents, pens and the like ready before the time arrives"

During February the following warnings and reminders were sent to those in charge. "February, March and April are the best months to build fences because the ground is soft. This is the month to plant fruit trees and vines. If there has been enough rain in the San Joaquin, the ground is in good condition, grain that is planted before the middle of February will do well.

"All those cattle that have not started for the hills should be sent now while there is plenty of feed and water. In front of the adobe, between the sheep headquarters and the last point of the Tulare Lake, there is usually a lot of water standing in pools on the high land, and they generally last until bout May. If cattle are sent now they will get some find feed; otherwise it will be lost, February is a good month to start breaking colts."

The month of March brought the following: "Get rid of all the fall lambs before this month is over, as spring lambs will soon be coming in and they are preferred. Cattle will not do well on cut feed after March. If there has not been much rain up to March, put everything under water that is possible. Take no chances of being scarce of feed later on. It is better to have a surplus than not enough. After the hay is off the grain lands, stock heavily and then irrigate, so as to bring up a volunteer crop. If feed conditions look very bad, save grain to mix with straw and hay for the poorest cattle."

For the month of April the following appeared in Henry Miller's letters. "Ship no more grown bull calves for veal, as time has passed for them. Look in the swamps for stock, as the water is liable to rise any day from now on. When the water gets high in the San Joaquin country, see that the uplands are flooded. Gather up the range cows most profitable for beef and put on pasture where they can get feed. Also all the blemished stock should be brought in the inside and fattened for the market, so as to get rid of them quickly.

"This is the month the superintendents should find out if they are going to have feed for their cattle on the outside. By finding out beforehand, it prevents looking for feed when the cattle have commenced to fall away."

May brought the following: "This is the best month to plant Egyptian corn. Look out for the levees and the high water in the San Joaquin. This is the month that the foxtail bothers the hogs, so examine them from now on.

Haying in the San Joaquin, 1880's

"Be prepared from now on to fight fires. Take out and examine all the rigging used to fight fires. How are the cattle doing? Is the feed coming up to expectations? May and June are the months to get good results irrigating for hay. During the high water in the San Joaquin is the time to catch the wild hogs, that have been driven out of the tules."

June was a busy month and Henry Miller was ever ready with his timely warnings and reminders.

"Break off the young sprouts of the fruit trees

and save cutting them next spring. Never work cattle in the San Joaquin from June to September, unless it is absolutely necessary. As soon as the haying is finished, plan for stocking and feeding of the stubble to the best advantage. See that the first crop of alfalfa is taken off in time. After the middle of this month it is best not to plant Egyptian corn; put in sorghum instead.

"Look out for the young trees and vines, as they need attention before the hot weather. Every effort should be made to irrigate this month, as in July the water is hot and can only be used on swampland. Be careful at this time in renting swamp feed. It is best to see how low the water is going down, because if there is much mud, the feed is absolutely worthless."

During the month of July the following reminders appeared in his letters.

"The latter part of July is the proper time to take cattle from the Cuyama for the stubble and alfalfa and the fore part of next are the months when the sickness is most prevalent in the lowlands of the San Joaquin. During this month and the next cultivate around young trees, especially those planted in the spring. This is the month to cut the firewood. Commencing with the latter part of this month, beef cattle will come from Oregon and Nevada. The California ranches should be prepared to hold back their beef for about two months, or as long as the northern beef keeps coming. Also have feed for the other northern stock."

In September we find the following: "All the bottom lands in the San Joaquin should be stocked heavily from now on before the frosts ruin the feed. During this month and the next have no cows or calves on the Carrisa ranch. Sort out the old cows and such stock as should be gotten rid of from the ranges, put them in pasture and hold them until the fall to fatten. Irrigate for fall feed alfalfa and foxtail fields. Arrange fields for the cattle that have to come out of the ranges during this month and next. Sorghum can be planted as late as the latter part of this month. Keep a close watch on the water flooding low lands, and draw it off before the tules commence to grow."

Bell Station, Pacheco Pass in 1880's, destroyed by fire in 1932

August was a busy month for the superintendents in the San Joaquin. "This month to insure plenty of pasture for the stock coming off the range and stubble. Find out from the ranches, how much the hay, grain and pasture they have. Be prepared to have suitable feed for those cattle that are on grass which will be ruined after the first rains. Now is the time to fix up the feeding stanchions. Clean the corrals and manger and fix the fences. Get the wood hauling done before it is too late."

October was a very important month throughout all the ranches. Henry Miller's instructions were more numerous during this month than at other times. "Irrigate in October and November so that the cattle which do not make any gains on cut feed will have suitable feed. After the middle of this month, cattle intended for cut feed should not be moved around until ready for the corrals. Use all the hay first that is in stacks exposed to the weather. Cattle can now be put in those swamp lands that are dangerous during the hot weather.

"In the last gathering of cattle in Nevada, be sure that all perishable stock is brought to the railroads and shipped to California. The latter part of this month put the colts on dry feed, as generally this time of the year, they are not doing very well. Commence to wean those calves that need it. When the rains start the volunteer, the cattle shrink and they should be taken off. Commence feeding sorghum as early as possible, before the frosts and rain strike it. Irrigate those fields, which are best suited for ewes and lambs. During the latter part of this month, the fall lambing takes place, so remove all stock from the fields where the ewes are. The early part of October is the time to strengthen levees, dams, and fix ditches.

"Commence to fatten the select turkeys for Thanksgiving. This is a good time of the year to buy bulls, as no one wants them. Be prepared to use the water on the high lands at the first rise. By the middle of this month all the cattle that are to go on cut feed and those that are to be turned out on grass, should be collected. There should be enough feed to last until late in the spring."

The November instructions were usually not so numerous. "This is a good month to clean up around the ranches. Haul out the manure; fix the barns, outer buildings and fences. After the stubble is eaten, taken out the veal calves and the beef cattle and wean the calves that are left. Now is the time to vaccinate the calves."

On November tenth the following was sent to those ranches where turkeys were raised. "Select the Thanksgiving turkeys."

December saw the end of the year and the reminders, instructions and notes although less numerous, were still obviously important. On December fourteenth we find the following; "Christmas cattle should be shipped about this time. No man should be allowed on the hay cutting machines with loose sleeves, as they are liable to catch in the knives. Every superintendent feeding cattle should send in a report, as to the number being fed, cows and steers; how many in fields, how many are to be fed later, and how many will not be taken up."

In the conduct of the various activities on his many ranches, Miller had his men as well as the work done thereon, thoroughly organized. Without some well-planned system of farm management the business would have failed. One example will suffice to give an idea of the thorough and complete organization of a ranch. The Santa Rita, which was Miller's pride, had about one hundred and fifty men divided into many groups. There was the haying gang, which kept going for eight months in the year on the irrigated alfalfa fields; the irrigation gang, the ditching and scraping gang, the shovel gang, the box and fence makers, the cattle feeders, the drivers of the big freight teams, and so on. A foreman, who had a sub-foreman for farming, another for cattle, another for the sheep, etc, managed each ranch. All the men worked under their own foreman, who hired and fired them without question. Other groups were the blacksmiths, the men in the harness shop, the saddle shop, the wheelwrights, the dairymen and the cheese makers, who labored under their own bosses, and these in turn reported to the general yard foreman. A general manager managed each group of ranches in turn. Then there was a superintendent of all the sheep interests of the state, etc. As near as possible, each ranch was under the personal supervision of Mr. Miller during his active life.

The vaqueros at headquarters were a group apart and worked under the orders of their own boss. These vaqueros were primarily of California-Mexican stock and a close corporation indeed. Once in a while an outsider was admitted into the circle but only after much deliberation and hesitation. To them was entrusted the working of the cattle, including the two yearly rodeos, the selection of the steers for feeders, the picking out for markets of the "topped off" steers, the watching of the bulls, the branding and marking of the calves, the shipping of the fat stock, the shifting to different pastures, the breaking in of the mule and horse colts, etc. They were likewise the line riders, who followed the barbed wire fences for hundreds of miles and made the necessary repairs; they were the men who watched the rise of the rivers and sloughs and brought in the news to the nearest ranch house, to be passed along down the line. They watched against horse and cattle thieves, against those duck and geese hunters who invaded the lower pastures and sometimes molested the livestock. It was they who had to carry orders from headquarter to camp and ranch foremen, often riding all night in order to deliver the orders before sunrise, so that the foremen in charge could start the hands early. This chore ended with the invention of the telephone. It was they who had to watch against the drifting of the cattle during storms, and had to patrol the range for suspicious looking characters. These men constituted the nerve system of the great ranch enterprise, where shipping of fat livestock from one of the five railroad stations within its boundaries was a daily occurrence.

Miller always kept a close watch on his employees. Not that he pinned them down by any means, for some of his trusted employees were permitted to draw checks on him for large sums in the duties of their business. Nevertheless, Miller, always on top of things, kept a close watch on all his men and often timed his visits when he was least expected. In each rancho use of one room was always reserved for Miller. No one slept in the room and it was kept in readiness for the cattleman. Miller often went to his room upon arriving at one of the ranches. If the room was spotlessly clean, well and good for the foreman's wife, but if the room was not in order he made trouble, big trouble.

Miller had the following to say to his bosses. "Each superintendent should know the exact locations of every section corner under his jurisdiction. The superintendent and all the lesser employees of the firm should go out of their way to maintain friendly relations with the neighbors. As friendly neighbors are a valuable asset, it is best when they are under some obligation. Never charge the neighbors for meals, but when ever any of the employees of the firm eat outside, they should insist on paying for the meal."

Miller was greatly concerned about fire, always a danger, and its resulting destruction. There were two subjects that came in for considerable attention: fires and squirrels. Regarding fires, Miller said, "Always be prepared to fight fires, as the best work is done before the fire makes any headway. At the commencement of each summer all the fire apparatus should be inspected to see if it is in the proper working condition. There should be a wagon especially rigged for firework, and it should be located in some handy place where it can be gotten out quickly. On this wagon should be barrels of water, rakes, long handled spades, blankets and coal sacks—not grain sacks are they are too light. Also have a barrel of drinking water and a tin cup attached, because men fighting fire need large quantities of water to drink. Whipping out a fire with wet sacks and long handled spades has proven itself a much more effective way to fight fire than plowing furrows or mowing down grass, unless the grass is very short. If the growth is rank the flames will jump the furrow easily."

Fires were often started on the firm's property by the carelessness of outsiders with their campfires. Miller would repeatedly tell his men, "At places where people are likely to camp, sheep should be held while the grass is green until they eat it up clean. This will remove some of the dangers of fire later on. Graze off closely the roads and lanes where people are in the habit of traveling. In putting up haystacks and barns it is always best not to place them too close together, so that when one catches on fire it can not spread to the others. In the barns and out-buildings there should be axes and buckets."

Miller observed the squirrel and was familiar with his ability as a pest. Miller is reputed to have said that he knew the San Joaquin when squirrels were unknown there. Once in coming over the Pacheco Pass road he observed a squirrel below the hills and in the valley. Upon arriving at the Los Banos division he instructed a man to go back and "get that squirrel." Later when the squirrels had become very destructive he said, "That man didn't do his job, he didn't get that squirrel." Miller undoubtedly believed in the policy of a squirrel in time saves killing nine. His instruction to the superintendents clearly indicates that he understood how to deal with this rodent. Miller would become incensed when he talked about squirrels. "Squirrels damage the firm thousands of dollars every year. The superintendents of the various ranches should insist that the neighbors kill their squirrels. Very little good comes of killing them on one ranch while the land just across the fence is a breeding ground for them to migrate from. All neighbors who do not comply with the new laws concerning the extermination of squirrels, should be reported and the officials will that the law is obeyed.

"The best season of the year to kill squirrels is during the winter and late fall when the ground is wet and feed scarce. Strychnine is the most expensive way of killing squirrels, and should be used with care. For poisoning always select a clear sunshine day when the animals are out. One of the most effective methods of handling poison for squirrels, is to have the riders attach a boot leg containing poisoned wheat to their saddles, and every time they see a squirrel around a hole, drop three or four grains into it. Water is by far the cheapest and most effectual method of killing squirrels, as it kills old and young, and ruins their store houses, so that they starve if not drowned. In flooding with water, the best way is to have two trained dogs that will catch and kill each squirrel as it comes out of the hole. Often the squirrels are only stunned when they come to the surface, and will revive if not killed."

In the conduct of a large business many men must be hired to carry out the necessary duties of that undertaking. Henry Miller, of necessity, had many men in his employment. He appreciated good men and did all in his power to help them better themselves. "One way of keeping good men is by having their sleeping quarters clean. They do better work and take more interest in everything, if the rooms are clean and comfortable. When possible have a room set apart where the men can talk or read at night. Such a room needs a lamp and a stove."

Miller understood cooks and their responsibility for running the kitchen. "When a new cook is hired his first duty should be to take down everything from the shelves and dust. In this way the kitchen will not only be clean, but the cook will be familiar with the provisions and cooking utensils at hand. It will also prevent ordering materials, of which there is a large supply on the shelves." He appreciated the benefits of vegetables in a workingman's diet. "The ranches should raise more vegetables. They have plenty of water and suitable land. Good vegetables make the men more contented and draw a better class of laborer." Miller also knew he had to keep peace with his cooks. He would sometime let his temper get the better of himself and rant at one of the cooks. However, he never left without making it up to the cook and doing whatever it took to keep the cook happy.

Regarding farming equipment and utensils Miller had the following to say; "Always keep the farming implements under cover when not in use, because if they are protected from the sun and rain, they will last twice as long. Before putting away any machinery that has been out in the damp, wipe it dry or otherwise it will rust. Every year the warehouses should be cleaned out. Dry lime should be put on the floors to kill the weevils, the bats should be smoked out, and if there are not cats, traps should be set for the mice and rats."

The Miller and Lux empire was not built and developed in a year or two. Rather slowly and gradually, year-by-year and decade-by-decade, the properties were developed and expanded. Only a man of vision and unique business acumen could have built such and enterprise and operated it successfully, through good, bad and indifferent years.

The following incident brings out, in bold relief, how Miller handled his men. Charles Warfield, then a

Miller & Lux teams hauling grain to Salt Slough warehouse, 1890's

young man, came to one of the Miller and Lux ranches in the San Joaquin Valley and asked for a job. Warfield was employed. At the time he knew little about stock raising. One day, during the spring rodeo, Warfield was given the job of keeping the fires of the branding irons burning. Miller was applying the irons. To an experienced hand it was easy enough to determine when an iron is sufficiently heated. Warfield, due to his inexperience, handed one of the irons to Miller, but it proved to be too cold. Miller thereupon threw the iron down and exploded in a barrage of curses that were not particularly complimentary to Warfield. Warfield became aroused, picked up the iron and struck Miller down with it. Warfield then left to his bunkhouse, signifying that he was quitting. Miller soon followed him to the bunkhouse and asked him what he was doing. Warfield replied that he was getting ready to leave. Miller came up to Warfield and calmly told him that he didn't want him to go, that he liked him, and ended by saying, "A man who will defend his rights will also defend mine as an employee." Warfield remained, and for many years was general superintendent in the San Joaquin.

Years afterwards Warfield decided to engage in other activities and notified Miller of his intentions. Miller pleaded with him, offered to pay him any salary he would name. Warfield had made up his mind and Miller's offers and pleading were to no avail. Shortly afterwards, Charles Warfield became sheriff of Merced County. Miller never forgave Warfield for leaving his employment.

The Miller temper sometimes was aroused and he often gave it free rein. He was at a cattle sale near Gilroy one day and was busy counting a herd of oxen he had just bought, when a rather impecunious looking person approached and, with respectful familiarity, said "Well Mr. Miller, it's a pleasure to meet you after all these years." The cattle were filing by in twos and threes. To reply meant to lose count. Miller glanced at the speaker and said nothing. There was a moment's silence. The speaker, imagining he had not been understood repeated his salutation in a louder tone. A mixed-up knot of oxen streamed by. Miller nodded but said not a word. His lips moved in the effort to keep tally. Two minutes of silence. Then the impecunious looking individual in a grieved tone said, "You use to answer me quick enough when you were padding small wares thirty years ago."

"And if I had been a damm fool like you, I'd be peddling them yet," was the answer, delivered in quick tones that fitted the sentence into a gap in the passing herd.

Miller refused to have anything to do with tramps and travelers. In the late sixties or early seventies about a dozen tramps came to one of the Miller & Lux ranches and demanded something to eat. They were told to get out; Miller & Lux could not feed everybody. The next morning the plains were on fire and Miller & Lux's haystacks were burning in every direction. Miller knew the cause and laid down a definite policy from this time on regarding the treatment of hobos and travelers; they were to be fed and then urged to move on. Because of this humane policy there was very little trouble from carelessly or maliciously set fires, gates left open or any similar ranch troubles until the time that Nickel took over.

What was picturesquely known as "the dirty plate route" was long an established institution on all the ranches. Men who "packed their blankets" from one ranch house to another were always sure of something to eat, at the second table. There were some whose orbits hardly extended beyond the company's ranches in the San Joaquin Valley. At one time, on the Poso Ranch, a record was kept of all the free meals handed out to transients, to be listed as given to the boss. During a February in the 1890's, the shortest month of the year, 1,028 free meals were given to "the boss." A six-month record for the same time frame showed that two free meals were handed out daily to two hundred and fifty-three individuals during that period. Not only were the tramps never turned away from the Miller ranches with empty stomachs, but also no money was to be taken by the foreman for food. Many a wealthy traveler who stopped at the Miller ranches for food was fed, but his offer to pay was always refused.

Miller hated waste. He could not stand to see anything that possessed any possible future use thrown away. This characteristic in Miller has often been misinterpreted as tight-fistedness. Actually, he was liberal and gave away much without hesitation and without expecting reward of any kind. This lifelong habit may have had its start from the time he left home with next to nothing. Miller's policy of "save everything and waste nothing" permeated his organization. Miller employees grasped the fundamental idea and soon realized its effectiveness in their own private affairs. It was not uncommon for Miller, upon arriving at a ranch-house, to immediately go to the trash can and dig into it to see if the cook was throwing away any meat on the bones or peeling the potatoes too thick, thus wasting the meat.

Economy was Miller's watchword. He practiced it in everything he did. Over this issue, Henry Miller and his superintendents occasionally got into difficulties. "When a man can save a dollar it learns him to save on a large scale," said Miller. "I would rather have a dollar saved than five dollars made" was another of his truly philosophical statements.

Henry Miller was almost continually on the move in the years of his health and activity. Although he did most of his work in the days before the automobile, he was one of the first to own a car. He came to dislike the machine, however, because of the rough roads he was generally compelled to travel. In rather short order he discarded the automobile, and again took to his favorite buggy or buckboard. Shannon from the Los Banos ranch was to meet Miller at Gilroy. The agreement was that Shannon was to come for him with a buckboard. Morning duties delayed Shannon's departure from Los Banos so that it was practically impossible to be at the Gilroy Station on time to meet Miller, if the horses and wagon were used. Miller's daughter, who was at Los Banos at the time with her big car, proposed that Shannon use the auto in order to meet her father on time. When Miller alighted from the train he was escorted to the car, while Shannon explained that his daughter was responsible for the change in the mode of transportation. There being no alternative, Miller got into the waiting car, and in very short time they arrived at the ranch. Shannon asked Miller how he liked the ride. "Umph! Like it, why I didn't see a damm thing," replied Miller.

The advantages of cheap freight transportation appealed to Henry Miller. As late as the 1890's Miller made use of the Salt Slough Warehouse to receive freight from the river crafts. The boats would always come up the river during the June rise. After unloading at Salt Slough the boat would proceed to Firebaugh. It required about half a day to unload the cargo. At Firebaugh the barge was loaded with wood and then returned down the San Joaquin River. The principal freight was salt and groceries. It usually required two or three eight mule teams an equal number of weeks to haul the freight from Salt Slough Warehouse to Los Banos and Canal Farm. The teams would come down and load on one day and the next day they would return loaded to Los Banos. The boats continued after the railroad came because of the much cheaper water transportation. Later too many dams in the river and the great quantities of water diverted from the river for irrigation purposes resulted in the boats ceasing to run to Firebaugh.

Miller would usually have a crew of about twenty or twenty-five men working down the river every summer fixing up openings in the levees and cutting wood. During this period of the year enough wood was cut to last the ranches a whole year. The wood would then be hauled to Canal Farm (about a mile east of Los Banos) and there split into stove lengths and taken to the other ranches. Sometimes four-foot lengths would be hauled direct from the river to the larger ranches and split there. Miller did not believe in paying fancy prices for fuel when he could make use of the timber growing along the river.

At one time an unusual situation developed with Miller's woodpiles as the center of attraction. It was discovered that some of the wood was mysteriously disappearing during the night. Miller took steps immediately to prevent further losses. Very carefully, holes were bored in a select number of sticks of wood. These holes were then filled with black powder and stopped up so as to look perfectly harmless. The treated sticks were then placed haphazardly on the pile. The next day there was a great commotion. One thief had the misfortune to have the entire top of his stove blown off and his breakfast blown all over the room. Eggs were found smeared over portions of the ceiling and walls of his kitchen. Another also had his stove explode. From that time wood was perfectly safe in piles outside.

Henry Miller taking his ease in a tree. Mt. Madonna, early 1900's

In conducting his extensive business enterprises, Miller had a large number of employees. It was the custom, during his active years, to occasionally hold councils with his foremen. At such times matters of business were discussed and Miller would listen to their plans and ideas. If he was convinced that it was a good move he allowed the foremen to proceed in their own way, but if he differed from them in any way he would tell them clearly and distinctly why he disagreed and would instruct them to proceed in his way. Miller was an aristocrat in the management of his men. In speaking to them he always spoke in a commanding tone of voice. If an employee was told to do a certain thing and failed in a single instance to carry out the instruction given, he was discharged at once. On the other hand, if by carrying out the instructions given by Henry Miller it should come to pass that the firm had lost money, he would commend the employee, telling him that he had done just right in obeying orders. A notable instance of this kind occurred on the Las Animas Rancho.

Once Miller had intended to write from his San Francisco office to his foreman at Gilroy to not gather up certain cattle, but in writing the letter the word "not" was omitted and the letter read "gather up certain cattle by a certain date." The foreman could not understand it, but, knowing the positive instructions that he had, put a large force of men to work on the ranch, in order to be ready by the date mentioned and to have the cattle on hand. Miller arrived on time, in a perfect passion of anger at finding that the cattle were being chased all over the ranch and corralled, contrary to his instructions. He met the foreman in a vociferous manner. The foreman claimed that he had carried out the instruction of the letter and that the letter said "to gather up the cattle." Miller said that the letter stated not to gather up the cattle, but upon being shown the letter he saw that the foreman was right, he had omitted the word "not." He at once turned to the foreman and told him that he had done just right in carrying out the instruction of the letter, and, while it would a serious loss in that the cattle had been run and lost much flesh and some had been killed, he wanted him to understand that at all times when he received a letter from him, the instructions contained in the letter should be carried out. Miller, while autocratic, was at all times responsible for his words and deeds.

In September of 1884, Miller was singled out as the victim of a kidnapping conspiracy. It appears that a man named William Ritchie, a man named Davis, his two sons, Thompson and David, and an employee of the Miller ranch near Sargent's Station put up a job to waylay Miller on the Pacheco Pass road. They planned to compel him to make out his check for $30,000 payable in San Francisco. After the check was cashed, word was to be sent to the confederates who were to release Miller and effect their escape. Thompson Davis lost his nerve and notified the authorities, who immediately put Ritchie under watch and communicated their discovery to Miller.

In the meantime the informer, for some reason or other, swore out a warrant against Ritchie and he was arrested at Gilroy. It was suggested to Miller that in order to entrap the other conspirators he should go through the Pass and allow himself to be captured. The officers assured him of their protection and that the check should never be cashed. Miller, however, did not propose to take the chance of losing his life should the desperadoes discover that their plot was detected. He went attended by a couple of well armed men, and at the appointed place where the attempt was to be made the party did actually meet the suspected men, who made no attempt whatever to impede their progress, doubtless owing to the apparent strength of the travelers.

In the absence of any evidence to corroborate the informer's testimony, no further arrests were made. On Ritchie's person documents were found which showed him to have been connected with a gang of Eastern bandits. At the trial the defense moved to dismiss the action on the ground that no attempt had been made to rob anybody, it having been conspired without the parties acting.

At another time Miller was stopped on the Pacheco Pass by a gang of bandits who proposed to put him to death on the grounds that he had notified the officers of their cattle rustling. Miller denied that he had notified anyone but admitted that one of his men had reported to the authorities certain facts. He also said that he had not given them permission to drive off his cattle but that they might appropriate them only for their own use. Some of the bandits were for putting a riata about Miller's neck but one of the leaders recognized the logic of Miller's statement and he was allowed to go on his way unmolested.

Once Miller was held up by a lone bandit and robbed of a considerable sum of money. Miller told the bandit he had a long way to go and would need some money before he got home. The desperado returned twenty dollars to Miller. Some time later, while Miller was in a saloon, he recognized the voice of the self-same bandit who had robbed him sometime before. Immediately he walked over to the bandit and said, "Here is the twenty dollars I borrowed from you some time ago." The man was nonplussed at first, but seeing that Miller was sincere, accepted the money and very quietly and quickly made his exit from the saloon. This tale has been told in many versions since. You will find many of them in later chapters.

Henry Miller was first married in 1858. Thirteen months later his first wife died. On June 10, 1860, he married his first wife's niece, Miss Sarah Wilmot Sheldon, after a whirlwind courtship of one day. Proposing to Henry Miller was a matter of fact as any regular business deal. After proposing to the lady of his choice he gave her an opportunity to think it over while he went out to look at some cattle. Upon returning from his business deal, he was accepted, and on the very same day the marriage took place. Three children were born into the Miller family, two girls and one boy. The youngest daughter, Sara Alice, was killed when she was thrown from her horse while out for a ride with her sister. The son, Henry Miller, Jr., died when a comparatively young man. Drinking, recklessness and irresponsible living sent him to his grave prematurely. Only his daughter Nellie, later Mrs. J. Leroy Nickel, survived him.

On October 14, 1916, after nearly four years of enforced inactivity because of a serious case of gout, Henry Miller passed away at his daughter's home in San Francisco. He was buried in Woodland Cemetery, where his wife and two children had preceded him.

The influx of populations, the shift to intensive agriculture, the rise in land values with its attendant rise in taxes, and the changing policies of a different type of management, made it impossible to hold intact and conduct profitably the far-flung properties that Miller had built up. Eventually, in order to pay taxes and meet other obligations the properties were mortgaged to the extent of twenty-five million dollars. Consequently, certain portions of the Miller & Lux lands were thrown on the market for sale. By 1923 regular land offices and agents were established throughout the state, with headquarters in San Francisco, in order to interest people in the Miller lands. Thousands of acres have been sold, and it was reported that in Kern County alone, where in 1916 Miller had nearly two hundred thousand acres, considerably over half this acreage had been sold. The depression retarded the sales activities by agents but the work of disposing of the remaining lands continued.

Dr. McClelland stated, "Henry Miller was born a ruler, he led, and pointed the way to others. He also made sure he got there first." When Miller died it was printed, "He has endowed no colleges but he has given millions as he went along without expecting any pledge or remembrance, or making any condition of publicity. Unthinking persons will perhaps censure him for his great wealth, but the fair-minded will think of the fact that in accumulating his wealth he developed and took advantage of opportunity, but did not crush and destroy other men along the way. When all is said and done, his was a life of intense usefulness, and his contribution to the future of California was large."

As I stated in the introduction, Ralph Milliken, David Lawrence and others worked on Henry Miller's story. As they worked back and forth, many changes were made within the work. I have edited it down to make it readable. After sixty years, it would be impossible to give one or the other total credit for the work. This is the combined work of both men.

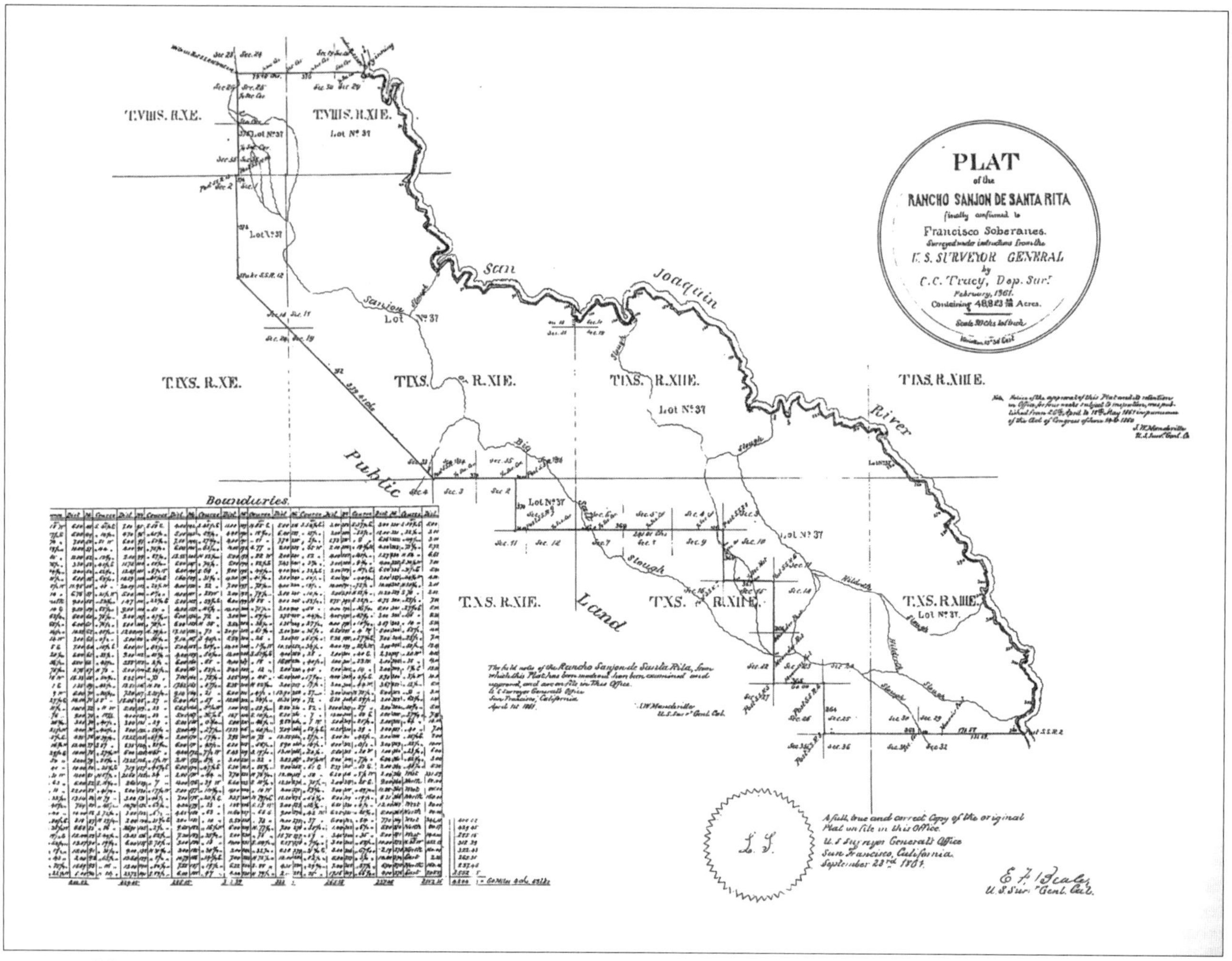

Map used to prove ownership of the Santa Rita Ranch before the Land Commission in 1861. Lux's clever use of such a map changed the size of the Buri Buri to increase his landholdings at the expense of its land grant owner.

Chapter Six

Bancroft Interview

Interviewer: Mr. Fowler of The History Company

The now famous Bancroft Library at the University of California started with Bancroft and his aid de camp, Mr. Stone. Bancroft endeavored to write a history of California and employed numerous writers to interview figures of interest, collecting news clippings, visiting sites and in general doing the research needed to complete his epic work. Mr. Fowler was assigned to do Henry Miller. Miller gave him short shrift by allowing only two interviews. It is somewhat difficult to follow at first because Fowler did not in the first dictation put down the question he had asked to elicit a response from Miller. This is not the exact transcription of the interview. The editor has made some corrections in the attempt to make it somewhat clearer to the reader.

(First dictation undated)

"I was born in the city of Brackenheim, Kingdom of Wurtemberg, July 21, 1828. My father's name was Christian Kreiser. My mother's name was Frederica. My grandmothers came from the stock call Buchwald and my grandfather on my mother's side was named John Fisher. My father was a butcher and cattle trafficker. My people on my mother's side were in the wine business, those were the two principal businesses there at the time.

"The cattle trade in America and the cattle trade in Germany have no connection at all. When I was a little boy, I would go around the town and hunt up calves three weeks old and bring them in for my father's shop. There was a law passed that calves under three weeks old could not be slaughtered and I would buy them three weeks old. I would buy one sheep or one goat and drive it to my place of business and then of course, it was cut up and retailed. There was no wholesale business there then. The industry was carried on a small scale then. I left there in 1842, but I went back in 1875 and the business had changed so—there was nothing hardly similar to what it was when I left. When I was at home, a man had to go through a certain state of performance to be admitted as a journeyman. From being the apprentice he had to perform a certain duty to show himself competent to become a journeyman, then he had to work seven years as a journeyman before he could be admitted as a boss or master butcher. Then he had to perform a duty more advanced to show himself competent to assume the mastership of the establishment. Then he was only allowed to be master, provided he could pass a certain examination and come up to the high grade, he was allowed an apprentice's low-grade if he could not pass the high grade.

"In order to be master in his profession, he had to be master of his business. The Jews were only allowed to live in certain towns, but now they are all over my country everywhere, and the business I was in has passed into the hands of the Jews mostly.

Henry Miller taken in the 1880's about the time of interview

"My father was a smaller man than I and his ideas and mine did not correspond at all. My idea was to leave the country as soon as I was able to make my own living. I had the nicest spot of Germany for home. Stuttgart is called the upper country and where I was raised is the lower country and there is no wasteland there—everybody knows how to make his own living and the people are all well-to-do people. I had very little intercourse with the Germans, for the simple reason that in that country people mostly live among themselves in their own town, but I have come in contact since with them. The men that came from my town has been more than ordinarily successful.

"In our country there is a gulf, an unsurpassable gulf between the Catholics and Protestants. I was born a Protestant. There is no such thing as their association together except in the ordinary everyday business. The people that come from my country are not very sociable. In that country they are very thickly settled, every mile or so there is a town. Every town runs to itself—has distinct ideas and the language is different—different dialects. A great many from our country are exiles from France and have towns of their own, but they become identified with the Germans.

"My folks were natives of that part of the country as far back as I can remember. The country where I was raised was depopulated in 1740 and 1750 by wars and pestilence. The country was almost put back in its original unpopulated state. The houses are large in that country, but rarely is it the case that only one family lives in a house. One family owns one story and another the other and so on, that is the rule in that county. One part of the house is owned by one family and the other part has no connection at all. I left there when I was only 13 years old. My father was well to do. The butchers and bakers and hotelkeepers and millers are mostly all well-to-do people. There was no man ever more tested than I was in that country. I can picture it out where I spent the early time in my young days. The people and I could never assimilate. What was pleasure for them was simply an annoyance for me and it was not a pleasant country for me to live in. My sympathies or attachments were never for the people but for the soil I was raised on. My course has been away from them from a very early time.

"Well, I saw people coming back from America that hadn't been there long that had made lots of money and I would think, there is a man been over to that country and if he can go there and make money so quickly, that's the country for me to go to. I thought if anyone could do that well, I surely could do proportionately well. I wanted more freedom and greater opportunities, home had no attractions for me. I had very little acquaintance of children of my age. So my mind was fixed to leave. My ambition was to have something to do that I could spend all my time in my business.

"No butcher there put in all of his time attending to his business and my great aim was to get into a place where my whole time was occupied with that business. The butcher business is different there now; it is carried on a large scale, it is more of a monopoly now. They do the same in regard to going around and picking up calves and stock to kill for meat now as they did then. They don't bring any from other countries, they raise more food generally than they can consume. Sheep are fattened for the French market. Each butcher selects out and feeds his stock and then sells them to the French market.

"The schools are compulsory there. A boy has to go to school from the time he is five or six until he is fourteen, then if he wants to go into the higher school, he can do so. Though the majority of them never aspire to higher schooling. They have the ordinary branches the same as they do here, no classic studies, languages or anything of that kind. Their public schools correspond to our common school. At the age of six I started to school and went till I was confirmed in April, I would be 14 in July.

"I don't think the surroundings of my boyhood had anything to do with my success in afterlife. I think it was my determination and pride, I wanted to be an expert in my business and in what I learned to do. There the butcher business comprises everything even down to making a fiddle string. He had to know how to clean guts of sheep and turn them into fiddle strings; he had to know how to cure meat and to make sausage and everything connected in that trade. Here one man kills the sheep and another stays in the shop and does the shop work, but there a man has to know how to do all those things. He has to learn the business in every branch, otherwise, he is not a butcher. It takes a man a long time to learn that.

"I was always trying to learn from men, which were successful in life seeking the best master, where I could imitate him to advantage. The manner of educating there can be very little improved upon. There is not any religious training in the public schools. They teach them to sing. The school children all turn out at a funeral and sing at the services. They would get paid for it according to the man's wealth they sang for.

"In every branch they learned as much as they needed in ordinary life. I had no particular study that I excelled in. My whole aim was to get out of school, I wasn't very fond of school. It was more of a compulsion than a pleasure for me. I was considered a very fair scholar, especially in figures.

"They have a great many questions in their educational training, for instance, the school teacher will go around and ask questions and even the preacher would stand up in the church and ask questions on religion, and in all these cases I was amongst the first to have an answer. Yes, I had a good memory, and thank the Lord I have it yet. No, I didn't have any friends, that is close friendship.

"A guardian or father contracts you out for some many years when he wants to apprentice you, say, some three or four years and he has got to pay so much for your support and they teach you; then, after that, you are supposed to earn your own living and get your wages. Some apprentices are more competent than journeymen but do not receive wages.

"When a man is a journeyman, he goes free and is allowed to shift for himself. Sometimes on account of sickness or something of that kind, he falls back on the old folks. I have made my living ever since I was 14 and I think I made it ever since I was 8 years old. My services rendered were simply worth my board and clothing. I had my face and ears frozen one time when I was about 10 years old. That shows what hardship I had to endure and exposure.

"I worked as a journeyman until I commenced here in this city in 1851. I worked at my trade. I thought I had been home long enough and I wanted to learn something of the world. The first wages I got I think was $10 a year, ten Prussian dollars. That included board but nothing else. I did my own washing and furnished my own clothes and did my own mending. The second year, I don't think I got much more—I got a trifle more of course—a man advances a little as he works along. The highest wages a man gets in Germany is about $2.50 of our money. The expense of living at that time was very small but a man don't figure that in, as he is always boarded. Clothing was made there of homespun cloth, no cashmere.

"I was a butcher in New York, but first I worked in Holland two years. I left home and went to Holland first and worked there three years. Yes I left home and went to Holland 3 years and then was in England 2 years after I was in Holland. I left home and went to Holland because my main aim was to work myself to another and new country. I thought I could work my way along and always have enough money, always to be independent. I left Holland to go to England for the same reason that I left home to go to Holland and from England, I came to New York.

"The first that turned my attention this way, was I saw men from here and they seemed to be successful, from the way they dressed and they seemed to have means, and I thought he was not so much smarter than I and if they could do so well, I, had confidence that I could do something.

"The next thing I did, I engaged with a man by the name of Halstead, in the pork business with two stalls in Washington market. My business was dressing hogs and working in the stall in the afternoon working 10 hours of the 24 regularly. I got $8 a month for that. I continued there until the 13th of April 1850. The wages I got were the offal from the hogs—the guts, heart, liver and lights. We had the privilege of selling them, which amount to almost a dollar a day; that was quite a premium on good work. Yes, I practiced economy from the first.

View towards bay from Kearny Street, 1850. Notice the forest of ships' masts at the top of the photo. Most were abandoned – *Courtesy Bancroft Library*

"I came to America in 1847. I was sixty days on the way. I was sick about fifty of them. I was entirely alone, no friends nor nothing with me. I cut loose from everything. I came with the intention of never going back there to make it my home. Yes, it was a sailing vessel that I came on, I don't remember the name.

"Landing in New York, I went to work in a garden so as not to be idle. I worked there three months, as in the summer season, the butchering business is very limited and then in the fall and winter there are a great many more men needed and so as to not be idle, I hired out to work in a garden. I think the wages I got in New York, was $4 a month and board included. Yes, I spoke English then. I was then 19 years old.

"The way I came to think of California was, I saw a man that used to come into the slaughterhouse in New York and buy meat of us. He was gone a year and a half and came back from California with money, and from the appearance of the man I judged he had made it by work. I thought where that man could go away a short time and come back so successful, I thought that was the country for me to go to. I landed here the 24th of September 1850. I left on the 13th of April.

"I came across the Isthmus of Panama. I came down to Panama by the steamer Georgia and then crossed over on mules and took the steamer Republic up to San Francisco. My ticket from New York to San Francisco cost me $350. When I got to Panama, there were some four or five thousand there waiting for the boat to come up and those that lived there, and the Panama Fever came on. Out of about five thousand, I was one of about 40 or 50, which lived through it.

"While I was in a private hospital in Panama, a gentleman whom you probably know, by the name of Zeile."

(It seems in getting the correct spelling of this name he was thrown off and didn't finish the incident—Fowler's notation)

"There must have been 1,500 persons on the steamer Republic. When I was in the Panama Hospital from May until August, I didn't have a visitor to see me, except a gentleman now living in San Jose who wanted to buy my through ticket. That was the sympathy I received from the people that came down in the Georgia. I had a steerage ticket. It took us 24 days to get here from Panama. About six dollars was all the money I had.

"I had a gold watch and some five or six dollars in coin. There were no associations formed on the trip that have been continued since. I saw a man last Saturday night who came over in the same steamer from Panama to this city. When we got to Acapulco, I went ashore and took my worldly goods and went out four or five miles in the country and washed them out and dried them and packed them back and went aboard the ship. We laid there two days on the steamer.

"If I hadn't got the fever when I was in Panama, I would have stayed there, because the opportunities were very great to make money. We had a great trip across, the Isthmus. Some of the men were robbed and left without anything. In order to get across, you had to hire a man to pack your things across—some men hired a mule. I hired, with others, a mule to pack our things across the Isthmus. We had a long sack with clothes in, at night the natives took their mules and went back and some of them took a good deal of the stuff belonging to the other men back with them. Of course you could not get any returns from them at all. I took the precaution to watch the mule we had, set up and watched him all night.

"I landed here the 24th of September 1850. I think it was at the wharf down between Commercial Street and the street north of it where we landed. The steamer anchored out in the stream and we got out into small boats and came ashore. I got here about 10 o'clock in the morning. San Francisco then consisted of about four or five streets. There was Commercial Street, Sacramento Street, and a few short streets what we call Clark's Point.

"I hoisted up calves on the corner of Clay and Battery. The population was very scattered and lived in small tents and the cheapest kind of shanties. I heard so much from here and had met people on the Chargres River and they looked as though they had come from a very wild unsettled country.

"After I landed I went and put my worldly goods in the Hillman House which was kept at the corner of Sacramento and Dupont streets, and then I went out to look for work. I went to various shops in town but I didn't know a single man. I went to a man by the name of Joe Bicken and he advised me to go down on Jackson Street to a French butcher shop which was kept by a man by the name of Scheurlante. He wanted a man to butcher sheep and I engaged that very hour and rode out on a cart to the slaughter house at the head of Dupont street and down near the bay.

"There I worked 2 months. I think I got $2 a day. I cooked my own victuals. I only stopped at the Hillman house about half an hour. I lived during the two months out at the slaughterhouse. When I wanted a square meal, I would come down to a

Chinese restaurant and get a potato, a steak, and a cup of coffee and put up a dollar for it.

"When I use to go down to save the guts from the sheep, the hide and the fat used to be thrown away and so I would get them too and sell them. I remember as I would go down there, I used to go past the gambling houses any number of them, but I never put down a sixpense. I stayed there something like ten minutes and then would go home. That was the year before the big fire. Everything was prosperous then. The business was scarce then because it was limited to a few men and they generally didn't have enough to do to need to employ help in the slaughter business. I had some five or six dollars when I landed. It cost me something on the steamer coming up from Panama. I was sick and weak and had to buy some little things, I suppose I spent a few dollars on the boat. I had something like five dollars when I got here. I don't think I ever spent that five dollars. I had that to commence on and I kept adding to it, so I was never broke.

"I worked for Salandy and Monrozhay, my object being to get acquainted around town. I would work in the shop and get acquainted with the people and in that way, carried on my aim to get into business myself. I worked for a man on Dupont Street for one month. I gave good satisfaction wherever I went. When the first two months were up, I had saved everything but what I had to spend for a square meal and I bought a pair of boots and a pair of socks and a woolen shirt. Then I worked for two brothers Adam and Fred Menches on the corner of Dupont and Jackson. That was at the expiration of the three months. I got something like $60.00 a month.

"Then an incident occurred. Mr. John Center, who kept the Jackson restaurant (a most elegant restaurant in San Francisco) down on Washington Street on the alley running in from Portsmouth Square. I butchered hogs for Center. He engaged me from (Menches Brothers ed.) to go and butcher hogs for him. They were hogs that had been brought from the Sandwich Islands—hogs were then .50 a pound. Their feed gave out and they wanted to butcher and sell them.

Lux's brother went to butcher for them, but he could not kill more than five or six a day and at that rate it would take him too long. He made inquiries about town for a man to butcher for him and someone told him there was a man over to Menches who could butcher all the hogs that he wanted killed. He brought in a mule and made an engagement with Menches for so much a head and I had to ride that mule. That was the first mule I have rode in my life. We had to go over the sand hills about all the way, but we got there about noon and I went to work and killed 30 that afternoon. The next day I killed some 60. I killed them all in five or six hours.

"Center offered to put me into business, I suppose because he thought I was an industrious man—I guess I surprised him by killing so many hogs and he offered to build me a shop at the corner of Mission and Dolores, but I declined to accept the offer. I declined because I wasn't acquainted enough and hadn't means enough to run my own business and I didn't want to take other men's money to commence on and so I refused and thanked Mr. Center kindly, but thought I could not accept it.

"That was the only benefaction that I ever had offered me, which of course I didn't accept. Then I worked for Menches until the June fire of 1851. By that time Mr. Menches thought he had money enough, and offered to give me the shop, if I stayed with him a certain time, and he would quit business and sell me the shop for what it was worth and let me carry on the business. Well, that fire came and burned him out, but I stayed until the new shop was built, then I didn't see any chance of any change being made and I thought I would branch out for myself.

"I went and rented a lot on Jackson Street between Dupont and California and built a house and commenced for myself. I think I commenced on the fifth day of July 1851, because on the 4th of July I send a young man over to the slaughterhouse to get some guts and they didn't have any and we had quite a time about it. The amount of capital I had after paying Mr. Pfam for building a house for me, butcher shop, and one story. I own that same lot today. I think I had a capital of $140 or $150 after paying him. I had no furniture at all hardly, just a block and a few hooks, everything was very crude then, not very attractive at first.

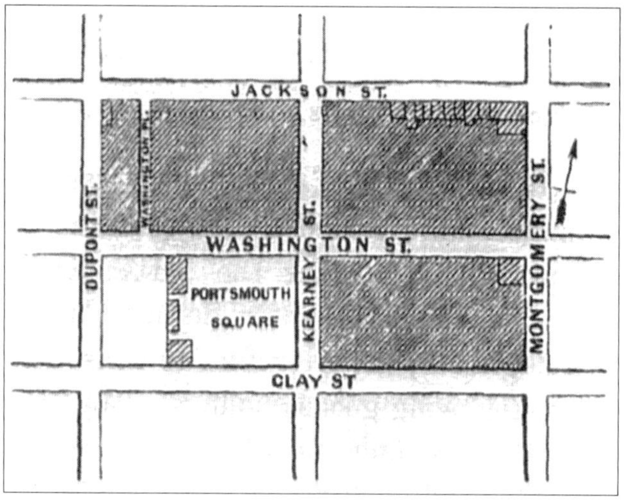

Miller's starting point, on the corner of Jackson and Dupont. Dupont is known today as Grant Ave. – *Annals of San Francisco*

"I opened the shop and went out to William Wood, butcher for Morris Smith; they were connected in the business some way, and had promised to see me about some stock to butcher. They had a iron house on California, you remember about those iron houses they were very few in number. Well, I killed some calves, which Menches and I had bought together, and some small stock that came in on sailboats. I was in the habit of going down to the wharf and buy what I wanted when they would bring them in these boats. When I was working for Menches, he saw I was handy I guess, and said whenever I started in business for myself, he would assist me. The way he assisted me was, he sent his collector around before I sold a pound of meat.

"I bought fourteen calves and when you bought beef you, had to take one good side and a poor one. You couldn't get all good beef. The calves I had to catch in the corral and tie them down and then get drays and haul them down to the shop. I did the first butchering right in the shop. I took those calves and packed them around to butchers on my back and so I commenced doing a wholesale business, on a small scale, right away. I packed some over to Clarks Point and some over to North Beach. It took about all the money I had for these beeves. I was then principally buying small stock off the boats and I would go around and buy a hog here and one there – I tell you they were scarce then and the only way you could get them was by going to these people that lived out of town and buying them one and two at a time all around, but I had pork and the other butchers didn't.

"I went to the slaughterhouse and got bulls heads brains which were a great rareity them times. All the cooking was done by French cooks then and they are great on brains you know I could not sell them meat without I had some brains and kidney to give them. And in that way I held their custom right along and in that little by little I kept growing. The first cattle I bought, I got of John Wilson, he used to be in the circus business. He brought the first elephants from Henderson that brought them here.

"I bought the remnants of the cattle out of the slaughterhouse and used some in my shop and sold the remainder out to the butchers. This business was built up on a very small scale, step by step. I was alone for two years, working my way, then I went and saw a man by the name of Zimmerman. I had then a shop and a slaughterhouse where I now own, open on the corner of 5th and Howard. The reason I took in anybody for a partner, was, I thought I could use a man around the shop while I was gone, attending to other business. I furnished the capital and did the work outside and he attended to things in the shop.

"I went over on the San Joaquin and bought a band of America cattle from Livingston & Kindman and I paid for that band of cattle $33,000 and paid down cash for them. That was more money than any other butcher had in town. These cattle were brought from Salt Lake. The partnership formed with Zimmerman was this way: I furnished the capital and he did the work around the office and looked after the business, for which he received a certain percentage, he put his labor against my capital.

"There were three hundred head of this band of cattle and they cost $110 a head, making $33,000. I brought them over here and butchered them and wholesaled them out. That was the first band of American cattle, to any extent, butchered in this city. They were prize beef. Over in the San Joaquin Valley there were wild horses and elk and you could see them running in every direction. I sold those cattle out for 18 and 20 cents a pound wholesale. The average carcass weighted over 800 pounds.

"At the expiration of my association with Zimmerman, I took another young man in with me, Westir by name. He is living yet. He put in his labor against my capital and got a certain percentage of the profits. We continued in partnership something like a year and then I quit having a man with me for a certain percentage and I got a man with me by the name of Lawler, Joseph Lawler.

"The second big lot of cattle we got we bought of the Spaniards in the summer and drove them eight miles to San Francisco. They came from the Southern part of California, Monterey, Los Angeles and San Diego all through that country. I just got them as I needed one, two and three hundred at a time. We kept them out here on the Guadeloupe Valley. We had them herded during the day and put in a corral at night. They varied in price sometimes they were high and sometimes they were cheap. I took advantage of the number in the market and when cattle were scarce, I got a good price for them.

"A gentleman by the name of Erskins brought a large number of Texas cattle up here to this country, something like 2,000 head and turned them loose on the Pacheco Pass. After the cattle became marketable, he engaged Lux to butcher them on what we call commission. Erskins asked a very large price for the cattle and he thought he could get more to have them butchered.

"I had very little acquaintance with Lux at that time, hardly a speaking acquaintance. Finally Erskins got tired of the slow way of disposing of the cattle and his business interests called him back to Texas and so he concluded to sell the cattle and he sold the whole band out and left the State. He came to me and proposed to sell the cattle to me. Lux heard of it, I suppose he was in the confidence of Erskins, and so Lux came to me and said to me, I am told Erskins wants to sell you his cattle and if you buy them I wish you would give me half of them. I consented, but it was because there was a great many of them and half was enough for me.

"We agreed to buy the cattle and Lux agreed to take half of them; I never lost a day in business from the time I bought those 14 calves up to the present time. We received the said cattle and held them in joint ownership feeding them on the stubble and other feed until the market was suitable and then put the cattle on the market. There was something like 1,000 head of them at $67.50 a head. Mr. Lux paid for his half and I paid for mine. We kept account of the cattle and when we sold them we divided the proceeds. That was in 1857.

"In the meantime I became acquainted with my first wife, who was a sister of Mrs. Lux. Her given name was Nancy Wilmont Sheldon. In the spring of 1858, Mr. Lux disposed of his business and went east with the determination to quit business.

He thought he had means enough and retired. I kept right along in business the same as before. There was no connection between us until Lux returned along in September 1858. He returned from the East and was in San Jose attending to some business and he wrote to me that he would like to go into business with me again and proposed that we buy cattle together and hold them for an advance.

"I wrote him to come up and look over the situation for himself and if he saw an opening, I would be willing. He came upon the 5th day of October and on the 6th of October I bought a large number of cattle from Thomas Dias of Milpitas nearly 2,000 head, and we drove them in.

"I rented a larger amount of feed and we kept those cattle for the winter market. I had my own stock of cattle outside of the firm of Miller and Lux. The buying of those cattle was the beginning of a very genial partnership. After those cattle came into the market I closed out my own cattle and what I had left I put into the partnership. There is no similarity between Lux and myself—not in sympathy nor in view nor nothing. He was a native of Germany and his stock is more German than mine. He comes from Alsace. Prior to going into partnership with Lux, I only owned town property.

"In the meantime I bought property in Gilroy on private account. Our capital was always separate. The property, from the beginning, was never mixed. This city property is mixed, to a certain extent, of course. The first partnership purchase was two leagues 8,830 acres from William Dumphy and Thomas Hildreth. That was called the Santa Rita Ranch. Before I had bought the Bloomfield Farm, that was my private property, something like 1700 acres.

"Lux owned the San Brun? (Buri-Buri) Ranch. (The Buri-Buri was actually in San Mateo but it appears Fowler at the time thought it was in San Bruno. Ed.) But long after that, Mr. Lux purchased property there. My idea of buying that Bloomfield farm was for a stopping place for the bands of cattle from the south. It was called the center place by stockmen. It was well-watered, nice country, nice grass, and a nice place to camp—stockmen have peculiar faculty for finding nice camping places—and that induced me to buy it. I thought it was a nice place and in the future would be valuable. I bought it for a stopping place for the cattle, not for my own use.

"We had a conversation one time about merging all our property into a partnership property and the only difference was a lot of brood mares that Lux thought I ought to pay for separate and I thought I hadn't and we couldn't agree, and so dropped it and have never mentioned it since. That was in 1860. I have since bought other interests to the amount of 13,000 odd acres – that 1700 grew up to that much more.

"No, I never had any intention of going into the raising of grains. Agriculture didn't amount to much in this state. I intended to use it for grazing purposes. I used it for, a stopping place for my cattle when I wanted to hold them for better market. We only owned cattle as we wanted to butcher them—sometimes as many as one thousand or two thousand—we never entered into breeding cattle. The purchase price of the Santa Rita Ranch was $10,000.

"I brought the cattle and horses and everything together and the whole purchase money amounted to $10,000. The cattle were $5 a head. That was the second land purchase. We bought another addition to the same ranch until it consisted of eleven leagues. We bought the two leagues and then we kept adding adjoining leagues—the first two were the extreme southern leagues and then the next two leagues joining and kept on until we had the eleven leagues. We made the purchases from time to time. I bought a property on the 20th of February 1863, and received it on the 3rd of May 1863.

"The whole ranch (48,400 acres) cost on average, $4 an acre. The Santa Rita Ranch was used for cattle—and in the meantime I have bought a good many cattle down in San Louis Obispo County—mostly beef cattle. I got so many cattle that I had to have some place to put them and I have paid James Dunn $100 an acre for land to keep cattle on. Sometimes I would buy a large number of cattle and sell the most of them, but have some 50 or 60 left and of course I would have to have some place to put them. When I bought cattle I had to buy the whole bunch and had to have some place to put what I couldn't dispose of. I bought a good many refused cattle. I thought if other people could make money out of these refused cattle I could.

"Always where I did the work, I wanted to make the money. When I saw this Santa Rita Ranch, I thought that is just the place I am looking for and didn't think I could find any better place than that. The grass is green there 9 months out of the year and a great many natural advantages and handy to all parts of the state—nice level country—and I thought that was the most favorable place that I had struck yet. It took no small amount of engineering though to get that property together, because if I had failed in getting some portion of it, I could not have had it together so nicely. I brought property in Monterey, then belonging to George W. Crane (he married a widow lady that inherited two Spanish Grants) and that we bought and then we bought what we call now Tequesquite and the Lomas Muertas. That consisted of 24,000 acres that joins on the south of the Gilroy property.

"That was purchased in 1865 and 1866. I had rented the property and paid more rent for it than the property cost me when I bought it. Yes, business changed somewhat there was a time when cattle wasn't being brought up from the lower country and in the fall they were scarce and if you did not have cattle on hand you had to pay pretty good prices for them. That property averages about $5 a acre, bought at that time. I only owned 1700 acres, but I kept on adding to it till I now have over 13,000 acres – private property."

Interviewer Fowler does asks a question at this point:

Wasn't 1864 very disastrous to you?

"Yes. In 1863 we purchased 7,500 cattle of Dumphy and before I received those cattle I had it arranged that I was to move over to that ranch to give me room. 1864 was very disastrous on account of the drought, because we didn't know how to take advantage of it. We hadn't any other resources to resort to and the consequence was we lost 2/3rds of the cattle. We could not drive them off for everybody's cattle were out there, and in moving my cattle I would have to move all the others. They

Canal Farm buckaroos, 1895

would run from one of the tidewater down by Stockton up to what we call Tule Plains, and it was a universally mixed up affair.

"There was a depression in everything. The land was cheaper in 1864 than it had been since and almost everybody wanted to be out of business. Lux wanted to go out of business, but I told him we had found it there and we had lost it and I told him we were not in any condition to sell out. I said we would wait until we had made some more money and then we will talk. From that time on, the people settled in around the country and utterly wiped out the free range and then the question arose what should we do, should we keep less cattle or buy more land. So we commenced to buy a little land and then a little more and then a few more cattle and we have kept that up until the present day. I bought 8,000 head from Abel Sterns. I paid $8 to $10 a head. The steers and the cattle I bought for $5 a head were all mixed cattle. Yes, the cattle business was successful in a measure. Cattle advanced from $10 to $20 a head and up to $40 and $50, and then they commenced to go down again.

"In 1871 we bought a lot of land in Kern County. We have now, in Oregon and Nevada and California, something like 750,000 acres. It is hard to tell how much we do own in Oregon, we are buying land there now. We probably own something like 25,000 or 30,000 acres—located in Malheur and Grant Counties. In one sense of the word we are alone in the cattle business; in Nevada, we are entirely alone and in Oregon, we have what we call a partnership connection with Mr. Mason. It is now formed into a joint stock company, which will be known as the Pacific Livestock Company.

"In Nevada, we must have something like 25,000 or 30,000 acres. The land is heavily stocked with cattle—so is the Oregon land. We control ten times the land we actually own. You could not buy land in Oregon as much as you needed, without going to a greater expense than the business would warrant. I do not know whether I am the largest cattle owner in the world or not—I never pay any attention to that. We know what cattle we have. I think take it all in all, we have a strong hundred thousand cattle. Yes, we are largely interested in sheep also. We have in this State and scattered around over the country, something like 75,000 or 85,000 head of sheep. We raise them for wool and partly for mutton. We are buying young wethers and letting them grow up and we buy ewes to have lambs and then we fatten the ewes—it is constantly a putting in and taking out process. We raise for the market exclusively. We have a few ewes that we use for breeding purposes but we generally kill them when they get fat and raise the lambs up. We buy a large number of wethers and keep them for the winter market. We have today between 18,000 and 20,000 wethers, holding them for the winter market.

"We raise hogs the same as we do cattle and sheep. All the ranches have hogs on them. It is a stock business right through, cattle, sheep, hogs, horses, poultry and dairying. We have been improving our cattle for years back. They are now highly improved shorthorn cattle. We have a few Devon's which we have exclusively for to run on the hills, because they are more suitable for the mountain range.

"We regard the San Joaquin property as the most expensive, and the best improved. Gilroy seems to be the headquarters, the most central place and the most suitable to communicate with all the ranches, and then another place on the San Joaquin we have which we call the home property, that is the Santa Rita. That is where the cattle interest is conducted. The superintendent lives here. Our business amounts to something like a million and a half annually.

"We raise draft horses principally, we have raised a few driving horses but not very many. We have quite a large number of horses. Of course we sell a great many though we raise them principally for our own use. Of late our business has increased so that we don't now raise enough horses for our own use.

"The San Joaquin Canal began six miles south of the Firebaugh City and extends through our own land principally and is 78 miles long. The canal is 68 feet wide at the top and 45 at the bottom. It is constructed in that condition for 38 miles and then drops down to 52 feet wide at the top and 35 feet wide at the bottom. We have a side canal which runs 28 miles which increases the capacity of the canal for a certain length. For nearly forty miles the fall is a foot to the mile and then it drops down to six inches.

"Captain Simmons and John D. Bensley were the originators of the canal. They made a financial failure of it. We were interested from the beginning. We were stockholders in 1871, but had no voice. We paid assessments to the amount of $120,000 before the canal was completed and the water run in. Then in 1876 we owned a controlling interest in the canal. The canal was not a success at all until we got a hold of it. The work begun in 1871 and completed in 1873. The expense over and above earnings was about $1,500,000.

"We have miles and miles—hundreds of miles of cross ditches running out of that. Some of the land that this waters is sandy loam, some it is black, some is alkali bog. The slope of the land from the ditch down to the river is from nothing to 13 or 14 feet to the mile. Our present plan of managing the canal combines economy with skill and speed. Labor being so high and irrigation so expensive that we had to adapt a plan which is the most speedy and cheapest, but if it comes to scientific irrigation, our plan is entirely out of full reason – wasteful and extravagant.

"We have a superintendent take charge of every district, which comprises from 5,000 to 10,000 acres where water can be used to advantage. We are perfecting it year by year. The canal would irrigate by proper management, by cutting the land into small tracts 100,000 acres. (This is in error, the tracts were 100 acres each. Ed.) A small fortune was expended by asking the government for a subsidy and the State for a subsidy and floating bonds on the English market. Something like $50,000 was spent for that purpose. That was before we controlled it.

"Since we got a hold of it, it does not pay very good dividends to the stockholders but it pays to the landowners. Some years it is just a burden on our hands but when a dry year comes along, we make out of it. The charges are from $1.25 to $2.50 an acre—depending upon what it is used for. The rule is to charge more for grain than for watering grasses. The land east of the Canal is only suitable for irrigation, because the west side is above the grade.

"In fact, there is more good land west of the canal than east of it. We have had some lawsuits with the company, but they broke a certain valid contract. The stockholders—John B. Bensley, Mora Morse, Isaac Friedlander, Mr. Ralston and William S. Chapman, tried to break our contract for the purpose of forcing us to buy the stock. We agreed to buy them out at a reasonable cost, or at a cost, which I could construct, a canal for, but we were not to pay for any wild schemes that we had no voice in. We were fighting for our rights and another thing, providence stood in with us in a great many instances. That is very valuable property.

"We have a wool-pulling establishment. That is out on 18th Street. That institution is carried on by Mr. Davis (F. A.) we furnish the capital and Davis does the work for a certain percentage. The property involves a capital of $250,000. We sell our sheepskins to them just the same as any other butcher shop, but we get the profits and of course he gets a certain percentage for managing the business.

"I have been engaged in farming to a great extent, raising grain, cutting our own hay and improving land. We have around the canal, something like 150,000 acres of land. That is more land than the canal will furnish water for. Some of the land we own is worth from $100 to $200 dollars an acre. The enterprise I have been engaged in are the wool pulling establishment, the canal improving land, taking wild land and putting it under a systematic cultivation, raising our own cattle, sheep, hogs, and horses and making the most out of them. We are fatting between six and eight thousand cattle by artificial means. Making steer weigh a third more than he would weigh otherwise by feeding him hay and grain.

"We do it to help our business, sometimes it is profitable and then again we lose money at it. We have to sell out grain fatted beef for what the butcher can buy other beef for, or else we can't get them to take it. I devote my whole time and thought to the business, studying how I can advance it and help it in a short time and how I can make those men do all possible, economizing and getting work out of the men and making a dollar go as far as possible.

"We have in our employ between six and seven hundred men in California alone and in Oregon we have nearly a hundred men. We are digging that compromise canal now, some 180 miles to Lake Tulare. There is a barrier between our canal and the lake a 14-foot cut and we are doing that now. I have got over 100 men to work at that.

"There is only one man out of 100 gets my confidence and then only partly. There are certain things, which I take men in on. We have men that we talk to freely in a social manner and in our business, but ordinarily sociability has not room to get in—it is business and nothing else. I superintend all the business myself.

"I was in Nevada and remained a month last year. Mr. Lux knows very little about the business. This land has largely increased since I bought it. It is probably worth on an average in California $20 a acre. The amount of land is limited and the population is unlimited—our resources are unlimited. The people of California know very little of the resources of their state. They expect too much and too little for it.

"Booms are only beneficial in one sense, they have developed a certain portion of the state and the people that have went in will be compelled to stay and improve their land in order to make it pay. There is no city in the Union has a better prospect than this City, it may be delayed but it will come sooner or later. Our state is very sparsely settled. There is 2/3 of the farming land that there is only about 1/10 of the 2/3 under cultivation. Some of the best land is just as nature made it.

"The cattle business is not profitable now. We are now on the decrease in prices have been so for the last two years and will be for a year or so to come, but from then on it will be profitable. Agriculture has not been profitable as a general thing, because grain has been worth less than it can be produced for. There have been years that you could not sell fruit but of late years, it has brought a better price in the market. They raise a better quality of fruit than they used to.

"I think, good land, there is no telling how high it will go. There is a time coming when people will find it out. I think there is no doubt but that California will be the most densely populated of any portion of the United States. I see nothing to hinder it. A man on ten acres can make a good living.

"I haven't taken any interest in politics only in a general way. I am a Republican. The first president I voted for was Lincoln. No, I haven't had no change of sentiment in that regard. I voted for Lincoln when it was worth a man's life to vote for him, that was in Gilroy.

"In regard to capital and labor, it looks sometimes as though there would be trouble, but I have got too much faith in the good sense of the good people who will take such precaution to keep that down. Even take the most radical and let them be successful and they will turn over with the good people and help make the others obey. If I had the say, I would do away with general suffrage and would compel a man to be worth a certain amount of property before he is entitled to vote. You let a man own a piece of property and he attends right to his interests.

"In regard to free trade and tariff, my idea is this; If it was kept to a certain limit and that limit could not be over-stepped, there would be no danger of a rise from it, but we do not know what might be the result and to guard against that, we better take the sure thing and let well enough alone. This country has been prosperous and people can afford to pay good prices for labor and they ought to be willing to do it; therefore, I am in favor of a protective tariff. As regards to emigration, the Chinese are a necessary evil. If all of our people would work, there would be no Chinamen among us, there would be no occasion for treaties and agreements. If every men that scorn the idea that would have the Chinamen abolished could get them done away with, then it would come to the test, who would work. It use to be when I wanted a white cook, I would send and get one, but now I don't get any. I suppose we have 50 cooks and there are 49 Chinamen, down at Gilroy."

In regard to benevolent organizations

(Fowler's notation)

"I belonged to one, but then resigned. When I see an institution is worthy of assistance, I have never refused them. But when I think it is for the good of only a few isolated persons, I refuse it. I never refuse a man who I know he is worthy of it. I may be giving him $5 to save him from jail or from doing some crime. While I might miss it 19 times out of twenty, having helped that 20th man would repay me for the missing of it on the 19.

H. G. Tanner, long time Miller head superintendent, 1890's

"I educated fifteen nieces and nephews and I have also helped educate a great many of my employees' children. I live on the corner of Eddy and Harrison streets. It is a place owned by a noted man before. I purchased it in 1875. I was married in 1858. My wife died in 1860. I was married 13 months to my first wife. Then I married her niece. My present wife's name was Sarah Wilmot Sheldon. I was married the second time, June 10, 1860, in San Francisco. I have 2 children. My daughter's name is Nellie Sara Miller, at present, the wife of Leroy Nickel. (His name was J. LeRoy Nickel. Ed.) My son's name is J. Leroy." (This is incorrect; his son's name was Henry Miller, Jr. Ed.)

In regard to disposition of children to take up the work were he leaves off—Fowler's notation

"There is no disposition that I know of circumstance and time will develop that (he is referring to his children's disposition). My son is 24 years old and my daughter is 22. They were educated here on the coast. I am 5 feet 8 inches tall. Square build, weigh 182 pounds, brown hair, light complexion, light brown eyes, I am quick, curt, incisive and forcible.

"People want to know why I take more to the business of raising cattle and quit slaughtering, because our country out here is too limited, you can't get any market. Mr. Armour, in Chicago, has a market all over the world. We are limited to our own consumption. There is no chance for me to better my condition now. I cannot let loose. Our business has increased since Mr. Lux died to almost double."

This concluded the first interview Mr. Fowler held with Miller. Fowler sadly did not, as I stated at the start of this interview, include all the questions he had asked Miller, and his comments appear somewhat disjointed. Fowler felt that he needed more information from Miller and after several letters Miller gave in and gave him an additional interview. Ed.

(*Second dictation Aug. 1st-no year*)

"I don't know in regard to that—I hardly know how to answer it. I never needed any urging—I don't recollect of ever being scolded for not working enough. (His natural qualities were sufficient of an incentive to work, without being compelled to work in those days. Fowler)

"Yes. I believe a practical education is better than a college education—what I have seen of it generally. I believe there is just so much in a man and if he used himself up doing one thing he cannot do anything else. I think that there is a good many men that have been over educated. Some men would more successful in one lien, if they only applied themselves to it; but they go over a good deal of ground and become jack of all trades and master of none.

"The education of these days is of a new style, I believe it tends to educate a man in a great many instances, to get along without hard work. I have heard boys say, 'I am not going to work hard for a living,' and I have seen mothers work hard to support their daughters in idleness, I think it is not right. I think children should expect to work for their living the same as their parents did. It seems to be the tendency to begin where their fathers left off and not to make their own start. There are a great many exceptions to this rule of course, but that is the rule.

"I left there with the expectation never to return to live. I went back for my health in 1875. I stayed at that time seven months. I was the youngest in the family, I had three sisters. I had to do work that was humiliating to me. I had to work in the stable and yoke up the cows and haul feed and I didn't like that. Home had no charms, because my work was on my business. I was born and brought up a butcher. My treatment in childhood by my parents was very severe, too exacting. I felt that there was no obstacle so great but what I could not overcome it.

"It was indomitable will and determination of

purpose that was the chief characteristic of my boyhood. It is more, the exacting will and disposition of my father that required me to do these things that was distasteful to me, my whole early life in fact. I never sowed any wild oats to speak of. (Bancroft) I was a staid, sober, hardworking, looking to the future kind of boy. My aim in life was to become master in my profession and then become independent."

Q. You had definite ideas even then about making money? Fowler

"My idea then and what is now, have changed to a very great extent. My idea was this: To accumulate a certain amount of wealth and then travel all over the world—without any fixed place. That desire comes on me once in a while now and I would like to go and see the world yet. That is the only aim I have now in life, that is the only bright spot I look to."

Q. Your surroundings at that time didn't influence you to any great extent then? Fowler

"My surroundings didn't influence me to any great extent. I had no adviser, I had no man to go to for advice. I never called upon any man for assistance and never accepted assistance. I never was without a day's work when I wanted it. At the age of fourteen, I went to work as a journeyman worker I never was apprenticed—I assumed the work of a journeyman without first being an apprentice for being trained up in that business, I was competent to assume the work of a journeyman at 14."

Q. Don't you think men today have principally come right from the workshops and from the hard places of work, instead of from homes of plenty and ease? Fowler

"Yes, mostly they have mostly educated themselves in business by imitating men who have been successful and who have worked their own fortunes out. I think that a man should try to improve himself in manner and knowledge. I think a man can always improve on his general life if he tries."

Q. What has been the key that has unlocked the door to fortune for you? Fowler

"The key to unlock my door of fortune was I had the power sufficient to leave off what I considered wrong and that power would direct me in what was to my interest. I don't think any man has seen darker days than I have. I have overcome difficulties with my determination and self-reliance by thinking it impossible to go back that I must go ahead.

"I think when a man takes care of his matters that it will bring him luck. I think everybody has trying times but if he looks far enough ahead he can overcome them. I believe a man's resources are unlimited a man don't know what he can do until he is put to the test. Lots of men never know their ability, they are never brought to the test to see what they can do. That is where the great secret lies. I believe as a man increases in business he should increase in capacity and determination one success adds to another and one acquirement requires another and so on till a man is forced to grow in his business and no matter how far up the ladder he gets, he can still do better one success begets another. Sometimes a man's business outgrows or grows faster than his ability to care for it and that is what causes a great many failures, the business gets too big.

"That should not be the case, a man should grow and learn more as his business grows, and if a man is determined and applies himself, he can do wonders. There are a great many men fail—just for want of force and resolution and self-reliance confidence in himself. A man that looks to others never succeeds—he must have self-reliance—that was always my strength, my fort—my guiding star."

Q. You never made large sums by accident or chance? Fowler

"No not in any way shape nor form. Our business has been conducted in a simple, slow way and accumulated little by little. Of course our land has increased in value, but the main investments are not equally as remunerative, we have made our money by saving. If we had invested in mining stock, we might have made a great deal more, and then again we might have lost it all. A great many men make money and accumulate a surplus and put it in places where it is saved, where they might have invested it where they would have lost it. Lux and I were very different in that respect. He wanted to invest in gas stock and collaterals and corner lots, government bonds and the like; but I told him we were not calculated for that business, we were calculated to put our money where we could not take it back again, where it had to stay, and at the same time, pay us some revenue, so we invested in country lands, they never rot on your hands nor waste away.

"I told him in regard to town property we might have property in one part of town that would be fine property and then again in another part it would be worthless on our hands, while country property would always be good and the city depended on the country to be prosperous and if the people settle in around the country and were prosperous, the city would grow with it."

(In regard to his habits) Fowler notation

"I only consider eating a necessity and I have many times wished I could do without eating. I eat to live, I eat as a necessity. I don't drink anything but water and milk sometimes a glass of wine. I never drink liquor except for medical purposes and I eat very little meat and very little fresh hot-baked bread. I have not smoked for fourteen years, haven't tasted tobacco of any kind. Sometimes I carry a cigar in my pocket and offer it to a friend but I never smoke myself.

"I am a great fault finder in regard to my eating. I like to have the food prepared and make the most of a little. I rather have a dollar saved than five dollars made. When a man can save one dollar, it learns him to save on a larger scale. I frequently say to other gentlemen—'I could live where you would starve.' I can make more out of a little and get along where other men would be discouraged. No I don't change in my habits as I get older—I think if I was to live for a hundred years that I wouldn't get old—I would not be an old man."

(About making out a daily program in advance)

Fowler notation

"My daily program is such. It would be entirely wrong as such to make a daily program—every day brings up something new. What may be beneficial to do today would be against our interests tomorrow. When we start in to do anything, we deliberate beforehand and then carry it through, but if we see a change for the better is necessary we do it. We don't allow a man to be a mere nothing. I say to him to discuss the matter and if I am wrong I want him to tell me of it. I don't like a man that will agree with me in everything. When I am right, then they have to submit.

"I get up when the business requires it—mostly 5 o'clock in the morning—I have always done that way. There is no certain number of hours to constitute a day's work. I retire when my business is done—if I don't sleep, I will rest anyhow. I am a poor sleeper. I suppose it is my destiny—I believe in destiny.

"It is impossible to tell how much I am worth, because our cattle and our land depend upon the prosperity of the country. If the country goes on prospering and the people come here as they have been doing there is no disease or any war, our land will be a great deal more valuable. I suppose it will be worth double what it is now. Our stock today are worth only about half what they were some five years ago. Cattle which I bought for $25 and $30 are only worth about $15 or $16 now. Still, they are worth to us exactly the same as they were then. Our canal cost us nearly half a million, and it will be worth every cent we paid for it and improving it, but if it were sold today, it wouldn't bring over two or three hundred thousand dollars.

"I have no objection to your guessing at what I am worth, but I would not try to put an estimate upon it. Some people call that Gilroy property worth a million and a half, but that is three times the value I place upon it. Last year, I sold ten 10 acre lots of that at $150 an acre to people from the East—because they will develop it and make what I have more valuable. I refused to sell some I have there for $400 an acre and I sold some for that price."

At this point the interview ends.

Fowler sent the following notes to Bancroft when he submitted the interview he had done for the Bancroft's history. The letter is several pages, which I have cut down. This is to give the reader a view of Fowler's thoughts of the man and impression of the overall interview. Ed.

"On the occasion of my first interview with Mr. Miller, he said a biography should be prepared to please him and the object should now be to please him absolutely. He attaches great importance to the fact that he is a self made man. He places great stress upon the fact that he was always a hard workingman. He says a day's work is 24 hours to him, if he gets the work done he rests, but the work must be done whether he rests or not. A happy contrast may be made of his life with the lives of other great men on this coast. He differentiates this himself very clearly and distinctly. He called particular attention to the fact that his success could not be compared to that of Stanford, Huntington, Hopkins, Fair and others for the reason that he did not have the help of the nation which contributed to the success of these men, while his success is the work of one man. Neither has he made anything by accident in the discovery of a mine or by fortuitous circumstances; nor does he admit that Mr. Lux, his partner, contributed anything to the success that has been achieved. In the course of the interviews it was developed that he regards himself as having been from the first partnership, the businessman of

Cattle being driven over the Pacheco Pass in the early days

the firm. While he speaks of Mr. Lux in the most pleasant terms at all times it is clear Mr. Lux should be kept in the background, from first to last, while holding Mr. Miller up to the gaze of the applauding multitude.

"I would also call to your attention the fact he is positively the largest stock raiser in the United States of America and possibly the largest in the world. Also note the firm of Miller & Lux represents between seven and eight hundred thousand acres of land in several states.

"The biography should be prepared upon the basis of not less than 15 pages and it may be found wise to expand it to 25 or 30 pages. Miller is a man who thinks for himself and lets no man, allows no man, to think for him. His indomitable will, his underlying determination and his resolution which are unlimited have placed him above other men.

"I doubt if there is another man on the Pacific Coast that furnishes a brighter example of the success that may be attained by an individual, unaided that the life of Henry Miller. The material herewith furnished affords the biographer an opportunity for holding up to the youth of the age an example worth of emulation.

"I should be glad to furnish further information desired regarding the character of the man and what is likely to please him. Believing the foregoing will be sufficient

I Remain, Edward Fowler."

This interview was done in the 1880's. Henry Miller ended up being mentioned in three paragraphs of Bancroft's history, which suited Miller, as he preferred to keep out of the limelight unlike many of his contemporaries.

Miller & Lux team getting ready to scrape the streets. H St. looking towards Miller & Lux Building on 6th St. 1890's

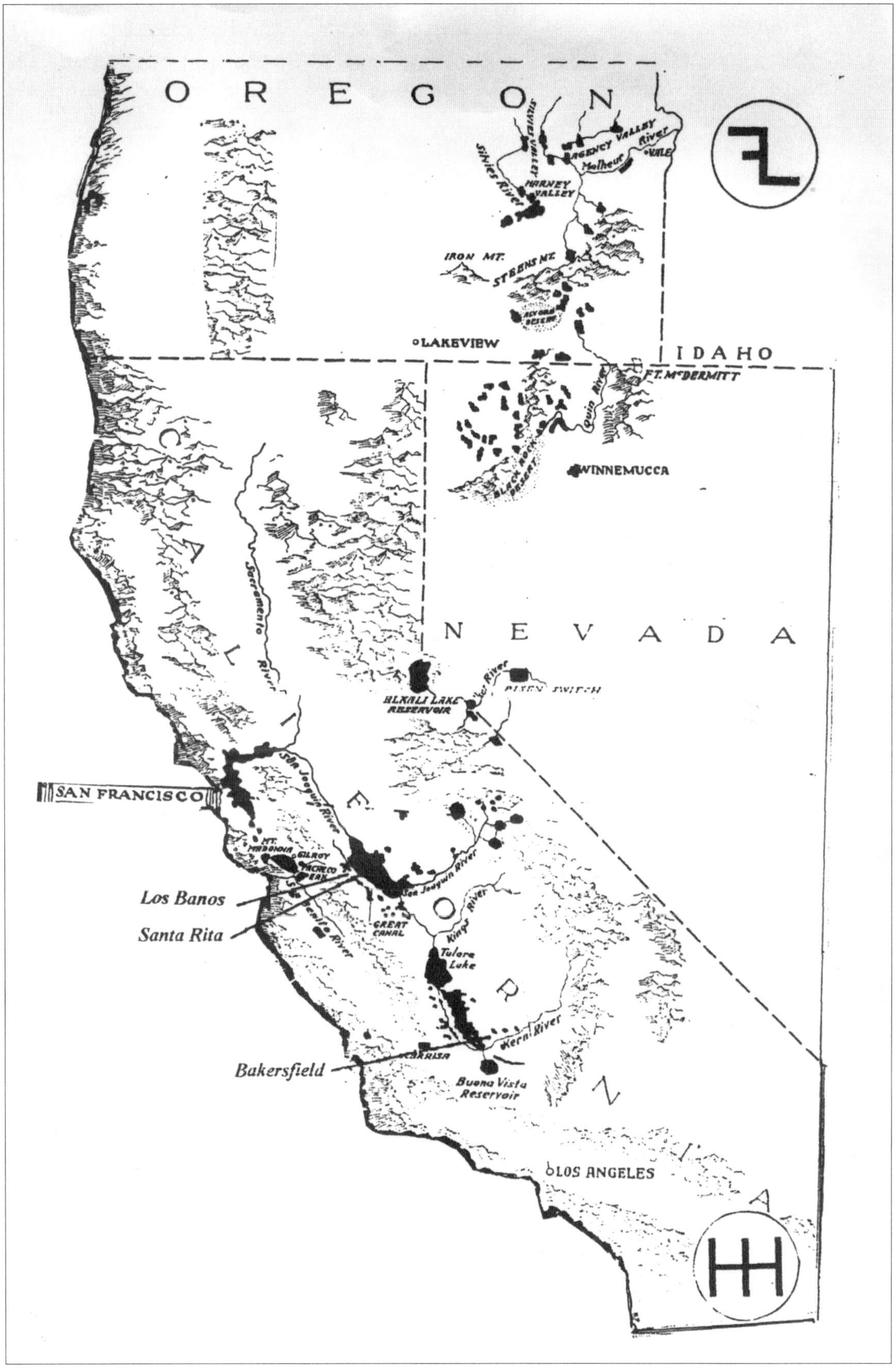

Map of Miller & Lux holdings, 1880's

Henry Miller – Sarah Wilmarth Sheldon Miller. Photos taken shortly after their marriage in 1860

Chapter Seven

Family Matters

Charles Lux was born in Germany in 1823. After working at his father's trade as a wheelwright, he migrated at the age of 15 to New York City where he found employment as a butcher's apprentice at six dollars a month.

The gold rush in 1850 lured him to California. He settled in San Francisco where he continued to work in the butcher business for Captain North. North considered Lux a treasure. Lux was indefatigable, never seemed to need sleep, never forgot anything, and was never in bad temper. The captain, who shared the roving disposition of most Californians of that time, decided the time had come to move on. They had worked together for about a year when the captain sold out to his assistant. For several years Lux carried on the business with the help of his brother, Henry. In about 1854 or 1855, Lux took into his partnership a man named Edmondson. They embarked in the wholesale cattle business in a small way and did well. After a year or two the partnership was dissolved.

In 1853 Isidro Sanchez sold his part of the Rancho Buri-Buri, which included the present site of South San Francisco, directly to Charles Lux for $10,000. Lux built a home on this property at Baden crossing. On his way to and from his Baden Ranch Mr. Lux passed the home of Homer Kings, which was located south of the city. It was here that he met and wooed the attractive widow, Mrs. Jesse Potter, who was a frequent visitor at the Kings'. Mrs. Potter's husband had been killed in the San Francisco Bay when the ferry San Rafael blew up. Lux courted Mrs. Potter three years and they married in 1857. Mrs. Potter was from the Rhode Island family named Sheldon and her sister Nancy would later marry Henry Miller. She had a son name Jesse Potter.

The year 1857 was also the year in which the first transaction of the famous partnership of Miller & Lux took place. A proposal was made by Lux to his competitor in the butcher business, Henry Miller, that he and Miller purchase a band of cattle consisting of 1600 head at $67.50 per head with each partner paying half the purchase price and the proceeds being equally divided. A year later in 1858 the partnership was consummated. In the spring of 1858, Lux disposed of his interest and went east but returned in September and proposed to Miller that they join in the purchase of cattle with a view to holding them for an advance in price. Miller agreed and on the following day bought nearly 2,000 head, which were pastured on rented land.

It was probably at the Lux home, which stood on the north bank of the creek near the present Chestnut Avenue in South San Francisco, where the true partnership was formed. At first the firm was called Lux and Miller, but by 1864 for unknown reasons it had changed to Miller & Lux. Charles Lux and Henry Miller, though strikingly different, were alike in many respects. Both were hardworking men of sturdy integrity, close habits of business and that power of concentration which can and did secure their fortunes.

The first Miller & Lux wholesale meat company was located on Sacramento Street, but it was later located on California Street near Montgomery.

From buying cattle the firm soon ventured into the ranching business. The first important purchase of land was part of the Santa Rita Ranch on the banks of the San Joaquin River in Merced County. This was bought in 1863 from Dumphy and Hildreth. The amount of land was 8,835 acres and the purchase price was $10,000. As the firm's stock of cattle increased and as neighbors desired to sell, Miller & Lux, who always kept ready money on hand, bought field after field and ranch after ranch until their property covered 600,000 acres.

At about this same time Miller & Lux bought two adjoining ranches known as the Tequesquite Ranch and the Lomas Muertas Ranch in what was then Monterey County and is now San Benito County. These were Old Spanish grants and covered an area of about 24,000 acres. The next major purchase was the Canada de San Lorenzo in the extreme southern portion of Monterey County (this was also called the Peach Tree Ranch). After the initial purchase they added to it as land came up for sale and soon they had 35,000 acres.

It was in 1869 that Miller & Lux began the purchases of land in Kern and Tulare counties. These purchases were follow-up additions, so that the Kern-Tulare ranch finally covered 114,000 acres. These lands did not include Charles Lux's private ranch of 2,000 acres in San Mateo County or the other pieces of land in other counties and cities. Miller also owned land outside the partnership, the Bloomfield Ranch outside of Gilroy in addition to lots and acreage around the state. Outside of California Miller & Lux owned the Glen Ranch, which covered two counties in Nevada, a tract of more than 25,000 acres. They also bought, in association with another cattleman named Overfeldt, a tract of 25,000 acres more or less in Baker and Grant counties in Oregon. This did not include the numerous leased lands, which cost the firm in excess of $20,000 a year for pasturage with one landowner receiving over $14,000.

This land was put to use by the raising of meat for the wholesale market. Charles Lux was the city man, the banker, the fundraiser the P.R. man. Miller was the man in the saddle, the land buyer, and the cattle broker. Miller claimed they were nothing alike but the partnership endured for 30 years without major problems, each man in his own sphere of responsibility.

Vast numbers of stock require vast sums of money. Charles Lux excelled at raising it. He was a social man and unlike Miller thrived in the world of big business, and dealing with banks and lenders. The administering of it successfully required a great deal of work. There were few men with large amounts of money who were willing to put in the work necessary to conduct such a business. Lux was such a man. He was always in good temper, always smiling, with a pleasant word for everyone. His serenity went below the surface. Miller, on the other hand, had somewhat of a volatile temperament. Lux would never break a butcher because he could not pay, but would carry him until the debt could be settled, and if he felt he would never be paid he would help the merchant to move on. He was also heavily involved with Ralston and the Bank of California. When Ralston drowned and the bank closed it was Lux who put the Miller & Lux funds, and got others behind the bank to keep it open. His wife was well known for her many charities and many a group or individual knew that in time of need Charles Lux could be called upon.

Lux's ranch in San Mateo County was about 1,500 acres. The barns and gardens spread out along

Charles Lux

Tents at Mt. Madonna, Miller's summer retreat, 1870's

the creek west of where the South San Francisco pumping station is located. His elaborate gardens on the banks of the ravine contained many ornamental trees and shrubs and were adorned with statuary and garden furniture. Tidewaters from the Bay reached inland as far as Lux's barn and schooners loaded with hay come up to the barn doors, there to unload their cargo. He also had a home at South Park near Harrison Street, but as industries were moving into the neighborhood the Lux family moved to the corner of Jackson and Gough Street. Charles Lux lived in this house until his death from pneumonia on March 16, 1887.

It was at Charles Lux's insistence that he and Henry Miller drew up an agreement that if one of the partners were to die, the other would take over total control of the business for a period of 7 years in order to ensure that it could be liquidated on the most favorable terms. Lux had seen the ups and downs of business in San Francisco, and had been closely involved with the Bank of California affair, and knew full well that either party could die at a time that was not good for the sale of such a large business. At first Miller was not in favor of this plan but he did sign and became the beneficiary of it, and in fact in many ways it made Miller.

Mrs. Lux preferred city life and society rather than the rough life on the ranch. She and her friends Mrs. Stanford and Mrs. Hearst were interested in the improvement of the community and were active in much charity work. Child welfare was a special interest of hers. She established the first kindergarten, the California Home for Feeble Minded Children, the Lux School of industrial training, Protestant and Hebrew orphans' asylums, public libraries and many others.

Due to the agreement between Charles Lux and Henry Miller enabling Miller to run the business as he saw fit for seven years, Mrs. Lux was forced to sue Miller to get what she thought was her share of the Charles Lux's estate. Miller bitterly resented Jesse Potter's indulgence in high living and society, so that when he had any opportunity he got all the Lux family out of the partnership before the seven years was up. Miller would lend the heirs money against their account, or he would totally reinvest income into the firm so that there were no actual dividends. The family was often at odds over Miller's handing of the business. Jesse had attended school with Henry Miller, Jr. and Miller always felt that Jesse had led his son on the wrong path. At that time Miller was married to Mrs. Lux's niece, his first wife, Mrs. Lux's sister, having died.

At Mrs. Lux's death, her son contested her will as she had given over a half a million to charity in it. She died seven years after her husband, just at the time the estate of Charles Lux was due to be settled. Jesse Potter claimed his mother was not capable of executing a valid will at the time the instrument was signed. A compromise was effected by Miller and family members with Potter whereby the will was admitted for probate and Potter gained considerably by taking one million dollars. However, he was not satisfied with that and soon engaged Miller and the firm of Miller & Lux in a lengthy court battle in which the heirs claimed Miller had robbed them in the foulest of ways. This case was later settled with the heirs receiving some funds and lands and Miller being paid to manage it for them as well as gaining unfettered control of the firm.

> "LUX—Charlie Lux, of the firm of MILLER & LUX, died in San Francisco, Tuesday morning." Source: The Hollister Free Lance (Hollister, San Benito Co., CA), 18 Mar 1887.

> "LUX—In this city, March 15, of pneumonia, Charles Lux, a native of Hatten, Germany, in his sixty-fourth year. Friends of the family are invited to attend the funeral services, TO-DAY (Thursday), at 2 o'clock P.M., at his late residence. Interment strictly private." Source: San Francisco Morning Call, March 17, 1887.

Charles Lux was buried in the family mausoleum, located in Oakland, California.

MILLER'S FAMILY LIFE

The Home Place was located at the Bloomfield Ranch in what is now Gilroy. Miller had a self-contained community. His initial purchase of the land, 1,700 acres, was in 1859. It later grew to 13,000 acres. Everything he needed was there: grazing for upwards of 3,000 head of cattle, feed mills and yards, dairy, a power supply, blacksmith shop, tannery, livery stable, hay and feed barns, slaughterhouse, chicken houses, hog pens, general store, machine shop, warehouses, his own rail spur and shipping station, as well as a reservoir of over a million gallons of water. The buildings covered more than ten acres of ground.

The house, really a mansion, was built in 1888. The cost was $50,000. At the time, Miller had 165 workers of whom 160 were boarders who lived on the top floors. The home with its 19 bedrooms could accommodate the workers as well as the family members and visitors. Mrs. Henry Miller, Jr. (Sara Onyon Miller), Miller's son's second wife, lived there until the home burned to the ground in 1923.

His personal life was not easy for Miller. His first wife, Nancy Sheldon, died just over a year after their marriage in childbirth along with the son she bore. Both were returned to the Bloomfield and buried on the hillside under a large oak tree. In 1860 he called on her niece, her brother Joseph's daughter, at the home of the Luxes and made an offer of marriage. The story is told that he called on Miss Sarah Sheldon at 11 am, asked for her hand in marriage and told her he would be back at four o'clock for her answer. She was 13 years younger than Miller.

The Millers had three children. Nellie, Henry, Jr. and Sarah Alice, whom he called Gussie. She was killed at the age of 8 by falling from her horse. Sarah Alice and Nellie had been riding on the eastern edge of the Bloomfield. The family was going into Gilroy by horseback and wagon and the horse stumbled and she was tossed off. The horse in attempting to regain its balance fell on her, crushing her skull. She lies buried at the Bloomfield Farm alongside her aunt and half brother.

Miller had never known a day without work since he had come to America and taking a vacation was totally outside his nature. Miller left the Bloomfield and returned to Germany for several months. There are several versions of why he went to Germany: he had suffered from a throat problem that his doctors could not seem to cure; business difficulties with the firm Miller & Lux; his temper was such that his doctors had warned him of the danger of apoplexy if he did not control it; the shock of Sarah Alice's death; any or all of these could be the real reason.

He always carried Sarah Alice's photo with him and one time offered to give a bandit whatever he wanted if he would only let Miller keep her photo from inside his watch. The town of Gustine is named in memory of his "Gussie" according to Mrs. Carrie Jameson, long time Westside pioneer. Others claimed that is just a fantasy.

Henry, Jr. was not the apple of his father's eye. Raised mainly in the city and attending private school with his step cousin Jesse Potter, he took full advantage of his position as Miller's son. He had little interest in the business and far preferred the pleasures of city life. Miller was seldom at home and spent little time with his family. He did not let his son have any responsibilities in connection with the firm and yet was unhappy that the boy showed no interest in its affairs. Miller disowned him for his first marriage to Miss Maud Oldham of Gilroy. Miller discovered the marriage quite by accident. He arrived home one evening wishing to discuss some business with his son only to be told he was out. Miller waited and the coach and coachman returned without young Miller. Miller continued to wait and quite late young Miller came in. As Henry Miller was reproving him for tarrying at the home of Miss Oldham so late, he received the answer from young Henry that she was his wife and he had a perfect right to stay as late as he pleased. Miller got madder than ever, called in Kane the coachman, and fired him. He then disowned his son in high tragedy and verse and then tore down to the office of the Mercury. The next issue contained the following notice:

> I hereby notify the public generally that my son Henry Miller Jr., has no longer any connection with my business in any way, either personal or mercantile, either in my own or in that of Miller & Lux. I will not be responsible for him in any way or recognize his acts in any manner whatsoever from this date. September 18, 1889 Henry Miller Sr.

The disowning father was interviewed at his home 34 Essex Street last night. "I do not know where he is or where he intends to live. Nor do I care. It is true that I intend to have nothing further to do with him. I made known my objections to his marrying, and he willfully disobeyed me, and therefore, I am done with him. It looks hard for a father to do as I have done; yet I did it with full knowledge of the facts, and I have not yet found any reason to change my decision. He has made his own way, now let him make the best of it," stated Henry Miller Sr.

"Young Mr. Miller and his bride are still living at Gilroy, and as far as can be ascertained are not at all bothered about his father's stern decree."

However, by 1898 young Miller was back in the fold and a house was built for him, one of the three Miller Sr. built on Mt. Madonna.

His oldest child was Nellie. She spent most of her life in San Francisco and though she often drove Miller in his later years, she was first and foremost a lady of fashion, what today would be considered a socialite. She married J. Leroy Nickel. Miller disliked Nickel and was often heard to say, "When he comes in a room, I's go out." Miller's judgment of Nickel proved somewhat correct in light of later events.

Miller's relationship with his wife is undocumented. They were married 45 years and none of his correspondence with her survives. He occasionally mentions that the "Family is well" in letters to Charles Lux but no mentions are made after Lux's death. Like most of his contemporaries his family lived in the nearest city. Sheep men favored Stockton, cattlemen Sacramento or San Francisco. The San Joaquin was considered too

House built at Mt. Madonna, 1890's, destroyed by fire in 1925

harsh for a wife and children. Mrs. Miller seemed to prefer the life of the City and Miller was seldom at home. They did have a summer camp, in tents at first and later houses at Mt. Madonna. A house fire in the 1920's took care of many of the personal items of the family and before her death Henry Miller, Jr.'s wife burned many of what she considered personal papers. People who knew Mrs. Miller said she was easy going, and would often stop at the local saloon for a beer with friends. She did visit the various ranches on occasions but it appears she never spent that much time in the valley. She preferred San Francisco, the Bloomfield or Mt. Madonna in summer. Like her aunt she was involved in charity and other civic causes.

Main house at Bloomfield Farm, 1880's

Miller realized that he could not continue to operate as he had in the 1860's and 1870's due to the changing cattle business and the influx of settlers. He held the most valuable asset of all land. As early as 1888 he had made the statement that "I think, good land, there is no telling how high it will go. There is a time coming when people will find it out. I think there is no doubt but that California will be the most densely populated of any portion of the United States. I see nothing to hinder it. A man on ten acres can make a good living."

Nellie had four children, none of which could benefit directly from the estate of Henry Miller. Miller had left his estate in trust for the third generation. Sally Nickel, the daughter of George W.

Nickel, was the first of this generation and was just six months old when Miller died at 89 in 1916.

Under Nickel's management many of the old time employees left the firm and it soon became mired in troubles. By the mid 1920's the cattle had gone and all that was left was selling the land. Albert Long, Miller's nephew and the one he had counted on to take over the cattle operations, did not see eye to eye with Nickel and left the firm after Nickel took over.

Miller's second wife, Sarah, had died in 1905 at the age of 65. Henry Miller, Jr. died at the age of 43 in 1907. It is through Nellie's children and grandchildren the remains of the Miller estate is ruled today, by the Bowles and Nickel families.

Henry Miller's three sisters had also come to America. Each made a life here and all were married. They came in stages from Mrs. Long's arrival in the late 1860's to Mrs. Pfeiffer's in 1881. The Pfeiffer family, as well as some of the Longs, are still in the area.

Elizabeth Johanne Krieser was born circa 1823 in Germany and died March 9, 1870, in San Francisco. She married Henry Long and was the mother of Albert Long, Henry Miller's chief cattle buyer.

Frederike Christine Kreiser was born in April of 1821 in Brackenheim, Germany, and died April 2, 1891, on the Bloomfield Farm, Miller's home place near Gilroy. She married Alfred Pfeiffer in 1846 in Germany. Mr. Pfeiffer died in Germany in 1877.

From the Gilroy *Advocate* on April 4, 1891: "Mrs. Pfeiffer, sister to Henry Miller of Bloomfield Farm, died of pneumonia on Thursday last at the home of her daughter, Mrs. Steve Sheldon. The funeral services will be conducted today, 1 p.m., by Rev. D. T. McClelland at the residence of Mr. and Mrs. Sheldon beyond the Carnadero School house"; and "Died - Pfeiffer - at Bloomfield Farm, April 2nd, Mrs. Pfeiffer, a sister of Henry Miller, aged 70 years. Funeral today at 1 o'clock." From the Gilroy *Advocate*, April 11, 1891: "The funeral of Mrs. Fredericke Pfeiffer took place at Bloomfield Farm Cemetery on Saturday last. There was a large following of friends and relatives, and the grave scene was made strikingly impressive by the solemn service conducted by Rev. D. T. McClelland of the Presbyterian Church, assisted by the church choir. Many floral mementoes were placed on the grave, one representing a pillow with the word "Sister" wrought in colored flowers, the gift of her only living brother, Mr. Henry Miller. There were other floral pieces, among them 'Gates Ajar,' 'The ripened wheat,' an anchor, etc.

"Mrs. Pfeiffer was 70 years old this month and until the past two months was a vigorous woman. Cold, la grippe, pneumonia and death, followed in succession and so passed away one of the brightest and interesting matrons of the German race in this township. It is ten years since she came to this country from Germany, accompanied by her youngest son, who died about four years since at the Sheldon residence. Mrs. Dalton, her daughter, made the voyage to bring her to be near her children, namely herself, Mrs. Sheldon and Albert Pfeiffer. This was in 1881. Her husband died in 1877, and four of her children also preceded her, two in Germany, and two since her residence in the Gilroy Valley. Many friends sympathize with Mr. Miller in the loss of his sister and also with the other bereaved members of the family, who naturally feel the loss of her counsel, friendship and love."

Karoline Christine Kreiser was born in 1825 in Germany and married a fellow German. She later came to America.

Outsider's View

"The Nickel element has never been just to Miller. It is the duty of Mrs. Nickel to get another person to write the biography of Miller properly. What is written is correct enough but the genuine merit of Old Man Miller, who was really a great man, was left out. The Miller element was where the genius lay and not the Nickel element. The

Sarah Alice Miller taken shortly before her death

In Memoriam.

Sarah Alice Maud Miller,

Summer's gold upon the sloping mountains.
   Summer's emeralds on the sheltering trees,
Summer's fragrance on the garden roses,
   Summer's bird-song on the passing breeze.

Life was at its fairest, childhood's beauty,
   Childhood's grace in heart, on lip, and brow,
Childhood's vista opening fair before her,
   She who lies so still and silent now.

Oh the glorious June day was not brighter
   Than the joyous spirit in her breast,
Where no shade of sorrow ever lingered,
Which no dark forebodings e'er possessed.

Hope with silver wand was pointing onward,
   Pleasure's voice rang out in sweet refrain,
When she went forth in her girlish gladness
   From the home she ne'er would see again.

Back they bore her, all of life and being
   Crushed in one brief moment from her heart,
Casting on that happy home a shadow
   Which can never, never more depart.

Veiling you her sorrowing, weeping parents,
   Stricken by this thunderbolt of woe
Which has pierced the circle your love guarded,
   Which has laid your cherished darling low.

Sunbeam of your house, the youngest, dearest,
   Loved by all who knew her, falling tears
Will keep green the memory of your daughter,
   She will dwell with you thro' coming years.

As your winsome child, love's light and treasure,
   Spared forever from earth's blighting touch,
She will be your Angel still to guard you
   For God blesseth those who sorrow much.

She will still be with you, her sweet memory
   Will be fondly linked with every place
Where in bygone days she gaily wandered.
   In her happy, care-free childish grace.

Ah! for you the pain, the loss, the anguish,
   And for her the sweet celestial rest
In His arms who loveth little children,
   In His home who doeth all things best.

MARCELLA A. FITZGERALD.

Mossy Woodland, Gilroy, June 1879.

Sarah Alice Miller's memorial card, 1879

element came in that was unfair to Miller. He was the great man and there were few men if any in his time that were his equals. I don't think Treadwell (author of *The Cattle King*) was fair to Miller. I heard Treadwell say that he had the book in hand a long time before Mrs. Nickel would let him print it.

"Miller took Nickel in as son-in-law and gradually let him take Lux's place after Lux was dead. Nickel got in a English lawyer by the name of Treadwell. His son became a lawyer and never worked for anyone but Miller & Lux. When Treadwell won the big case for the Miller & Lux inheritance, he put in a bill for $600,000. It was paid with company bonds. It made Treadwell rich and independent. After that, he became a big lawyer for big concerns. Mrs. Nickel couldn't see things as Treadwell did, and so she didn't want his life of Miller published. She made him cut out many things.

"Miller knew nothing of fashion and he cared even less. It was the English element that wanted to be fashionable. I used to know Mrs. Miller. She was very common and thought nothing of stopping in a saloon and getting a glass of beer. Miller had a fine house but never got anywhere fashionably. Mrs. Nickel went to fashionable schools and aimed to be fashionable. She would come here as a young girl and was very democratic. Dunbar used to like her. She would take on with all the fellows but seemed to have no favorite. She was very sociable and was acquainted with everybody."

Interview Billy Stockton, undated. RLM

Newspaper reports and clippings

Gilroy *Advocate* June 21, 1879

The remains of Sarah Alice, daughter of Mr. and Mrs. Henry Miller, will be brought on the two o'clock train today from San Francisco and conveyed to the Bloomfield Farm for burial. The friends and acquaintances of the family are invited to join the funeral procession and to be present at the service at the grave. The place of burial is to be on the plateau situated on the west side of the Monterey Road.

Gilroy *Advocate* June 28, 1879

The mortal remains of Sarah Alice Miller, beloved daughter of Mr. & Mrs. Henry Miller were conveyed here on Saturday last by train for interment in a private cemetery lot at the Home Farm. The day after the late lamentable accident, which deprived this blooming and affectionate girl of life, her body was sent to San Francisco and placed in the family vault at Laurel Hill Cemetery. The dismal precincts of the cemetery however seemed to be no place for the repose of her young form and this beautiful rural spot as the Bloomfield ranch where she had gamboled and spent many happy hours in play, seemed to her fond parents to be the most appropriate place for their buried love to rest. The afflicted parents and members of the family in mournful sadness, with hundreds of sympathizing citizens in carriages followed the hearse bearing the cherished form that only a few days before bloomed in youth and beauty, and moved in gaiety and smiles in the presence of those who are now heartbroken over her early and sudden death. The funeral procession finally reached the grave where the solemn Christian services were conducted by Reverends J. J. Drum and T. M. Oviatt of the Presbyterian Church. Friends gathered round with flowers, "emblems of the beauty and purity of the deceased and the crow of glory, which she has received in heaven." The scene was impressive as the beauties of nature were strew about the grave in the presence of uncovered heads while the plaintive hope-inspiring words of the song "The Sweet By and By" echoed on the hillsides. The crowd left the melancholy scene touched with sympathy over the sad event.

August 23, 1889- The Hollister *Free Lance*, Hollister, Ca. Married — MILLER-OLDHAM — in Gilroy, Aug 16th, by Rev. J. R. GRIMSTEAD, at the residence of the bride's parents; Henry MILLER, Jr., to Miss Maude OLDHAM.

2/25/1898 Gilroy *Advocate*\Merced *Express*

Died: Mrs. Henry Miller Jr. died suddenly Saturday last at her home in Gilroy. She was 50 years of age and had a number of friends in Gilroy and Merced who will regret to hear of her death.

Gilroy *Advocate* clipping undated

Mrs. Henry Miller Dead.

Monday evening, August 30, 1906

Sarah Wilmath, wife of Henry Miller, Sr., died at her home in San Francisco. Service conducted by Episcopal minister. Mrs. Miller was 66 years of age. Came to California from Providence, Rhode Island. Leaves two children: Mrs. Leroy Nickel of San Francisco and Henry Miller, Jr. of Bloomfield. A sister Mrs. Perry Robinson and Stephen W. Sheldon, both of Gilroy. Mr. Miller is nearly 80 years of age.

Merced *Express*, no date

DEAD: Mrs. Henry Miller

Monday evening, August 30, 1906 Sarah Wilmath, wife of Henry Miller, Sr., died at her home in San Francisco. Mrs. Miller was 66 years of age. Leaves two children: Mrs. Leroy Nickel of San Francisco and Henry Miller, Jr. of Bloomfield. The sister of Mrs. Perry Robinson and Stephen W. Sheldon, both of Gilroy.

Gilroy *Advocate* clipping, January 7, 1916

Word came today of the sudden death of Albert Long, the well known cattleman who was for so many years connected with Miller & Lux interests in Los Banos and other parts of the coast.

A telegram to the Gilroy *Advocate* states that he met his death at some point in the state of Utah by being struck by an engine. He was engaged in shipping cattle at the time of the accident.

Messrs. H. J. Raven and James Vandiveer of Dos Palos and other old time friends went over to Gilroy the fore part of the week to be present at the funeral. "This news is a shock to Mr. Long's old friends in Gilroy and all over the state."

Mr. Long was for over a quarter of a century the head cattle buyer for Miller & Lux a firm owned by Mr. Long's uncle Henry Miller. He was on the go all the time and a more faithful man could not have been found in the entire west. He was a recognized authority on cattle and it was generally conceded that he was the best man in the state. He was looked upon as the logical successor of Henry Miller the cattle king, whose interest he guarded as his own. After the death of Mrs. Long a few years ago he severed his connection with the firm of Miller & Lux and accepted a more lucrative position with the firm of Woodward & Bennett now in the meat business in Los Angeles. Albert Long was a native of Gilroy, aged 53 years. He was up to four years ago Henry Miller's right hand man. He leaves one brother Henry Long, three sisters, Mrs. Lizzie Berry, Mrs. David Brown and Mrs. Grant Whittington and two children Mary and Raymond who live with their grandparents in Pacific Grove. They are 16 and 17 respectively. Mr. Long was a member of the Gilroy lodge I.O.O.F. and was buried under its auspices.

MLLLER'S WILL

CATTLE KING LEAVES $20,000,000 IN TRUST TO DAUGHTER AND HER FAMILY Substantial Sums Given to Relatives, friends, Old Employees, Charity and Cause of Education. Many gifts in long document.

Poor of Gilroy and birthplace in German are remembered; daughter is trustee.

The will of the late Henry Miller, the pioneer cattle king whose fortune is estimated at over $20,000,000, was filed today by Attorney Edward F. Treadwell. All of Miller's land has for many years been held by corporations and his only property constituted of stock therein. This is all bequeathed to his daughter Nellie Miller Nickel, and her husband J. LeRoy Nickel, as trustees.

The trustees are directed to make a suitable burial place for the decedent at Mount Madonna, Santa Clara County Ca. and give $25,000 for that purpose.

CHARITIES REMEMBERED:

Many, many charities are remembered in the bequest of the late cattle king and endowments for education purposes to become effective several decades hence are also outlined. The poor of the City of Gilroy will share $15,000. The poor of Henry Miller's birthplace the city of Brackenheim, Kingdom of Wurtemburg, Germany are bequeathed $25,000. Bequests ranging from $50,000 to a monthly income of $50 are left to Mr. Miller's relatives and those of his wife, and to many of his old employees who grew gray in the cattle business with him. The San Francisco Home for Incurables is left $10,000. The First English Evangelical Lutheran Church of San Francisco is left $5,000. A gift of $300,000 for a hospital in Gilroy and $500,000 for a medical research hospital in connection with either the University of California or Stanford becomes effective only after the death of Mr. & Mrs. Nickel and their descendants. The bulk of the estate goes to the Nickel family.

The following amounts are directed to be paid to the relatives of Mr. Miller and his deceased wife:

Sara E. Miller Gilroy $50,000.00
Minnie Pfeiffer Sheldon Gilroy $30,000.00
Sara Long Brown Alameda $20,000.00
Elizabeth Long Barry, San Jose $20,000.00
Albert Long (deceased) $25,000.00
Martha Rembold, Stockton in trust
For children $10,000.00
Minnie Elisabeth Pfeiffer Turner
Gilroy $10,000.00
Henry John Pfeiffer, Los Banos $10,000.00
Albert Pfeiffer Jr. Los Banos $10,000.00
Gustav Rembold Maas, San Francisco $5,000.00
Gustav Rembold, Burns Ore. $5,000.00
Phillip Rembold Maas, San Francisco $5,000.00
Nancy Probasco, Burlingame $10,000.00
Anna Sibley Robinson, Gilroy $10,000.00
Joseph Place, Los Banos $5,000.00

The trustees were also directed pay certain employees of Mr. Miller and of Miller & Lux, Inc. the following amounts.

Charles Z. Merritt, Oakland $2500
Edwin G. Rudolph, San Francisco $2500
David Brown, Alameda $2500
John Dillon, San Francisco $2500
William Mitchell, Alameda $2500
John Gilcrest, Livermore $2000
Charles Reddy, San Francisco $2000
James Ogden, Buttonwillow $1000
D. A. Leonard, Dos Palos $1000
D. W. Wallis, Los Banos $1000
E. F. Ogle, Buttonwillow $1000
Hugh T. Gower, San Francisco $1000
Charles Robinson, Bolinas Beach $1000
Edwin M Jordan, San Francisco $1000
Peter Arbios, (deceased) $500
Henry Bader, Firebaugh $500
Mrs. Henry Bader, Firebaugh $500
H. C. Safstrom, Los Banos $500
Cyrus B Swift, Madera $500
Benjamin Miller, San Francisco $500
Gus A Lastsine, San Francisco $300
Henry Pfeiffer, San Francisco $300

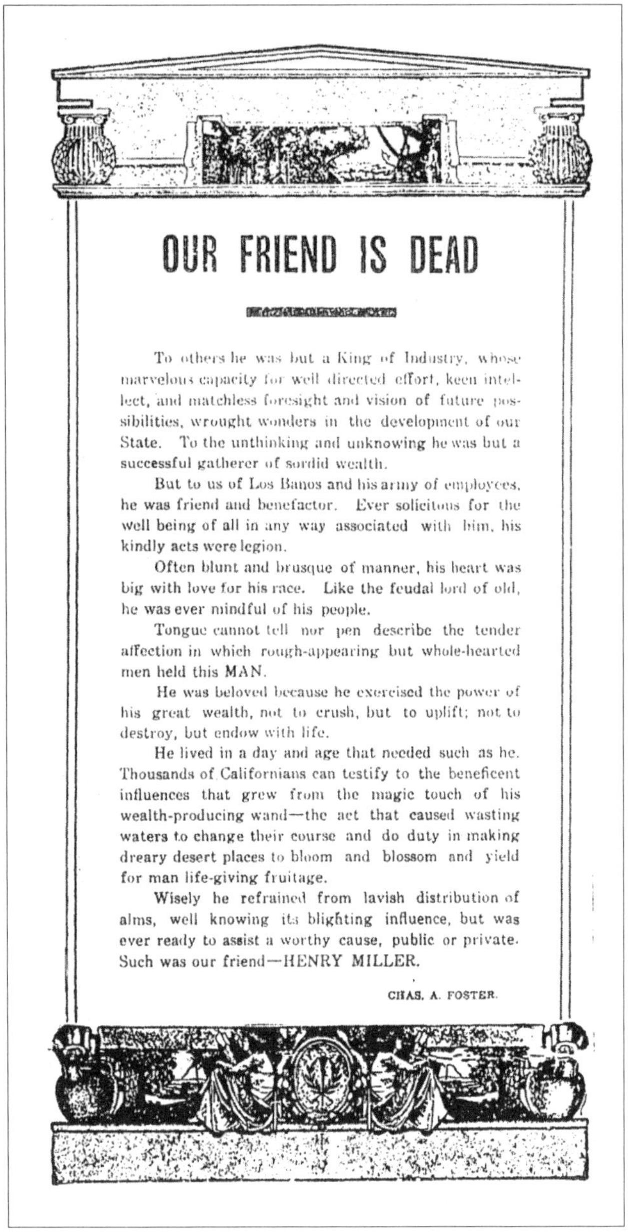

OUR FRIEND IS DEAD

To others he was but a King of Industry, whose marvelous capacity for well directed effort, keen intellect, and matchless foresight and vision of future possibilities, wrought wonders in the development of our State. To the unthinking and unknowing he was but a successful gatherer of sordid wealth.

But to us of Los Banos and his army of employees, he was friend and benefactor. Ever solicitous for the well being of all in any way associated with him, his kindly acts were legion.

Often blunt and brusque of manner, his heart was big with love for his race. Like the feudal lord of old, he was ever mindful of his people.

Tongue cannot tell nor pen describe the tender affection in which rough-appearing but whole-hearted men held this MAN.

He was beloved because he exercised the power of his great wealth, not to crush, but to uplift; not to destroy, but endow with life.

He lived in a day and age that needed such as he. Thousands of Californians can testify to the beneficent influences that grew from the magic touch of his wealth-producing wand—the act that caused wasting waters to change their course and do duty in making dreary desert places to bloom and blossom and yield for man life-giving fruitage.

Wisely he refrained from lavish distribution of alms, well knowing its blighting influence, but was ever ready to assist a worthy cause, public or private. Such was our friend—HENRY MILLER.

CHAS. A. FOSTER.

Memorial Card, Henry Miller, Los Banos *Enterprise*, 1916

Mrs. Emil Marquardt, Gilroy $250
Narcisio Castro, Gilroy $250
J. Rock Tofalo, Firebaugh $250
Jesus Lopes, Firebaugh $250
Louis Gonzales, Gilroy $200
Mary Taylor, San Francisco $200
William Beddell, San Francisco $200
M. M. Thomas, San Francisco $200
M. V. Deeney, San Francisco $250
M. C. Wilson, San Francisco $250
Pachuca Higuerra, Firebaugh home for life
Richard Brem, in trust for children $5000.00
Dorothy Otto, Gilroy $1000

These amounts are only to be paid if those parties are in the employ of Mr. Miller and the company at the time of Mr. Miller's death.

INCOME TO NICKELS

The income from the balance of the property is to be paid jointly to Nellie Miller Nickel, daughter of Henry Miller and her husband J. LeRoy Nickel during their lives. Upon their death the sum of $50,000 shall be divided among such of the San Francisco charities as my be selected by the trustees for that purpose and the balance of the income to be paid to the children of Mr. and Mrs. Nickel and their descendants during their lives. Those children are George W. Nickel, Beatrice Sara Nickel and J. Leroy Nickel.

FUTURE BEQUESTS

Upon their death the sum of $300,000 is to be devoted to the establishment of a hospital at the city of Gilroy, Santa Clara County and also the sum of $500,000 is to be devoted to a research hospital under the University of California or Stanford University and then the balance of the property is then to go to the descendants of Mr. and Mrs. J. LeRoy Nickel, if any and if not then one quarter of the property is to go to the descendant of Mr. Miller's sisters and the balance is to be applied to charitable institutions in the State of California, particularly to the support and assistance of charitable schools and institutions devoted to the practical education of boys and girls in mechanical, agricultural and domestic pursuits without regard to creed.

Nellie Miller Nickel and her husband J. LeRoy Nickel are named as executors of the will. While the property is held in trust this will in no way prevent the sale of land, which is all held by the corporations and the sale and colonization of the land will be in no way affected by the trust created.

Call Bulletin, October 19, 1916,
Published three days after Miller's death.

Merced *Express*
Oct. 21, 1916
PASSING OF HENRY MILLER
FAMOUS CALIFORNIA LAND AND CATTLE KING DIES AT HIS SAN FRANISCO HOME

Henry Miller, one of California's notable men, who rose from butcher's boy to be known as the cattle king of the West, and founder of the famous Miller & Lux Company, died in San Francisco last Saturday in his 90th year at the residence of his daughter, Mrs. J. LeRoy Nickel. He had been bedridden for over two years and was unconscious for two days when death came.

Henry Miller, once a poor butcher boy of San Francisco owned at the time of his death an empire in the West described as "twice the size of Belgium," and because of his vast holdings of sheep and cattle, he was known along the Pacific slope as "The Cattle King."

He never knew himself exactly how much land he held. Several years ago experts estimated approximately 22,717 square miles or 14, 539, 200 acres were under his control mostly in California, Nevada and Oregon. Miller it was said could drive a herd of cattle from Arizona to Oregon through the Central California Valley and each night camp on his own land. At no stage of his trip would he be out of sight of one of his own ranches. More land was held on the Pacific Coast by Miller than by any other individual. At times 150,000 cattle and 100,000 sheep grazed on his western pastures and bore the Miller & Lux brand. The firm of Miller & Lux in which Miller was heavily interested was the cattle business and operated a chain of slaughter-houses, banks, stores and hotels, in addition to ranches and ranges. His managers, clerks and workers were numbered in the 100's, and vaqueros were employed by the scores. Miller reasoned in miles not acres.

Friends said Henry Miller's policy was always buy land and never to sell. As a boy he had visions of wealth. He bought land when the Spanish and

American governments sold it cheap and hoarded his property and lived its dream. Miller was born in Brackenheim in Germany July 21, 1827. He emigrated to America penniless and worked as a butcher boy in the Washington market in New York City. The discovery of gold in California set him on the path to our state. In 1850 he was still a butcher boy in San Francisco. By 1851 he launched into business for himself. He met Charles Lux and six years later the two formed the partnership that later helped make history on Western ranges.

Miller was first married to Nancy Sheldon who passed shortly after the marriage. In 1860 he married Sara Wilmarth Sheldon her niece. In 1887 Charles Lux died and Miller took control of the firm. The years that followed were ones of expansion and history making. Range land swept from the Sierras to the Pacific Coast.

Gradually Miller watched the settlers come, saw their ranches marked off by barbwire and orchards grow up where cattle had formerly roamed. He changed with the times keeping Miller & Lux at the forefront of cattle production. It is thought that the dismemberment of his domains probably will come soon for conservations policies, high land taxes and a clamoring for land from settlers makes the cattle empires of old outdated. He was the last of the great land barons. In his later years he lived in San Francisco with his surviving child Mrs. J. LeRoy Nickel and became somewhat of a recluse due to ill health. His wife Sara and two children Sara Alice and Henry Jr. preceded him in death. Private Services were held in San Francisco.

A press dispatch from Los Banos reads: This city which owes much to the generosity of Henry Miller closed this afternoon during the time of funeral services in San Francisco. The name of Henry Miller will long live in the history of the Westside where he spent the greater part of his life. Los Banos has always been known as his pet. The village was founded on his land and his gifts to the city have been many and substantial ones. The city park and was inaugurated at the May Day Barbecue and for more than a quarter of a century he contributed beeves in abundance so that all might come, enjoy and eat their fill on that day. Pioneer residents always referred to him as "Uncle Henry" and he is known to nearly every man, woman and child on the Westside of the San Joaquin River. Crepe was hung on the door of every business house on the Westside yesterday in memory of the one whose name have been a household word to residents of this and other Westside Cities.

June 8, 1937, unknown San Francisco paper

Miller & Lux Executive Dies

Funeral services were held at Colma Saturday for James LeRoy Nickel Sr., 86, former president of Miller & Lux. He died Thursday night of last week at Palo Alto hospital from injuries suffered in a fall following a heart attack. Services were held at the crematory in Woodlawn Cemetery where the ashes were interred.

Nickel was born in Collomsoville, Pa. In 1851, came to California in 1878, he joined Miller & Lux in 1888 and married Nellie Miller, daughter of Henry Miller, founder of the big land company that same year. He retired in 1925 as general manager of the company and became chairman of the board of directors, holding that office until his death. He was president of the former Bank of Los Banos at Los Banos and at Firebaugh and the Bank of Gustine.

He is survived by two sons, George W. Nickel of Palo Alto and J. LeRoy Nickel of San Francisco, and a daughter, Mrs. Beatrice Bowles of San Francisco. The Miller & Lux office here was closed Friday and Saturday in observance of his death.

August 1, 1944, unknown San Francisco paper

HENRY MILLER'S DAUGHTER DIES

Mrs. James Leroy Nickel, 78 Succumbs at Atherton

Mrs. James Leroy Nickel, daughter of Henry Miller, one of the California's great land barons, died yesterday at her home in Atherton after two weeks illness. She was 78.

The former Nellie Miller was the widow of James LeRoy Nickel, who was president of the great Miller & Lux Land Company, which owned more than $70,000,000 worth of farmland in the San Joaquin and Sacramento valleys and which was founded by Henry Miller.

HUGE ESTATE

Mrs. Nickel at the time of her death lived on one of the last of the huge San Mateo County estates located on Middlefield road near the James L. Flood land which in recent years have been cut into subdivisions. She leaves two sons, George W. Nickel, and James LeRoy Nickel Jr., and a daughter, Mrs. Beatrice N Bowles of San Francisco.

Funeral services will be held today at 2 p.m. at Woodlawn Chapel in Colma, followed by burial in Woodlawn Cemetery.

Los Banos *Enterprise* – undated September 1954

MILLER & LUX HEIR DIES IN PLANE CRASH

John Beverly Nickel, one of the heirs to the multimillion dollar Miller & Lux estate, was killed yesterday afternoon when his small plane crashed and exploded in an alfalfa field seven miles south of Los Banos. Killed with him was his sister-in-law Charlotte Nickel, whose husband George Nickel Jr. is another of the three heirs who filed the $100,000,000 Miller & Lux fraud and conspiracy suit.

The two took off from Los Banos airport, near the George Nickels Delta Ranch about 4 pm to fly to the Bay Area for a family Thanksgiving dinner.

PLANE IN FOG

About five minutes later three men working in a field heard the single engine Cessna 170 circling overhead in thick fog. The men told Deputy Sheriff Rex McBride the plane nosed down under the low overcast, zoomed up and then came down again. It ploughed head on into the field and buried its engine. An explosion followed and the searing flames prevented the workers from attempting a rescue. The bodies were taken to the Whitehurst Funeral Chapel in Los Banos where funeral arrangements are pending.

KIN WAITING

George Nickel Jr. was waiting at the Oakland Airport to meet his wife and brother-in-law, John Nickel's wife Nancy and their two small children were waiting at the home of her mother, Mrs. Charles Watson Burkett, at 870 Camino del Mar in San Francisco. Mrs. Charlotte Nickel was the mother of six children ranging in age from 2 to 14.

C. Ray Robinson, attorney for the Nickels, called the accident "a terrible, terrible tragedy." Asked if it would have any effect on the vastly complex lawsuit now in the courts, he said it would not.

BACKGROUND

John and George Nickel Jr. filed the suit earlier this year along with another great-grandchild of cattle baron Henry Miller, Mary Lombardi of San Rafael. They accused the trustees of the Miller & Lux estate – principally their uncle J. LeRoy Nickel, Jr. of mismanagement.

Just last week, the State District Court of Appeal voided all the complicated litigation that had taken place, because the other six heirs to the estate had not been parties to the suit.

SAN FRANCISCO *EXAMINER*

September 12, 1954

This is only a starting point

Lawyer says!

A $100,000,000 fraud suit filed in Federal court here Friday was only the starting point for a long range attempt to recount every dollar of the indeterminate millions allegedly stolen from Miller & Lux, Incorporated by its former officers and agents, Attorney C. Ray Robinson said yesterday.

Robinson asserted that a multiplicity of previous undetected frauds against the company are being uncovered daily. He said new charges and new defendants will be added to the Federal court complaint as rapidly as the facts can be assembled.

ALLEGED FRAUDS

And he noted that the company is attempting to recoup from those accused of perpetrating he frauds and the heirs of those who are dead but also from oil companies and other corporations that have profited from the alleged frauds during the past twenty-five years.

In this connection, one of the defendants named Friday is The Texas Company, but the complaint also joined as defendants 100 oil companies and 100 individuals not identified by name.

Among the other named defendants are the Crocker First National Bank, which was sued only in its capacity as executor and trustee of the estate of James E. Fickett, president of Miller & Lux from 1925 to 1935 and chairman of its managerial committee for the next five years. Fickett who died in 1950, allegedly participated in the frauds.

DAUGHTERS NAMED

Fickett's daughters, Mrs. Loretta Codington and Mrs. Alice June Lindsay, although not accused of fraud, also were named as defendants. Heirs of C. E. Houchin and F. C. Noel, former Miller & Lux land agents, and W. W. Colm, another official of the company were similarly named. The major named defendants all accused of defrauding the company are J. LeRoy Nickel Jr., Attorney J. E. Wollery and A. R. Olsen, former trustees of the multi-million dollar Henry Miller trust, which includes all stock in Miller & Lux, R. H. Anderson, millionaire Bakersfield oilman, and former Miller & Lux land agent; H & F Farming Company, R.H. Anderson Inc., and the Kern Property Corporation.

OTHER DEFENDANTS

Other defendants are J. LeRoy Nickel Jr.'s wife Gladys; Attorney Vincent J. McGovern of San Francisco, E. K. Potter and Lon V. Smith former Miller & Lux land agents; and George L. Bradford, identified as a partner of Houchin from 1940 until Houchin's death last year.

The complaint charges that during the Nickel-Wooley-Olsen-Fickett regime they and various other defendants formed a conspiracy to defraud the company. The complain charges that the conspiracy was carried out through dummy corporations, secret trusts and secret profits and fraudulent conveyances of fabulously rich Miller & Lux oil lands. The complaint cites, in this latter connection, sales and leases of land in the Buena Vista Lake gas field, the Coles, Levee, North Midway, Paloma, Cymric and McKittrick oil fields.

Merced *Sun Star* 9/14/1954

MILLER ESTATE SUIT IS HALTED

San Francisco – The California District Court of Appeals has ordered proceedings in the Miller & Lux fraud suit halted until it can determine three legal side issues. Two of the side issues concerned petitions for intervention filed in behalf of various heirs to the estate of the late Henry Miller.

In these petitions, attorneys for the heirs asked that they be given a voice in any action, which the Superior Court might take in trying to reorganize Miller & Lux, the chief asset of the estate.

The third petition was filed in behalf of A. R. Olsen, who was ousted as a Miller & Lux trustee in June after he was named in a fraud suit filed by three of Miller's great grandchildren. Olsen's petition asked that all heirs to the case be included as plaintiffs.

Presiding Justice John P. Nourse set Nov. 9 for legal arguments on the writs. Last week, court appointed attorneys for Miller & Lux filed a suit for 100 million dollars damages in federal court here against a number of company officers and former officers who benefited from allegedly fraudulent operations. They charge that valuable lands belonging to Miller & Lux were sold at bargain prices. Some of the land is presently yielding millions of dollars in oil (UP)

Clipping – dated 1954

NICKEL OIL LANDS DEAL UNDER FIRE

San Francisco – Attorney for the Miller & Lux Corporation tried to establish yesterday that its former president knew he was buying potential oil land when he purchased Kern County land from the company in 1928.

They introduced a document while the former president and board chairman, J. LeRoy Nickel Jr., was giving his deposition in connection with a 100 million dollar suit by Miller & Lux against him and other defendants. The document was a copy of a letter, supposedly dictated by Nickel more than four months before he bought the land, indicating the company was then "considering" leasing the property for oil development.

Attorney Thomas Greer, one of those representing the company, produced the letter after Nickel had denied he considered a deal with an oil company prior to the time his father suggested that he buy the 166 acres. The letter was not signed, but Nickel's initials were typed in the lower left corner, as though he had dictated it. Nickel said he did not recall it. The company contention is that as a trustee and officer, Nickel had no legal right to acquire for personal use land of the corporation. The land was sold at a large discount. It was appraised at $11 an acre, but may be worth $300-$400 an acre.

Merced *Sun Star* – 11/18/1954

SETTLEMENT MAY REACH $10,000,000

San Francisco- A second group of heirs of the

late cattle baron Henry Miller today forced into abeyance a compromise settlement in a fraud-conspiracy suit reached by another group of heirs last night.

If eventually culminated, the settlement may total as much as ten million dollars. Today's delaying action came as a direct result of move made in the state District Court of Appeals designed to halt the appointment of new trustees to manage the 100 million dollar estate. The action was taken by attorney John Neylan, representing five heirs. The appeal court rules that the previous Superior Court appointment of a new trustee, and most other actions in the lower court were invalid.

TWO GROUPS OF HEIRS

The plaintiffs in the original action are four great-grandchildren of Miller. They are George W. Nickel Jr., John Beverly Nickel, Mary Nickel Lombardi, and Sally Nickel Mein. Making up the second group of heirs are two grandchildren of Miller, George W. Nickel Sr. and Mrs. Beatrice Morse, and three great grandchildren, Henry Miller Bowles, George McNear Bowles and Amy Bowles Lawrence.

One phase of the fraud-conspiracy suite was believed ended last night when the plaintiffs agreed to accept property appraised at 4 ½ million dollars from the estate of the late C. Elmer Houchin. The Bakersfield millionaire and one-time leasing agent for Miller & Lux, Inc. had been sold valuable oil bearing lands at ridiculously low prices by former Miller & Lux officers, the plaintiffs charged.

APPEAL PROMISED

C. Ray Robinson, Merced attorney for the plaintiff heirs in the original state suit attacking the administration of the estate and attorney for the trustees in the attendant 100 million dollar federal action said: " Mr. John Neylan, representing the other five heirs has temporarily succeeded in blocking recovery by the trust of money due the heirs. However, we are confident the California Supreme Court will reverse today's ruling." The court ruled that Superior Court actions on and after June 15th are void because all heirs were not before the court. The ruling also prohibited any future action by the Superior Court until all heirs are included in the suit. Today's ruling did not however affect the appointment of a temporary trustee for purposed of filling claims against the Houchine estate. Therefore, the claims filed by the trustee before the filing deadline are admissible in the action.

FEDERAL COURT ACTION

In the suit that is still pending in federal court the plaintiff heirs are charging a number of officers, former officers and other individuals with fraud and mismanagement demanding restitution.

Miller & Lux Make Start To Sell Store

What might be considered the forerunner to dispose of their general merchandise business in Los Banos is the closing out sale of their shoe stock as advertised on page three of this issue. The sale is final as the store room has been leased to other parties commencing the first of the year.

It is thought that the entire general store stock will go on the block sometime during the next few months.

The firm has already disposed of all their Gustine holdings, having sold their butcher shop, lumber yard, banking interests and store stock.

12/10/1926

Miller & Lux Land Sales Total Million

3/26/ —— 1937

Land sales totalling more than $1,000,000 have been reported by the Miller & Lux corporation for the first two and one-half months of 1937. Though some of this acreage is from the Buttonwill holdings of the corporation, the majority is located mostly east and south of Los Banos and almost entirely represents newly developed agriculutral land. Water rights are sold with the land in nearly all instances.

Contrary to the trend of a few years ago for large acreages by a few holders, the average sale now is between 200 and 300 acres and means that many new farmers are moving into this community to make their future home.

Miller & Lux are still selling their agricultural land at an average of $90 an acre, which was established during the depression. However, in the face of a rising price for all farm products and increased value of farm lands, it is understood that a substantial increase in the price of M & L lands will be made in the near future.

MILLER & LUX HAVE BIG LAND SALES YEAR

12/31 —— 1926

Real estate and financial operations of Miller & Lux Incorporated, have reached major proportions during the past twelve months, according to announcement by company officials in this city. Land sales have totaled $3,500,000 during the year and negotiations in progress represent an approximate expenditure of $2,000,000 for additional land.

Proposal of the directors of the San Joaquin River Water Storage district to purchase canals and water rights from Miller & Lux involves a consideration of more than $10,000,000, it is stated. Water rights in Kern river are also to be sold to the Buena Vista water storage district for $1,500,000 it is announced.

Favorable prices for cattle are helping the firm's current position materially and as a matter of fact, operating costs for the current year promise to be better than those for past years. Cotton lands of the company have been yielding this year on an average of one bale per acre. During the past year the firm has reduced its bonds and notes outstanding in the hands of the public by more than $2,000,000.

Land Sale articles from the Los Banos Enterprise 1920-1930's

SAFSTROM COLLECTION

MILLERS HEIRS HINTED NEAR AGREEMENT

Lose Move to Delay Court OK of Trustees' Actions

Hints that heirs to the fabulous Henry Miller estate are trying to settle their differences out of court were voiced yesterday in Superior Court.

Attorneys, saying they spoke for all the beneficiaries of the estate, argued bitterly for two hours, asking Judge Herman A. van der Zee to delay action one week on the first report of the estate's new trustees.

Judge van der Zee rejected every legal move of the attorneys, however, and concluded the hearing by accepting the trustees' report, and with it, a contract for Merced attorney C. Ray Robinson whom the trustees have hired to recover approximately $50,000,000 he claims the estate was defrauded of by former trustees.

CHAT—George W. Nickel Sr., left, and his brother, J. Leroy Nickel, two of the principals in the Miller estate battle here, hold a conversation at the hearing yesterday. J. Leroy resigned in June as head of the Miller and Lux firm.

MRS. AMY B. LAWRENCE
A Miller heir

CONTRACT PRAISED.

Robinson's contract, with fee schedules that could make him a millionaire, was praised by van der Zee as **"a fair one, and one that no lawyer could find fault with."**

The Judge went on to praise the three trustees whom he appointed June 30 as having **"my fullest confidence."** In an apparent reference to the action of the new trustees in dropping banker Jerd F. Sullivan from a post as director of Miller and Lux Corporation, the estate holding company, van der Zee asserted **"it was their right to make changes in the directorate. They have a right to do that at any time. In fact, it may be their duty."**

J. Leroy Nickel, who resigned in June as president and chairman of the board of Miller and Lux, and is a key figure in the dispute, occupied a front row spectators' seat next to three of the eventual heirs of the estate as the hearing began.

ASKED POSTPONEMENT.

At the outset, Attorneys John Francis Neylan, Harold Faulkner, James D. Adams and Varnum Paul asked van der Zee for a week's postponement.

Neylan, representing Henry Miller Bowles, George McNear Bowles and Mrs. Amy Bowles Lawrence, asked permission to intervene in the case.

The Bowles family members are great grandchildren of the late land and cattle baron. They will share in the eventual distribution of his wealth with four other great grandchildren — John Beverly Nickel, George W. Nickel Jr., of Los Banos, Sally Nickel (Mrs. William Wallace) Mein, San Francisco and Mrs. Mary Nickel Lombardi, San Francisco.

Leroy Nickel is an uncle of the other Nickels and receives, along with his brother, George Nickel Sr., of Pebble Beach, and Mrs. Beatrice Bowles Morse, of Berkeley, annual benefits from the estate. Mrs. Morse and her son, Henry, are members of the board of Miller and Lux.

PETITION REFUSED.

Van der Zee refused to accept Neylan's petition to enter the case, advising him to file it with the clerk for a hearing at a later date.

"All the beneficiaries should be in this case and I believe the acts of the court will be void unless they are included," Neylan declared.

He said further that George W. Nickel Jr. had telephoned him earlier in the day, saying that he spoke for Robinson in requesting a delay in approving the trustees' report. Nickel and his brothers and sisters are the ones who started the court fight to oust their uncle from Miller and Lux.

PROCEDURE HIT.

Faulkner, representing J. Leroy Nickel, attacked the procedure by which the trustees, headed by Attorney Louis Ferrari, had hired Robinson. He said a delay would give the beneficiaries a chance to reach **"some agreement on this."**

Attorney Morse Erskine, for the trustees, then told the court "a delay is not justified, and I also feel that if these parties want to get together and arrive at a composition of some sort this report will not prejudice that action.

"The trustees do not want to stand in the way and will not stand in the way of any compromise of this matter. The corporation and not Robinson has the right to make the decision."

Van der Zee recessed the hearing at one point to take a telephone call. He announced afterward that the call was from Robinson in San Diego, advising the judge that he did not ask a continuance and had not authorized anyone to speak for him.

Neylan asserted that Robinson's contract was "railroaded through the trustees" at the July 26 meeting where Sullivan was ousted.

"Why are the trustees so anxious to rush this through?" He said. "The people who are going to pay are here and are asking a delay."

Robinson's contract provides a $25,000 basic fee and would give him twenty per cent of all funds be recovers up to $10,000,000 and fifteen per cent of all monies beyond that.

REPORT CRITICIZED.

Faulkner pointed to a misinterpretation in the trustees' audit report of Miller and Lux. He said that $2,930,937 listed as advanced to J. Leroy Nickel and former trustees A. R. Olsen and J. E. Woolley was money given them in 192[illegible] to pay inheritance taxes on the estate.

Attorney Adams in behalf of Olsen, submitted a statement of opposition to the estate suit. He argued that the trustees were not properly appointed and Van der Zee denied the objection.

Paul spoke for Mrs. Morse and George W. Nickel Sr., in requesting a delay.

8/3/1954

Clipping from unknown San Francisco paper, donated by Mrs. Safstrom

Henry Miller's Family

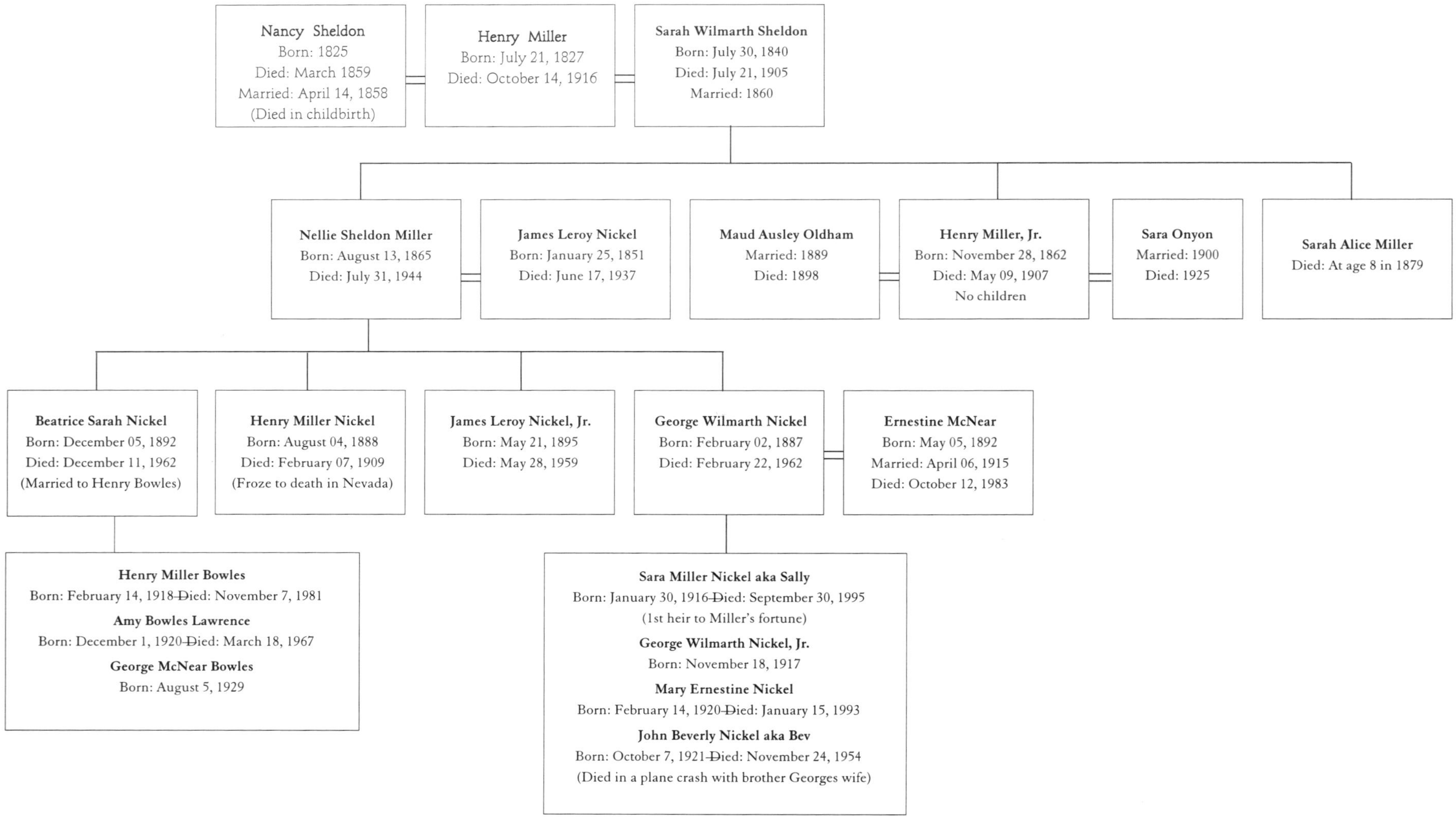

Miller's estate was left to the third generation. Sarah Miller Nickel was the first born of that generation. There are now heirs too numerous to list though the Bowels and Nickel families.

Sheldon Family of Rhode Island

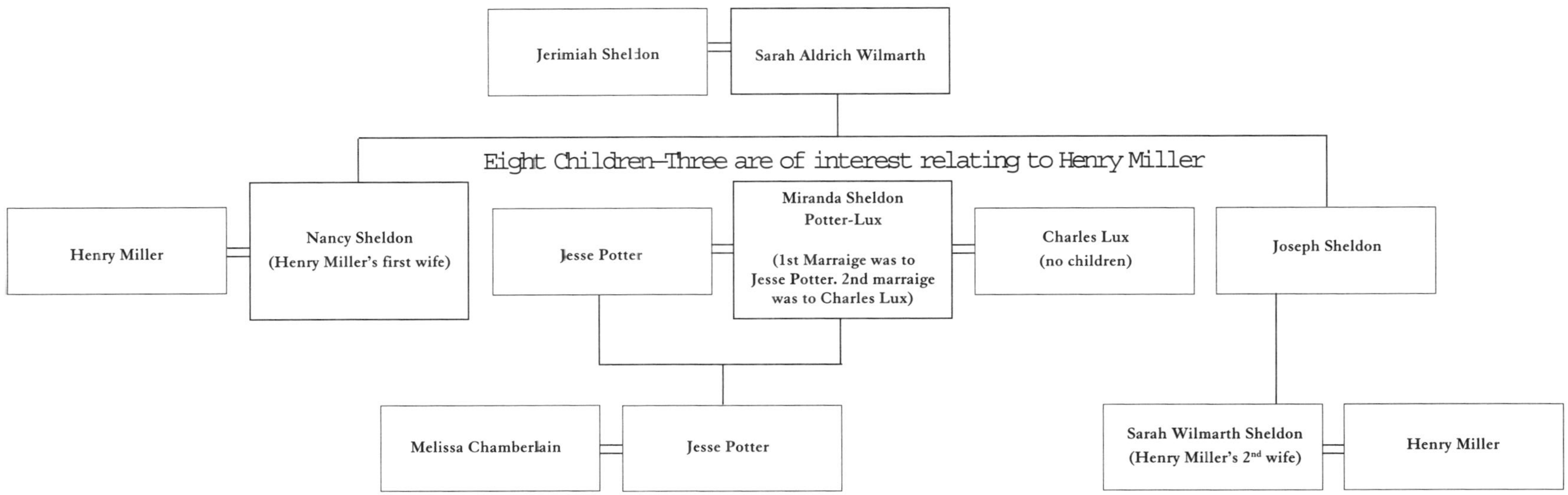

Henry Miller, Jr., as a cadet, age 13

Part II

Chapter Eight

Letters - 1860 to 1890

Henry Miller wrote hundreds of letters per month, sometimes 20 to 25 a day. Most that remain were to his partner Lux who was also his brother-in-law at one point and later uncle by marriage to his second wife.

He and Lux seldom changed their formality towards each other. Miller's early letters are full of misspellings, strange terms and odd phrases. I have left them to give the reader the flavor of the man and times. Miller also made liberal use of XXX in his letters. Why no one knows. What his meaning and reason for this usage is unknown. The letters are from various sources. Ralph Milliken Museum, Bancroft Library, Fresno State Library, ones held by individuals, and a body of these letters were found pasted in a notebook with a notation that they were written by Henry Miller. They appear to be in his handwriting. It was a common practice for letters of well-known persons to have the signature cut off as it was thought that the signature was more important than the actual letter. Other letters are intact and all are typed as written. Some have been worked over in a prior attempt to make them more understandable to the reader. Later letters became more fluent as his knowledge of written English became better. Miller was a man on the move and these letters reflect what was happening during the early years, including the loss of open range; the need of fencing; the battles and bills of the Canal Company; the purchase both legal and illegal of the land script Civil War veterans had been paid with; taxes, supplies, and day in and day out dealings of the operation.

In the 1876-77 period Miller started to cut back on handwritten letters from his note pad and started dictating them to clerks at the ranches. After Lux's death his letters were for the most part dictated to clerks at the various ranches who cleaned them up. A story is told of Miller, as he called for a certain letter. The clerk who had written it for him brought it to Miller. Miller looking at it said, "I's write good letters." Included here are various letters somewhat rustic in nature and style. These letters were written over several years and will give the reader insight into Miller and Lux's working relationship.

April 12, 1864
Lone Willow
Witness John R Means
Witness Robert Pebley –R. B. Hard
Papers belonging to the Lone Willow
Gilroy

Received of Henry Miller & Charles Lux, Five Hundred dollars being payment in full for rent of land located on the San Joaquin River Fresno Co. from the first day of January 1868 up to the first day of Jan. 1869.

J. C. Crocker

Enclosed was James Crocker's receipt for land rented

XXXX

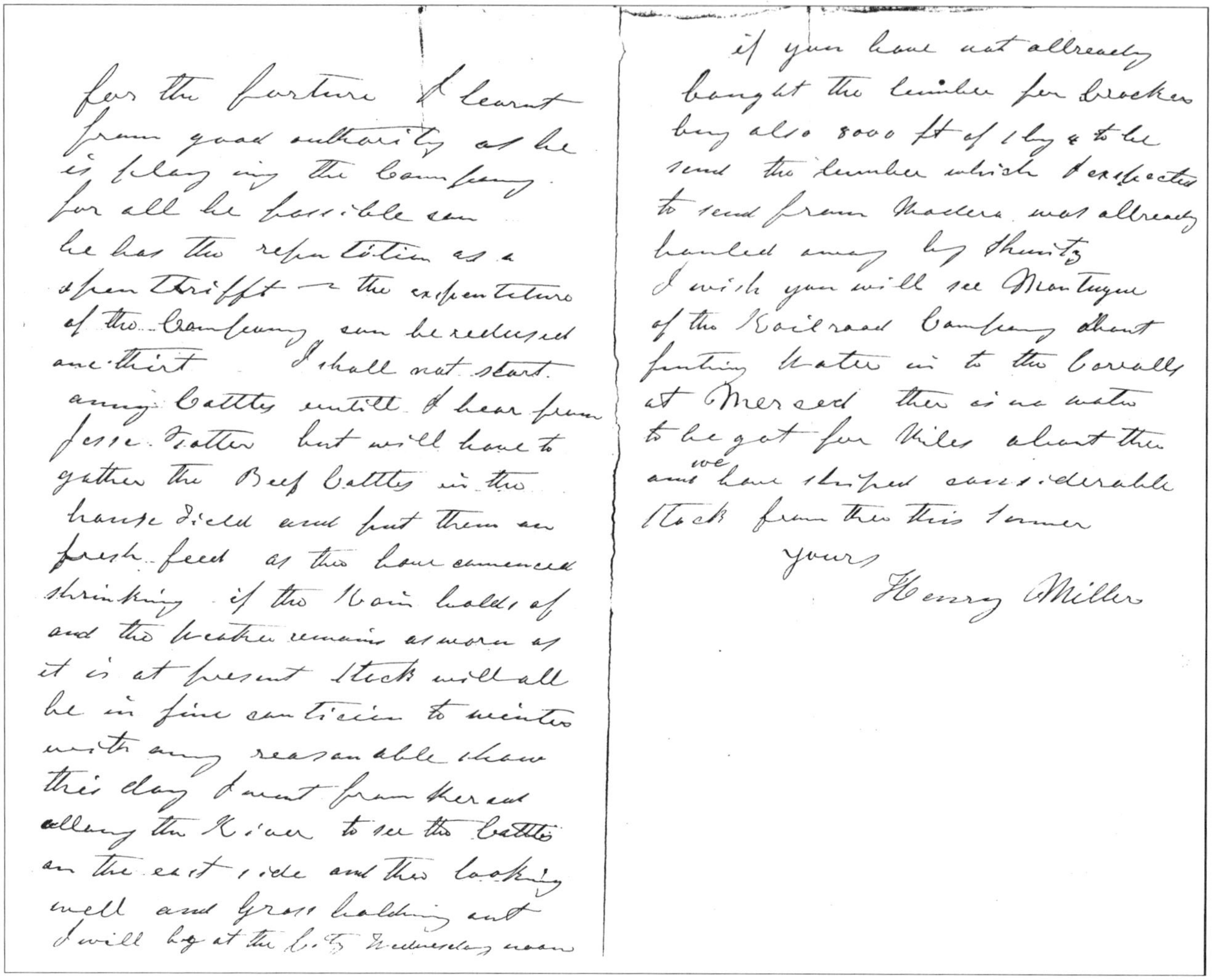
for the fortune I learnt
from good authority as he
is playing the Company
for all he possible can
he has the reputation as a
spen thrifft — the expenditure
of the Company can be reduced
one third   I shall not start
any Cattle untill I hear from
Jesse Potter but will have to
gather the Beef Cattle in the
house field and put them on
fresh feed as the have commenced
shrinking if the Rain holds of
and the weather remains as warm as
it is at present stock will all
be in fine condition to winter
with any reasonable show
this day I went from here out
allong the River to see the Cattle
on the east side and the looking
well and Grass holding out
I will be at the City Wednesday noon

if you have not allready
bought the lumber for Brocker
buy also 8000 ft of 1 by 6 to be
send the lumber which I expected
to send from Madera was allready
hauled away by thirty
I wish you will see Montague
of the Railroad Company about
putting Water in to the Corralls
at Merced there is no water
to be got for Miles about there
and we have shipped considerable
stock from there this summer
yours
Henry Miller

Sample Miller letter. Most likely done on the road 1870's

Lone Willow Station, Merced Co., State of Calif

This indenture made this the 3rd day of June 1864AD between R. B. Hard party of the first part, Henry Miller & Charles Lux of the City of San Francisco – Witnessed that the party of the first part for an in consideration of the sum of Two Thousand ($2000) dollars to him in hand paid by the party of the second part, does demise release quite claim all of his right to the claim passes on and in to all that tract or parcel of land known as the Lone Willow Station on the road leading from Pacheco Pass to Visalia in Merced Co., together with its improvements appurtenances there into belonging with all the right grazing with the above mentioned location and the place above this as the Tomas Deacon's with the improvements and grazing prevails there unto belonging together with all the improvements also and further more all the cattle branded (H) Also the brand (H) belonging to the party of the first part, the delivery of the home and improvements as above mentioned to E. Whiting to hold for the party of the second part. The receipt of the specified sum is herby acknowledge by the party of the first part on the date and day above mentioned.

In witness hereof the party of the first part unto set his hand and seal in the presence of B. B. Baker, R.B. Hard (seal) for in consideration of the sum of Two Thousand ($2000) dollars I sell to Henry Miller & Charles Lux all of my right and title to the

Lone Willow property together with all the cattle branded thus (H) and I also sell all my right to a certain claim bought of Tomas Deacon on improvements expected. Said improvements to remove from said claim between this and the Santa Rita Ranch.

Dated April 16, 1864

April 16, 1864

Received of Henry Miller & Lux the sum of sixty dollars in part payment for eighty acres situated on Stinson Slough.

Condition of contract as Follows:

J Hawthorn does herby agree to establish a genuine title to the above-mentioned land. J. J. Hawthorn does hereby agree to deed the same to H. Miller & Lux. Said Miller & Lux are to pay any further expenses which may occur in paying for said land and establishing title hereof satisfactory to said Miller & Lux.

Witness J. C. Crocker Jas Hawthorn

Endorsed Hawthorn Bill of Sale for water hole

March 7 1868

This agreement made and entered to this Seventh day of March A.D. 1868 by and between Miller & Lux of the first part and John Sutherland and James Sutherland of the other part.

Witnessed – That said John & James Sutherland agree to sell to said Miller & Lux all their steers from years old and upwards, and all their three year old steers that are fat and Choice. Also their fat barren cows, however said cattle may be branded at the price of Forty dollars ($40.) per head in gold coin – The said cattle to be counted and delivered at Murphy's Slough in the vicinity of Elkhorn Ranch at any time Miller & Lux may desire between this date and the Fifteenth of August Next, but all to be delivered and received in not more that three deliveries.

Said Sutherland agrees to use their best effort to get together as many cattle above kind as possible within the time specified.

Said Miller & Lux agree to pay said John and James Sutherland Ten Thousand dollars cash on the execution of this agreement and to pay for said cattle as fast as delivered and received said Ten Thousand dollars to remain and be applied as the last payment for said cattle. It is expressly understood that said Miller & Lux and not to be compelled to take any cattle that are cripple or that are sound and healthy.

Miller & Lux agree that after said cattle are received by them they will not turn any of them loose of the east side of the Pacheco mountains.

Miller & Lux (seal)
John Sutherland (seal)
James Sutherland
(seal)

XXXXX

Received San Francisco

Of Miller & Lux the sum of Ten Thousand dollars being the first payment in the above written agreement.

John Sutherland
James Sutherland

Miller & Lux
Bill for toll over Pacheco Turnpike
July 5th. 1867
Receipted by L. F. Bell

6 ox wagon	*$3.00*
4 horse wagon	*2.00*
130 mules	*5.20*
Cattle	*10.00*
1 saddle horse	*.25*
1 wagon	*1.00*
60 horses	*2.40*

Bill for toll over turnpike Jan. 3 1866

To O. A. Thoms
San Luis Gonzaga
1868
Charles Lux, Esq.

Send me up by next steamer 500# of cheap sugar and 500# of San Francisco refined No 2 to be left at Chamberlin Landing

Henry Miller

San Luis Gonzaga
Mar. 17,1868
Charles Lux Esq.
Dear Sir:
The cattles from the Joaquin are starting here this evening and will be in the city next Monday xxxxx
Henry Miller

Lone Willow
Sept 7, 1868
Charles Lux San Francisco
I am writing you these few lines letting you know I have started last evening 518 head of Cattles including 148 Cowes & Calves and 27 Oxen. There will be at your place Sunday next
Henry Miller

Firebaugh Ferry
Dec. 6 1868
Charles Lux San Francisco
About the Hogs ther have about 500 head gathert and there will gid up watt hogs there is outsite of the fenz and will start about a week from the Day.
Henry Miller

Santa Rita Ranch
January 7, 1869
Charles Lux, San Francisco
Dear Sir:
I am writing you these few lines letting you know as I have arrived here with Potter and showed him the class of cattle I will give them, and he concluded to take them and receive them in two lots. In the next band I will send half the number. Will write you how many oxen and how many stags. I shall commence giding up a band of beef cattle the last of this week. When I arrived yesterday at Kragenhagens I found the men wath were working at the fenz drunk and not having been to work for some time. There for you will write to Hassel at the Lauterwasser place not to cook for them. I wish you could find some men to send over. If you would send them, also write to Hassel so he can make his arrangement accordingle. There is only Joseph Bush and one more left and the work unfinished, and I can't spare them. The carpenter you sent Monday last has not come yet and I am afraid he will not come. The weather has been very cold here this last few days and the cattle are looking that ruffe but doing well. Write me by Thomas Dias or White. Let me know about the market and see Mr. Williams and Thornton about my case in the Los Animas.
Yours
Henry Miller

Auzerais House
I. M. Staples, Proprietor
San Jose, January 12, 1869
Charles Lux
Dear Sir
I went out with Carlos Carrut Sunday afternoon & bought 10 oxen for 640 dollars and 8 bulls at 30 dollars per head. He will call on you for the money. Yesterday morning I went to Mountain View and bought of Charles Davidson 17 head of cattle amounting to 706 dollars. Enclosed in this you find statement. I could not leave for Gilroy yesterday on account of suit is coming up about the Chewago Land. It seems there is nothing ready yet to try the case. Judge Peklam requested me to remain until this afternoon. If you make any arrangement with the butcher from the Almaden about delivering him some beef cattle I wish you will write me to Gilroy and let me know the condition. I shall remain at Gilroy until Friday. I wish you will see Green in regard of he wants to come down. Don't encorage him about coming without it is his free will. We owed Davidson for two bull and two calves for sometime amounting $130 dollars for wich amount I gave him a order on you. Let me know in regard to the post watt were reported loosed.
Yours
Henry Miller

Gilroy
January 15, 1869
Charles Lux
Dear Sir:
Bearer of this letter is Ward White wich I told you to send over to Black to gid the

horses. I wish to see to send Reynolds over with him. I wish you will try so ther be no delay. Give White money for expenses. Also give him five hundred dollars in small coin to bring to the Joaquin. Let Thomas Dias and the man go back with Ward White along with the horses. The cattles will be at your place Sunday evening 400 head including whatever cattles left at the Jose, also 106 calves. I shall start with Potter for the Joaquin this morning.

Yours

Henry Miller

P.S. Henry Fisher at Peachtree has been murdered the 9th of this month at his own house.

Santa Rita Ranch

January 19, 1869

Charles Lux

San Francisco

Dear Sir:

I am sending you those few lines in regard to the post. The way the county is plowed up and settling we are complet to fenz as soon as possible from San Luis Camp down. I wish you will try and gid post to make up for watt were lost of them were among and have them shipped as soon as possible. The price we overt Captain Shreck I think it is enouge, we paid them for the last lumber more than necessary. It has cost us about $200 dollars to remove the lumber were they landed the same. We want the post landed at the same slough, but at a better place. I will put up a flag wer to leave them. I shall commence Thursday next to gid up beef cattles and will have them at your place about one week from next Sunday. Write me at Firebaugh Ferry next Monday.

Yours

Henry Miller

Santa Rita Ranch

March, 1869

Dear Sir

The way the Country is plowed up and settling we are compelt to fenz as soon as possible from San Luis Camp down.

Yours Henry Miller

Firebaugh Ferry

March 6, 1869

Mr. Sandborn

Dear Sir:

Me and Mr. Lux returned this evening from the east site of the river and shall go to the ranch tomorrow. I expected a letter from you by stage this evening, but none came. I think you must put the letter at Wells Fargo Express, and that is wrong, because they do not open the box, as there is no office here. You must give your letter for this place to Robert the Agent for the Visalia Stage Line ten it will come in the waybill. I think Mr. Lux will start Wednesday morning for the City and will gid ther Thursday. I send enclosed in this an order for $500 dollars for to pay for some land Mr. Chapman purchased for us. In case he will call on you for the payment, give him the amount. It me be more, if it is, pay him the ballens out of your cash at hand. Tell Gadlile to have the cattles (blank line)___in way of having them killt and Robert to bring in the worse cattle fast, as I think there will be plenty middling cattles in the market soon. Have Jesse Potter tell Thomas Dias and have him to start Tuesday morning so he will be at the San Luis Camp Wednesday evening. I will try and have a band of cattles at the City by two weeks from the day or tomorrow so Gadlile can make his calculation in regard of killing with them is at hand. If the Weirer I bought is not shipt yet, I wish you will go and have it sent Monday next so it will go Tuesday up river. In case it comes before, I will telegraph from here. Wen the man we bought the Patent Right of to start for here, send letter by Thomas Dias. Also any letters connected with our Business. Also give him $500 Dollars in cash.

Yours

Henry Miller

Mr. Roberts is stoppen at Rufs House

Gilroy

March 18, 1869

Charles Lux

Dear Sir:

I received your letter in regard to the Post. Let Captain Shreck know as we want the Post and 40,000 feet of Lumber left at the same Slough were the Lumber was left last Fall but higher up the Slough. The other 20,000 feet to be left at Chamberlain Landing. I have telegraph to you in regard to the cattles. Kill all wath will be fitt out your field as we will have plenty of nice cattles after this. The band going down are better tan I expect from the outside. There are 400 head and 140 calves. The Crocker cattles are at Firebaugh Ferry by this time and com right along. I will be in the City Saturday morning.

Yours
Henry Miller

P.S. The hogs will start tomorrow and will be at the City next Wednesday or Thursday.

San Luis Rancho
March 21, 1869
Mr. Miller
Dear Sir

I drop you this line to let you no how things are about three o'clock this morning I had one hell of a stampede when daylight came I was in front of Chas Woods sheep camp I had five men on gard all the men this side of hell could not of stopped them. I am six head short xxxxx I had good luck whole trip until this morning

Yours in haist E. M. Crocker

San Francisco
March 27, 1869
John McPike
Dear Sir

I suppose you are aware at you stock is on ourer Land and at we shall commence to fenz from the San Luis Camp down connecting with your line fenz et will be necessary for you to find Pasturage for them.

Yours truly Henry Miller

Bloomfield Farm
April 5, 1869
Charles Lux
Dear Sir:

I shall start out this morning surveying at the Soap Ranch and shall remain about here till next Friday. I wish you will try and send the sulky you spoke to me about buying, also the articles I bought of Booley & Johnson such as Blankets and Harness & Books has not arrived yet. Send Jesse Potter to see about it. I want it before leaving for the Joaquin. Also wirte me before next Friday.

Yours
Henry Miller

Bloomfield Farm
April 6, 1869
Charles Lux
Dear Sir:

I am writing these lines in regard to the man build up the Range at my Kitchen. He wished me to inform Mr. Elis at Washington Street as he is going to the Joaquin tomorrow and he will return Saturday or Sunday next. I wish you will send down Jesse Potter letting him know also to have Elis sent out to the man wife his name is Pollard and let her know as he will not be home untill next week, I will go out tomorrow and meet the cattles and if I can make arangment about sending up some calves by railroad I will send them up tomorrow night. I will telegraph if I send them and how man ther is. Ther is about 500 beef cattles and 150 calves.

Yours
Henry Miller

Gilroy
April 7, 1869
Charles Lux
Dear Sir:

I send by Cars this evening 70 head of calves the price for freight is $40 dollars for the lot. The cattles will be at the Fellon field the evening and will gid to the Danitron field Sunday evening. There is only about 30 calves more among the cattles there is 509 head of beef cattles. Ther are as a lot, nice cattles. In regard to my Jackson Street Property I wish Mr. Sandborn will attend. I wish you will see Mr. Gellar about the oxen and if ther not sold buy them. I shall not be able to leave for the Joaquin until Sunday next. I shall send for some more calves ther will be here the last of next week.

Yours
Henry Miller

San Luis Ranch
April 12, 1869
Charles Lux
Dear Sir:

I have youst returned from Sam Clark Ranch and bought his land at $2.50 cents per acres and I think I done well. I give him a order of $1000 dollars on you and the bellens to be paid on deliver of Deed at Stockton. I shall gid to the Ranch this evening, will write from ther about how soon I shall start some cattles, but shall this week start a band of calves and some cattles.

I wish you will not tell any body the price I paid to Clark for his land. In case Mr. Williams want anything in way of money, let him have what he want, as I have to atent in person. Let me know emitadle. I wish you will see Mastick about my Tannery Property and tell him to leave nothing undone in regard toward bringen the matter to some end.

Yours
Henry Miller

I wish you will try and see this man Fleming or his agent about the bellens of the land. He owns about 800 acres if you could buy the same at a fair price.

Santa Rita Ranch
April 13, 1869
Charles Lux
Dear Sir:

I arrived here last evening and as there is some lumber wanted, wich amount I enclosed herewith. The five doors which Mastick send up with the lumber are only one inch. Them will not suit. Send up 5 doors in exchange according to order we wanted to come up with return boat. Al Lewis arrived last night with 60 head of oxen. I brandet them this morning. I shall commence surveying about next Friday.

Yours
Henry Miller

P.S. I want this lumber go by the T.C. Walker, and from there transferred to the Steamer Hariet ship to Chamberlains Landing if not ordered otherwise

Santa Rita Ranch
April 19, 1869
Charles Lux
Dear Sir:

I want to inform you as Charles Sheppard quit work and as I don't know watt amount you have paid to his wife I wish you will see watt amount you have paid. The total amount for work is $210 dollars. I haved reducted $60 dollars you paid him and gave him by order on you $130.00 dollars and $20 by cash wich above will make the total amount.

Yours San Luis Camp

April 25, 1869
Charles Lux Esq.
Dear Sir

I am writing these lines in regard to wanting Provision there started to work at the Fenz yesterday we want the following articles

One or two barl of fresh corn beef but not to be bought cheap I want good Beef one box of Bacon the hartest Grain fett is the cheapest about 250#, one meat saw, half dozen Knife with steel and Carving Knife & Fork, Cod Fish two half barls of secont qualite of sugar all this I want send by the T.C. Walker and transferred to the connecting steamer and be left at Hills Ferry in care of W. Wilson.

The Fenz we are building will be about 14 mile and ¾

Yours Henry Miller

May 8 1869
San Luis Gonzaga
Charles Lux, Esq.
Dear Sir:

About the renting of the land Malarin we can dig along without and for that I want to give him a cowe. I think et police.

Yours,Henry Miller

Bloomfield Farm
Columbia Ranch
May 18, 1869
Charles Lux San Francisco

About the Hogs there will leave the Lone Willow Sunday at be at Gilroy about Friday.

Henry Miller

Santa Rita Ranch
May 25, 21869
The water is over most of the bottom land and et is worth a Man Lieve to gid in and out at present.
Henry Miller

Bloomfield Farm
June 2, 1869
Charles Lux Esq.
Dear Sir
I want Thomas Dias to start Friday morning for San Luis Camp he want to bring along $1000 silver tell him to take the Visalia stage as far as San Luis Gonzaga and ourer men will meet him there with the Bugge. xxxxx
Yours Henry Miller

Santa Rita Ranch
June 14, 1869
Charles Lux
Dear Sir
We hate last week an Man drowned at San Luis Slough.
Yours Henry Miller

San Luis Ranch
Sept 3. 1869
Charles Lux San Francisco
I meet the Cattles on the Top of the Mountain there will be at the Soap Ranch Saturday. Ter is 500 beef Cattles 130 calves and will gid to the Denistson field Wednesday.
Henry Miller

Oct 25, 69
Henry Miller
Sir
I sent by last load 3,000 lbs of Flower 700 pounds Beans Some apples 6 kegs of nails the other wagon loaded with Pickets
Grinter

Western Union Telegraph Co.
San Francisco 1870 Jany 26
Received at San Luis Ranch 6.10PM
To Mr. Malarin
Take Mann and Ralphs message over to Millers Ranch immediately. He will pay expenses delivering. Signed Chas Lux
Miller will pay $10 for delivery Paid Paid

Santa Rita Ranch
May 1, 1870
Ef Thomas Dias hass not started yet wen you receive this give him $500 dollars part small coin and part silver encludded quarter and 10 cent pieces.
Henry Miller

Lone Willow Station
May 9, 1870
I want the same day the Cattles gids to the Soap Ranch 75 Calves sent by cars as I want the Cows to come back with the ones which are at the soap ranch now.
Henry Miller

San Luis Camp
June 27, 1870
The Cattles will start from here Wednesday evening and will be at Gilroy Saturday.
Henry Miller

Gilroy
June 30, 1870
Charles Lux San Francisco
Thomas Dias left the Lone Willow yesterday evening with about 650 head of Beef Cattles and 160 Calves will gid to the Soap Ranch next Saturday.
Henry Miller

Western Union Telegraph Company
San Luis Ranch, Cal. Aug 8, 1870
Received at San Francisco 8 12.20P.M.
Your cattle passed yesterday
I. Malarin collect
Kreyenhagen Store

Nov. 14 1870
Charles Lux San Francisco
I am my way to Gilroy will take stage here and be there tomorrow morning. I started this day about 150 head of Beef Cattle 130 Oxen and 150 Calves
Miller

Santa Rita Ranch
April 26, 1871
Charles Lux Esq.
Dear Sir

I arrived at San Luis Camp Monday morning and found the Hogs not started and everything in a muddle ther left Monday evening and will take them all week to gid to Gilroy ther have about 300 head.

Yours
Henry Miller

Bloomfield Farm
June lst 1871

I started this morning 745 head of Cattles 350 head to be left at the 15 Mile House and the bellens to go to the City. They will gid to your ranch Sunday.

Henry Miller

Bloomfield Farm
July 9, 1871
Charles Lux Esq
Dear Sir

I don't need Regle after his month is up His judgment don't suit mexx You can calculate about watt time you want calves from the tule and write to James Maxwell in care of Charles Warfield to have them ready in the Kimble field adjoining the Warf tan you can gid at any time there can gather about 200 cowes and calves wich make two shipments and let one man come down withxxx

Yours
Henry Miller

Firebaugh Ferry
August 20, 1871
Charles Lux Esq.
Dear Sir

I arrived here this morning left snelling Friday noon I found we were assest on 20,000 head of cattle improvements 40,000—land 165,000 acres at $2.00 per acre. I have reduced the cattles to 6,000 and improvements to 11,000 and land.

I shall have to send Collier up next week an witness in regard to the quality. I find the work on the Chapman canal gidding alllong better tan we could do the work the leaving machinery wich we have not god. I therfor put ourer man unter Mr. Overhauser wich hase charge of the work and I find at Chapman is working with econume and good judgment.

I Shall go down at the Bensley Canal and will be at the Ranch two eveningsxxx

Yours
Henry Miller

Gilroy
August 23, 1871
Charles Lux Esq.
Dear Sir

Art Ward will be at Modesto with 200 head cattles Friday evening-you better go to the office and see if there is a train Saturdayxx

Yours
Henry Miller

Bloomfield Farm
September 3, 1871
Charles Lux, Esq.
Dear Sir

xxxWill start the man back to the Joaquin with cowesxxx

Yours
Henry Miller

Firebaugh Ferry
September 21, 1871
Charles Lux
Dear Sir

xxxxPaid G. L. Hoffman for 10 ton barley and 20 ton of hayxxx

Yours Henry Miller

Gilroy
October 15, 1871
Charles Lux Esq,
Dear Sir

I don't know yet if I can ship cattles tomorrow night on account the cars ar all engaged by Hildreath and Dumphy and there will not be back in time for tomorrow nightxxx

Yours
Henry Miller

Firebaugh Ferry
October 25, 1871
Charles Lux Esq
Dear Sir
I drived fum the canal this morning there is a great deal of work to be done yet and I have to buy a bill of goods to be sed up ernitiall ther som to be out of every think Chapman is working on his onen account noth worth of his Fenz but the partner ship work will take 5 weeks yet before et be finished by sites the head gates at the river I will either be home tomorrow night or Wedsnesday nightxx
Yours
Henry Miller

Bloomfield Farm
November 15, 1871
Charles Lux
Dear Sir
I received tespatch from J. B. Bensley wanting to know even I was going to the Joaquin. I answered back to him I was going Saturday noon by stagexxx
Yours Henry Miller

Bell Station
November 18, 1871
Charles Lux Esq
Dear Sir
I engage Henry Shafer to work for me at 125 dollar fer month.
J.B. Bensley is on the stage with me herexxx
Yours
Henry Miller

Gilroy
December 29, 1871
Charles Lux Esq
Dear Sir
It rained verry hart here last night and wind blowing south east strong.
Strangerson the head enginee is here and want to go up to the Joaquin there is no stage this day in case I go with him I will telegraph to you—write Briggerman to use all possible means to finish the stables to put back the oxen.
Yours
Henry Miller

Chico
February 8, 1872
Charles Lux Esq.
Dear Sir
I want you to write to John Geer Stockton to hurry up ourer lumber as there is no excuse now we want him to take lumber and posts and want to be landed above the Muddy Bend ourer man Schmits is there to work
Yours
Henry Miller

Kreyenhagen store
March 2, 1872
Charles Lux Esq
Dear Sir
I arrived here youst now and shall gid to the ranch the night I learnt by Kreyenhagen at Maples & Ducan have finished and have turned over all the teams to Briggerman therefore I want you to send me up som more Barley about 20 tons which you art to bouy for 1.40 et will be time enove wen Jack Gier coms after the Fenzig also as much more hay as I bought about 25 tons because after Lone Willow Slought gid to room we will not be able to have over. Also enquire about the alfalfa see from Chille there art to know at watt tim et will gid xx let me if Harrold and Harrison sent you the lease signed by both. Send by steamer som Bacon & shoulder which you can buy cheap not ovr 10cent per #.
Yours Henry Miller

Santa Rita Ranch
March 6, 1872
Charles Lux Esq.
Dear Sire
xxx The grase is verry fine but cattles will not be beef before first of April I think by that time we can gid beef by picking out from all the cattles about calves I shall start tomorrow and lock enomgs the cattles an the out site and ef ther is enove to pay to gather them will put the man to work gathering them next Monday the man working emongs the hogs will be ready with a band the last of next week and will ship then from Merced

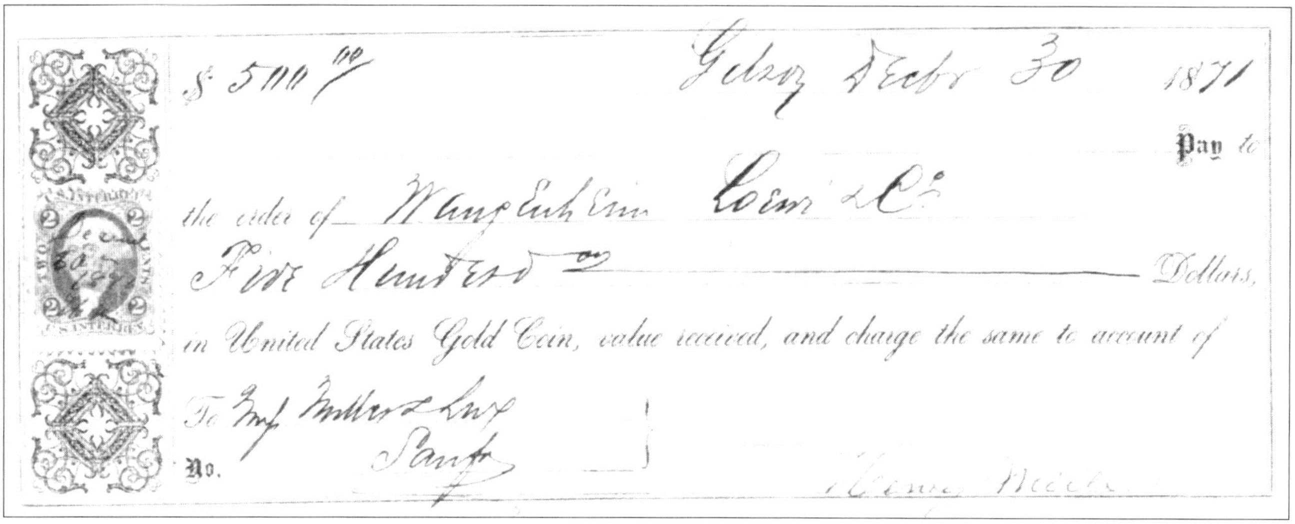

Gold draft signed by Miller in 1871

Cityxxx. The steamer with the lumber his not mad his apirens yet and etmaked verry badxxx

About the plowing there giding along every well ther is about 450 accres plowed and 300 acres on out the same sowedxxx

Yours

Henry Miller

Santa Rita Ranch

March 10, 1872

Charles Lux Esq

Dear Sir

The man started this morning and will commence gathering calves and watt beef cattles there will find ther have to up as far as Posso Chinnee about 40 miles south west of Firebaugh Ferryxxx About Firebaugh Ferry and all way to the foothills I never have seen the grase better and the cattle doing fine the cattles will be big fatt by the midle of Aprilxxx

The steamer Caroline arrived night before last with the hay and ground barley all in good order and about 60,000 feet of Fenzing and 800 post could not bring more on account of the low state of water but promosed to return to wer ther unloaded the trip before last and take up enove to finish the fenzxxx Also aks him Chapman ef he will take the Herrold & J+Harrison land wich is in site of his field according to ourer Lease with them for the same terms of yearsxxxSend me no letter to read but let me know the contends about yoou gid this before Thomas Diaz start for up here give him $500 in silver and send him by way of Merced City or as far as the railroad is complet ther the stage to Firebaugh connect and will send him a saddle horsexxxx

Please tell Millisia Chamberlain my familie are well and send regardsxx

Love to Henry and Nellie also Mrs. Lux Sarah

Yours

Henry Miller

Santa Rita Ranch

March 18,1872

Charles Lux Esq

San Francisco

Dear Sir

Captain Geir returned yesterday—unloaded the lumber about 8 miles north of Firebaugh—could not go up. The boat drawes to much watter he is now on heis way to the city to load up but he request ne to write to you at he has to exchange boat I promise your will go and respond for hin in way of not rooning the boatr en port deptxx

Yours Henry Miller

Santa Rita Ranch

April 17, 1872

Charles Lux Esq

Dear Sir

I arrived here Sunday evening stop at the Farming Camp ourer men have two ditches nearle completed Maples and Duncan ar also to work with ourer teams making site for iregation with out water there will be no crop even the earle sowed grain is beginning to dry up there had bin blowing a north west wind blasted the grase as well an the grain and we hath a great deale of heave frostxxx

Yours

Henry Miller

Gilroy

May 17, 1872

Charles Lux Esq

Dear Sir

Your letter I received in regard to the Ward land ef et yoins ourer land on the east side takeed ef et don't yoins anny of ourers we don't want to used-but ef it river front er is sheepxxx I am sending to you tomorrow 70 calves becase I wan the cowes to go back to the Joaquin tomorrow the Vagerosxxx

Yours

Henry Miller

Bloomfield Farm

June 20, 1872

Charles Lux, Esq.

Dear Sir

Thomas Dias arrived from the Joaquin with 350 head of beef cattle and 200 calves. Yesterday I selected 60 head of Texas cattle and put with the band. they will be at the Danstien Field Monday evening.

Yours, Henry Miller

Santa Rita Ranch

June 27, 1872

Charles Lux, Esq.

Dear Sir

The water is very high here yet

Yours Henry Miller

Gilroy

July 13, 1872

Charles Lux, Esq.

Dear Sir

Yesterday I sent you 174 head of cattle, 25 head of Oregon cows, they are very nice.

Yours, Henry Miller

Gilroy

July 22, 1872

Charles Lux San Francisco

Thomas Dias started with 354 head of cattle this morning for the city and 145 head of cattle were shipped by cars.

Your, Henry Miller

Visalia

August 17, 1872

Charles Lux Esq

Dear Sir

I hope you don't fee so poor as you generally doxx write mexx ef the Canal company signed the contrachtxx

Yours

Henry Miller

Visalia

Aug. 22, '1872

Charles Lux Esq.

Dear Sir

Your dispatch I received also a letter returned from Buena Vista Slough yesterday noon. Meet Crocker and Art Ward Crocker will not drive any more beef this year xxx I had a very (unreadable word)

Yours Henry Miller

Visalia

Sept 11, 1872

Charles Lux, Esq.

Dear Sir

I want Schafer to come to the Joaquin as soon as he returns from Sergant and collect out of the Middle Field a band of mixed Beefs. Davis can help him if he is needed. Tell him to make a small rodeo and don't hold the cattle long. Takes about 300 head. If you want Texas cattles he can collect in the House Field as many as you want without getting up all the cattle.

Yours, Henry Miller

September 16, 1872

Visalia:

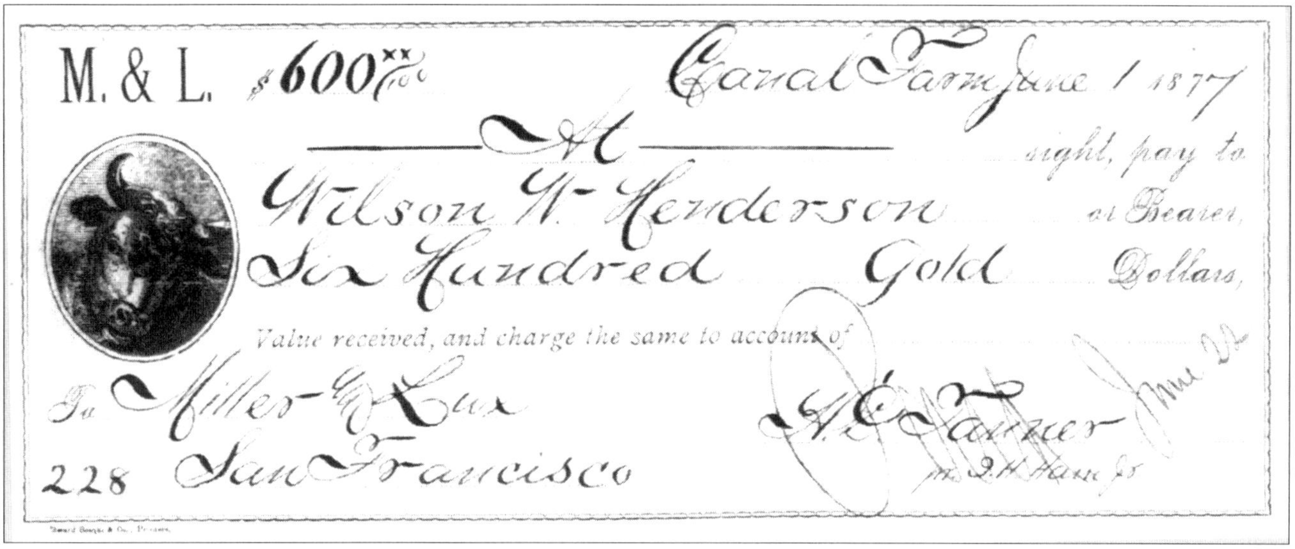

Miller & Lux's famous Bullhead check, 1877

Charles Lux Esq.

Dear Sir:

I have ben buying some genteel stiers most of them is good beef from 18 to 25 dollars per head wich ar verry cheap either to kill or to kip over…them is the cheapest cattles ever left Tulare. I know you will fret about money matter but ter ar large margin in them et will be 3 years before ther have anny other such cattles to sell and we can well afford to pay interest xxx. I want you to try and be at Merced City Sunday morning and I will have John Goodwin ther to receive you I shall start for Cross creek in the morning to clean the band and will gid to King river tomorrow night-will gid ready to start for the Joquin about Thursday evening to make your arrangements. So you be away aobut three or four days xxx

Yours

Henry Miller

Gilroy

October 2, 1872

Charles Lux Esq

Dear Sir

Pacheco Hegera wic ben working for us on the Joaquin for a long time is coming to you with his two boys he wants to put them in the Orphan Asilum as he is not able to support them and school them I promised him to assist him. No I wish Mr. Bolton will go in my place and see watt he can do. Ther father can pay something toward ther support and I will give $100.00 per year previted ther will conduct themselves properxxx

Yours

Henry Miller

Gilroy

October 15, 1872

Charles Lux Esq

Dear Sir

I received your letter in regard to Chapman now you must stop writing me all the tronble ef you don't I have to quit. I think I have trouble enove without gidding more

Yours

Henry Miller

Los Banos

Oct. 18 1872

I meet the Cattles Shafer started 200 Beef 220 Cowes and Calves xxxxx I want the Man to return with a band of Calve Cowes after Wednesday give Shafer $500 Dollars in small coin and silver to bring over here.

Henry Miller

Kreyenhagen Store

Nov 30, 1872

Charles Lux, Esq.

Dear Sir

A Ward will start this noon for the Tule

and gather up what cattle is left along the river.

Yours Henry Miller

Santa Rita Ranch
Dec 1, 1872
Charles Lux, Esq.
Dear Sir

Your letter of the 27 instant I received yesterday evening.

Yours Henry Miller

Bloomfield Ranch
1873
Gilroy, Ca.
Jan 9, 1873
Charles Lux, Esq.
Dear Sir

In regards to the card Tim Page brought. I am very sorry he got the land. He bought it on purpose to speculate on us.

Yours Henry Miller

Santa Rita Ranch
Jan 14, 1873
Charles Lux, Esq.
Dear Sir

Mills is getting along nicely. The grass is some places is almost a foot.

Yours Henry Miller

Santa Rita Ranch
Jan 19th, 1873
Charles Lux, Esq.
Dear Sir

J. Schmitz is coming down to spend a few days. I am sending to the Coal Vein to bring some down to be treated. You better go somewhere where there is someone which knows what is coal and what is not. I have been collecting stock cattle to take to the Panoche.

Yours, Henry Miller

Firebaugh's Ferry
Jan. 20th, 1873
Charles Lux, Esq.
Dear Sir

I left the Ranch this morning and cut across from Temple Field over the Plains. Grass is growing nicely and the cattle are doing well. Tomorrow I shall go down on West Side of the canal land and see what they are doing on the Canal Farm. About the man Case I will see him and he could get plenty of firewood.

Yours Henry Miller

Gilroy
Feb 2nd, 1873
Charles Lux, Esq.
Dear Sir

Find out if there is a steamer going up the Joaquin soon.

Yours Henry Miller

Firebaugh Ferry
Mar 5, 1873
Charles Lux, Esq.
Dear Sir

Yesterday noon I went up above Watson's Ferry and what cattles I have seen are looking well, I shall have to work over a good many cattle to get a drive. I will ship a few carloads from Berenda Station the last of the week.

Yours Henry Miller

Firebaugh Ferry
Mar. 5, 1873
Charles Lux, Esq.
Dear Sir

The cattle in Chapman's Field are not as good as they were a month ago, there getting just enough green grass to make them poor and eating old feed. As soon as Schafer returns from Pech Tree I will start him with another band of cattles. They will be at Gilroy one week from next Monday. I have not yet seen the man which come over to take Mills place. Will go down the canal this evening after I return from Watson's Ferry. Write to me at Los Banos after this.

Yours Henry Miller

Capitan Hotel Merced City
Mar 24, 1873
Charles Lux, Esq.
Dear Sir

I arrived here this morning met Stony

road and bought his cattle and 50 head of horses at $41,500. Don't stop for price but sell beef

Yours Henry Miller

Santa Rita Ranch

April 7, 1873

Charles Lux, Esq.

Dear Sir

We had a very heavy frost so much is to kill all the vegetables in the garden and part of the fruit. I have no had time to look around much yet. We leave here today and go to Farmers Camp and San Luis Camp.

Yours Henry Miller

Firebaugh Ferry

April 10, 1873

Charles Lux, sq.

Dear Sir

Monday night I stopped at Farming Camp have commenced irrigating the grain without irrigation the grain wouldn't amount to much. The country needs rain very bad. We are gathering along building ditches very satisfactorily. I think we will succeed with the old alfa which is sowed on about 1200 acres. All of our horses having the prevailing horse disease in a mild form, but it makes it inconvenient for fast driving.

Yours Henry Miller

San Jose

April 20, 1873

Charles Lux Esq.

Dear Sir

I have seen Mr. Houghton He will be at the City tomorrow noon. I told him go and see Mr. Mastick and he could tell him all about the Grant case and Sabrante.

Yours Henry Miller

Gilroy

April 22, 1873

Charles Lux, Esq.

The lumber I bought of Smith I want sent down the Dos Palos Slough or the place where a Mexican was farming last year and will have a flag put up by the time the lumber gets there. It will be about ten miles above Temple and eight miles below Firebaugh. There should also be 6000 posts left.

Henry Miller

Notation at bottom: Notified Capt. Moore April 23

(This is the first written mention of Dos Palos. RLM)

Bloomfield Farm

April 23, 1873

Charles Lux, Esq.

Dear Sir

I have sent for calves this morning to the San Joaquin but it may take them until the middle of next week to come here.

Yours Henry Miller

Notation April 23, 1873

I wish you would get me two men $30. per month wages. Steady work, but don't send anyone that may not suit.

Bloomfield Farm

April 24, 1873

Charles Lux, Esq.

Dear Sir

Now about the horse disease, you wanted oxen. Now there is no driver which can drive oxen. Not one out of 100 can drive oxen especially in the city.

All of our horses got the disease. We are working them now. Will have enough next week which has recovered their health which I can sent you. Those you have and had should have bee green grass instead of hay. If you have green barley at your ranch use that on some. Don't give the horses medicine. Don't depend on eastern people knowing how to treat western horses.

Yours Henry Miller

Bloomfield Ranch

April 29, 1873

Charles Lux, Esq.

Dear Sir

I have seen Dan Harris of San Juan last week. He tells me he and Carayago are in the process of buying a ranch in Santa Barbara so therefore Carayago would be interested in having his part divided and fenced separately *into 1000 acres. I made him an offer of $12,000 dollars as it stands allowing the renters taking their share of the crop per agreement with Carayago. Harris told me that Pat Breen offered him the same amount some time ago but he wanted $15,000. Now that he is making the Santa Barbara purchase, he may take my offer.*

Yours Henry Miller

Firebaugh Ferry
May 23, 1873
Charles Lux, Esq.

I find we need more lumber to fence in the Santanella Ranch so you better purchase 150,000 ft of 1x6 feet and about 6000 posts and contract them to be sent with the fencing to San Luis Slough or Salt Slough which is the same place. There is a party living in Stockton, Mr. Hamilton who owns the steamer Clara Bell he is taking the lumber which you contracted to the steamer Alfarab ? San Luis Slough is only about 12 miles from Hill's Ferry. They aught to bring up for $7 per thousand from the city.

Yours Henry Miller

Merced City
June 18, 1873
Charles Lux, Esq.
Dear Sir

I wish you will get Ackerson and find out if his brother can come up and build us a warehouse. He could commence by the first of the month.

Yours Henry Miller

Firebaugh Ferry
June 21, 1873

There is plenty of coarse grass but no overflow this season no prospects for any, it seems. The surveying party was sent by Mr. Bereton to survey our land in accordance to the system of irrigation. They have commenced already at Farmers Camp. I saw Capt Henden who works for Mr. Hamilton. Now about my remaining here at the ranch. It will take me three or four days to regulate the cattle.

Unsigned letter in Miller's writing

Santa Rita Ranch
June 23, 1873
Charles Lux, Esq.
Dear Sir

About Hildreth wanting you to kill less, Don't you do it. Let each take out for themselves. If they have cheap cattle they would not come to you and expect you to kill less and us to join them.

Yours Henry Miller

Santa Rita Ranch
Aug.1, 1873
Charles Lux
Dear Sir

The warehouse will be ready next Tuesday or Wednesday. Mr. Ackerman will bring you a letter from Edward Roehme. I agreed to pay Ackerson $10 per day for himself and he made two wagons with his men.

Yours Henry Miller

Santa Rita Ranch
Aug. 25,1873
Charles Lux, Esq.
Dear Sir

Today I am going to McPikes Field and Farmers Camp

Yours Henry Miller

Bloomfield Farm
Oct. 10, 1873
Charles Lux, Esq.
Dear Sir

I see by your letter the Canal Co has levied another assessment. Now we owe them about $3000 for water. They owe us about that much for beef, pork and hay. I think it will amount to close onto $4000. Now the account for building the fence will not be less that four to five thousand dollars. You can arrange so it will get against our assessment. Try and see W. G. Roberts about the Stage Barn as we can use it now and it is worth $100. I have made arrangements with Mr. Ackerson to put up a house for us on the Farm and he promised to start about the 15th. If he want to go by way of Hill's Ferry I will send out team to meet them there it will be

the quickest. Have him telegraph to Los Banos. James Mills will send for him.
Yours Henry Miller

Farmers Camp
Oct. 14th 1873
Charles Lux, Esq.
Dear Sir
I arrived here yesterday and have seen some of our cattle. They's holding up very well, but they have to come off the stubble by the first of November. If you see Schafer, tell him to lose no time to come back as soon as possible and come to Millo Camp. About W. Roberts wanting $125 for the old Barn, I am surprised. To take the barn and move it is nothing. But if we can let it remain and used there I made him the offer. Two hundred dollars to build a new one. The barn was built 14 years ago.
Yours Henry Miller

Santa Rita Ranch
Oct 17th 1873
Charles Lux Esq.
Dear Sir
We have lost 4 sections of range pasture this season because of a man named Shaw. He is one of the meanest men living.
Yours Henry Miller

Farmers Camp
Oct. 23, 1873
Charles Lux, Esq.
Dear Sir
Ackerson's is to work here
Yours
Henry Miller

Santa Rita Ranch
Nov 11, 1873
Charles Lux, Esq.
Dear Sir
I arrived her Sunday noon. Found old Jim Graham very sick he died Sunday night. We buried him this morning Sunday. I saw Ackerson his getting along well with the house and will be finished in about three more weeks.
Yours Henry Miller

Santa Rita Ranch
Dec 2, 1873
Charles Lux, Esq.
Dear Sir
Hogs will leave Farmers Camp about Saturday. You can expect them at the City one week from then, Write to me at Los Banos.

Dutch Corners Hills Ferry
Dec. 3, 1873
Charles Lux, Esq.
Dear Sir
I come from San Luis Ranch last night could not trade with Malaren. This morning at 9 o'clock it commenced snowing and storming so hard that I thought it best to remain here until tomorrow morning. The prospect is winter, with plenty of rain. Its raining now and wind is from the southeast. Cattles will suffer for a while and I think fat beef will be better in a short time. Let me know if you can buy Mr. Chapman barley its badly needed. Everything will be for the better after this.
Yours Henry Miller

Santa Rita Ranch
Dec 6, 1873
Charles Lux
Dear Sir
Your letter about the coalmine. I received we better let it rest until spring. We have enough at hand for the present. Hogs were started this morning. In all about 400 head also 100 head of cattles were started from the alfalfa field yesterday.
Yours Henry Miller

Kreyenhagen Store
Dec 7, 1873
Charles Lux, Esq.
Dear Sir
I wrote you yesterday by way of Merced City but you may not have gotten my letter. I came here last night to go on to Santanella. It has rained very hard, yesterday by spells and now its raining harder. The rain we make plenty of grass. The roads are very bad an no hauling can be done at the present.

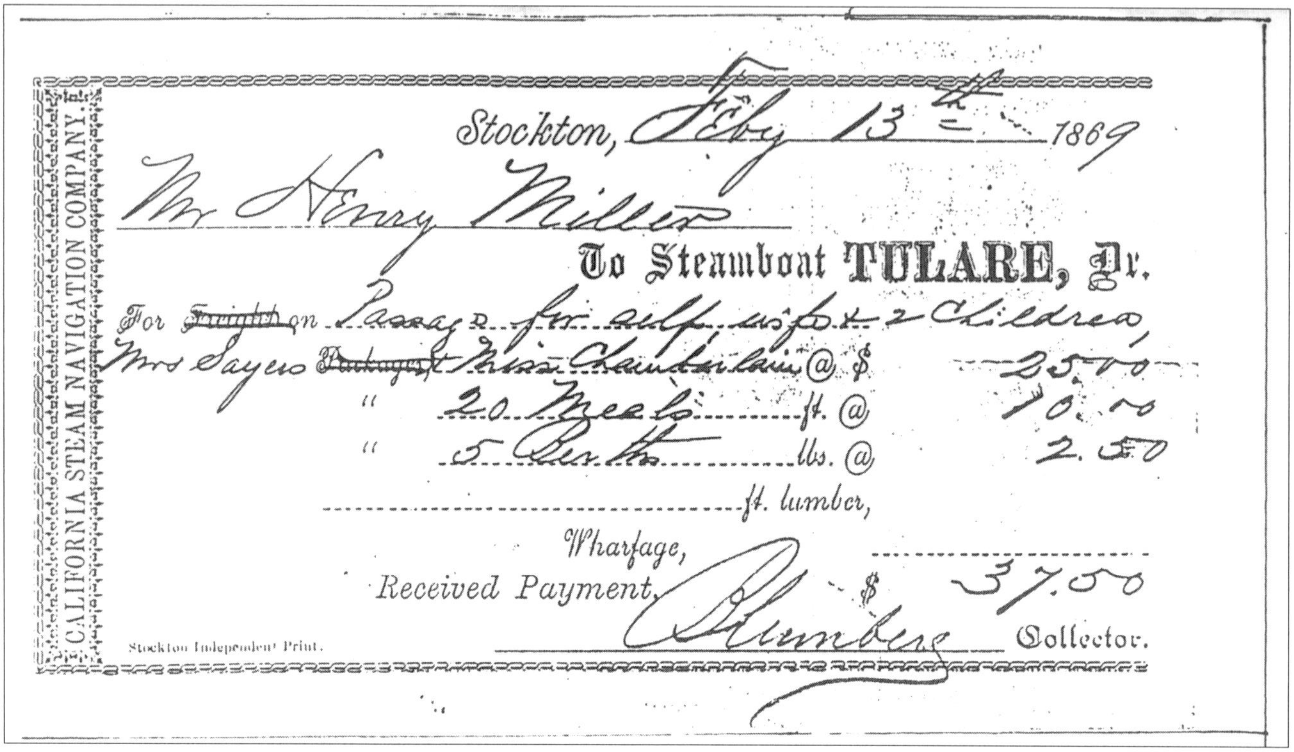

CALIFORNIA STEAM NAVIGATION COMPANY.

Stockton, Feby 13th 1869

Mr Henry Miller

To Steamboat TULARE, Dr.

For ~~Freight~~ on Passage for self wife & 2 Children,
Mrs Sayers ~~Packages~~ Miss Chamberlain @ $ 25.00
" 20 Meals ft. @ 10.00
" 5 Berths lbs. @ 2.50
ft. lumber,
Wharfage,
Received Payment, $ 37.50
Collector.

Stockton Independent Print.

Bill of passage on steamboat Tulare for Miller and family, 1869

Schafer will get to Gilroy with the hogs Tues or Wed. Among them are 150 wild hogs the balance are good gentle ones. About 200 out of the whole are pork hogs. I will instruct Schafer about selecting out all the small gentle hogs to be left at my place and ship all the balance as soon as they can get cars. Should weather remain for a day or two as if is not there will be water enough in the river to ship one.

Yours Henry Miller

Dec 11, 1873
Charles Lux San Francisco
Dear Sir

Let me know about shipping the hay to Alvarado. I want 10 coils of baling rope sent down immediately.

Yours Henry Miller

1874
Santa Rita Ranch
Jan 9th 1874
Charles Lux, Esq.
Dear Sir

I arrived from Hills Ferry last night on the steamer Clara Crow. Brought our goods to Salt Slough and also 100 sacks of barley, which I bought. The water in the River has falled, and it will be hard work for a steamer to come up to our landing but it looks like rain this morning. The cattles in the lower San Luis Field are doing well and are very cattles. Don't worry that I won't have cattle enough to send you for butchering.

Yours Henry Miller

Santa Rita Ranch
Jan. 15, 1874
Charles Lux, Esq.
Dear Sir

In regard to the articles shipped from Stockton which Capt. Hamilton was about we want them bad, and let him bring it up to Salt Slough to our warehouse. Should there be a sudden rise in the river let him come to the home landing.

Yours Henry Miller

Postscript: Sir the water is very low in the River and there is no hope for a boat to come up in about two days.

Santa Rita Ranch
Jan. 17, 74

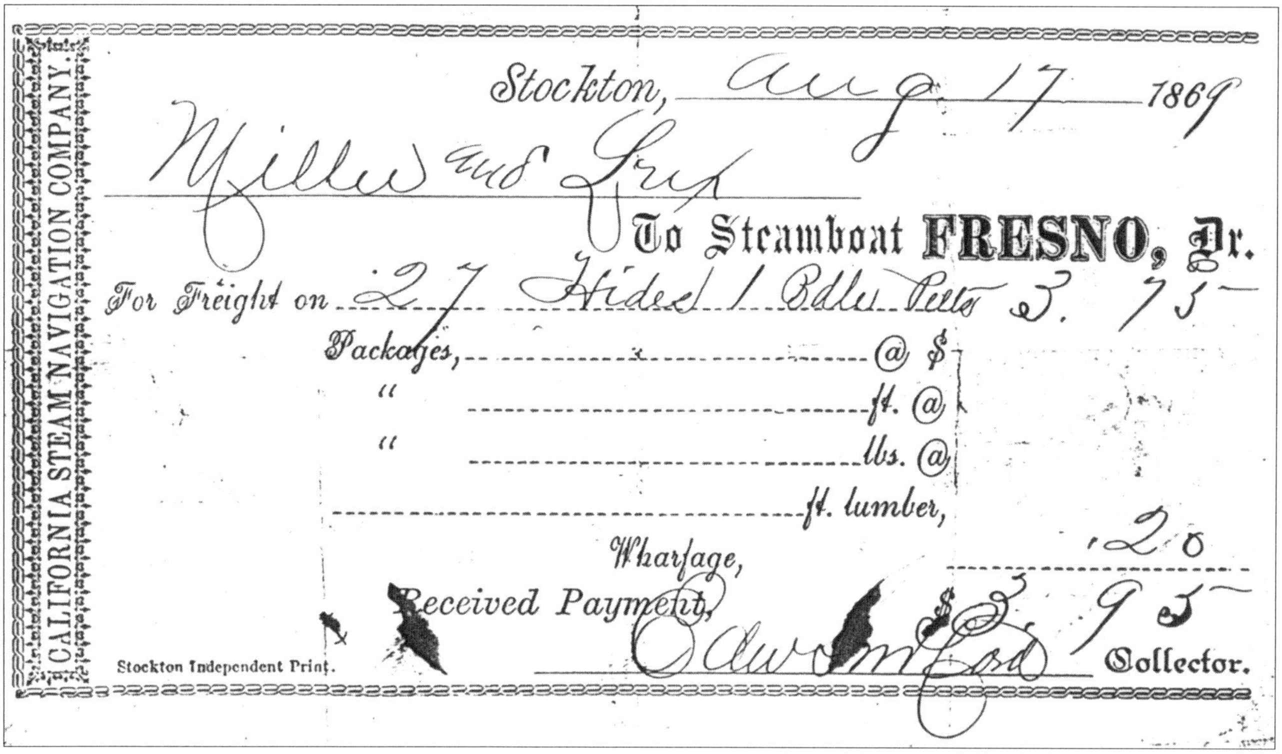
CALIFORNIA STEAM NAVIGATION COMPANY.

Stockton, Aug 17 1869

Miller and Lux

To Steamboat **FRESNO**, Dr.

For Freight on 27 Hides 1 Bdle Pelts 3.75

Packages, @ $

" ft. @

" lbs. @

ft. lumber,

Wharfage, .20

Received Payment, $3.95

Collector.

Stockton Independent Print.

Bill of lading for hides on steamboat Fresno, 1869

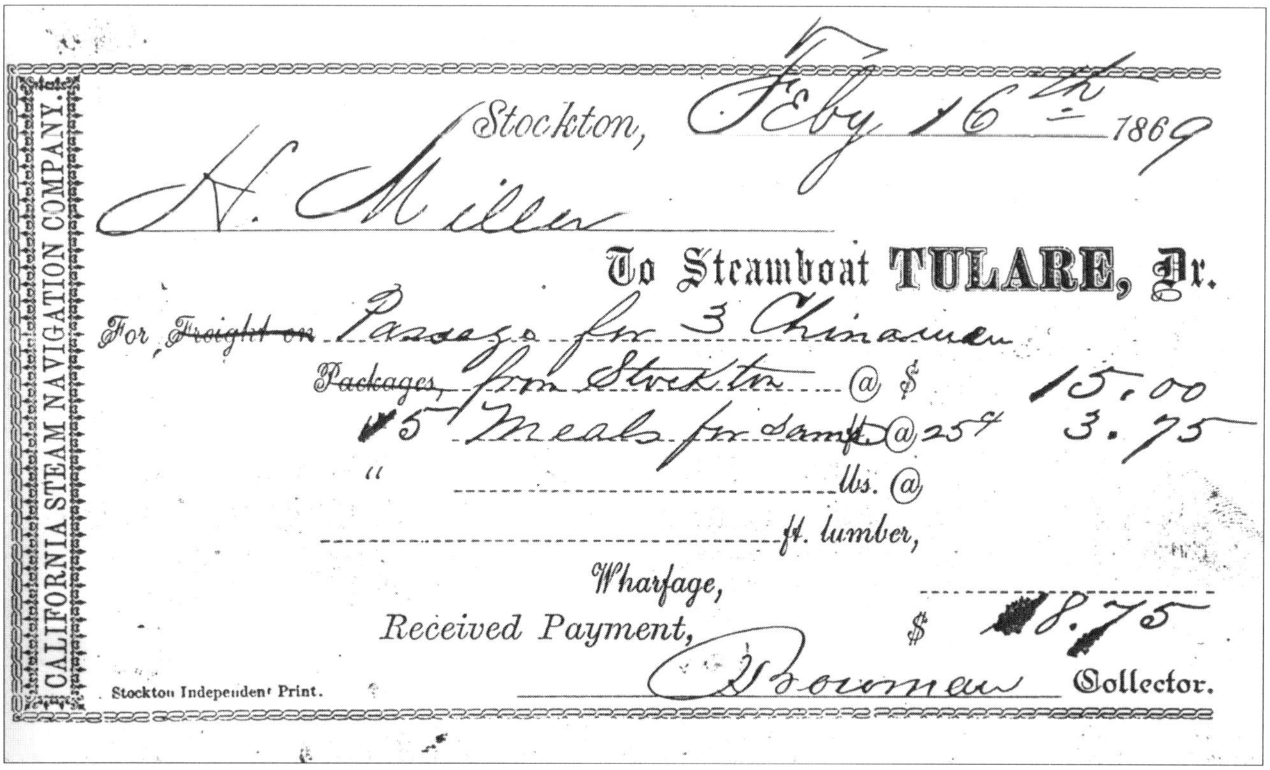
CALIFORNIA STEAM NAVIGATION COMPANY.

Stockton, Feby 16th 1869

H. Miller

To Steamboat **TULARE**, Dr.

For ~~Freight on~~ Passage for 3 Chinamen

~~Packages,~~ from Stockton @ $ 15.00

15 Meals for same @ 25¢ 3.75

" lbs. @

ft. lumber,

Wharfage,

Received Payment, $ 18.75

Bowman Collector.

Stockton Independent Print.

Bill of passage for three Chinamen on steamboat Tulare, 1869, most likely cooks for Miller & Lux ranches

Charles Lux, Esq.
Dear Sir
I have just returned from the other side of the river. There is no water in the river yet, I forded the river this noon. Its not over knee deep. As to our farming we are getting along very well.
Yours Henry Miller

Santa Rita Ranch
Jan. 22, 1874
Charles Lux Esq.
Dear Sir
Tomorrow we will collect the beef in the lower end of the San Luis Island, where we have not take away many cattle since last spring. Last Sunday the slough came down banks full but it is falling again quite low. The band I am starting will get to Gilroy next Tuesday and will arrange for them to get to your place the Sunday following. We have about 3000 bags of wheat to spare, which we want to sell. See Fred Lavder and ask him what he will pay. The wheat is good. S. Newman has seen it.
Yours Henry Miller

Canal Farm
Feb 24, 1874
Charles Lux, Esq.
Dear Sir
Yesterday morning I met the cattle band at the top of the mountains – now about buying barley, we must furnish our renters feed otherwise they cant put in the ditches with George Davis and told him he should find a place somewhere else as the ranch needed another foreman. I have written to Mr. Hamilton in Stockton to send the articles from Gilroy to our warehouse on San Luis Slough.
Yours Henry Miller

Santa Rita Ranch
Feb. 25, 1874
Charles Lux, Esq.
Dear Sir
The steamer Clara Bell brought up part of our freight to the Home Landing and Charles Warfield put on board 45 salted hides, four dried and barrel of tallow weighing 37.1/2 pound. It looks rain now. I am in hopes shall get out one band of good beef cattles from the lower San Luis Field. Afterwards I will manage so we get enough to kill from the east side of the river. We ought to have some nice cattles in Chapman Field. We will have good cattle in the Lone Willow by the middle of next month.
Yours Henry Miller
Postscript: This morning we are preparing to start for San Luis Camp and tomorrow night hope to reach Hills Ferry

Hills Ferry
Feb 27, 1874
Charles Lux, Esq.
Dear Sir
I shall go up on the east side of the San Joaquin River Sunday and will have to make on enough cattles to last until March 20 after that date would be able to get cattle from most anywhere. Choice cattles will be scare for one month yet. Telegraph me at Firebaugh if anything important If need any gentle cattle provided you get the steamer contract telegraph me at Los Banos.
Yours Henry Miller

Canal Farm
Mar 9, 1874
Charles Lux, Esq.
Dear Sir
Included in the band of cattle I started is a steer belonging to Sellig. Jeff James also has a brindled steer 5 years old. You will see that they get the cattles. They are not of the San Luis Island and I think they butcher better than they look. Today I am going up the Canal and will be at Home Ranch Tonight!
Yours Henry Miller

Bloomfield Farm
March 16, 1874
Charles Lux, Esq.
Dear Sir
We started three teams with 20 head of horses to the San Joaquin
Yours Henry Miller

Bloomfield Farm
March 18, 1874
Charles Lux
Dear Sir

I shall start Thomas Dias and all his men for the Joaquin this morning, this noon. I shall start Friday. I will get up a band of cattles out of the lower San Luis Field Saturday, will start them. They will be at Soap Lake Tuesday morning.

Yours Henry Miller

Top of the Mountain
Mar 19, 1874
Charles Lux, Esq.

If there is a change of getting ready a good many cattle, don't lose the opportunity. I shall meet Schmitz this morning at Santanella, from there will go up to San Luis Camp.

Yours Henry Miller

San Luis Camp
Mar 21, 1874
Charles Lux Esq.
Dear Sir

I am starting to gather calves to meet the band at Cottonwood.

Yours Henry Miller

Mar 24, 1874
Los Banos
Charles Lux Esq.
Dear Sir

Your letter of March 21 just delivered this morning on my arrival at Canal farm. About the cottonseed. I do not have any bought nor plan to until further order.

Yours Henry Miller

Postscript There are still here 17 head of stall fed cattle, the best in the state, and 18 head of fat cows. We will have to start a band by Schafer a few days after Thomas Dias has started

March 26, 1874
San Luis Camp
Charles Lux Esq. & Jesse Potter Esq.
Dear Sir

I wish you will try and take a general interest to see that everything is running smoothly. Your father arrived here yesterday from Merced.

Yours Truly Henry Miller

San Luis Ranch
March 31, 1874
John H Bolton Esq.

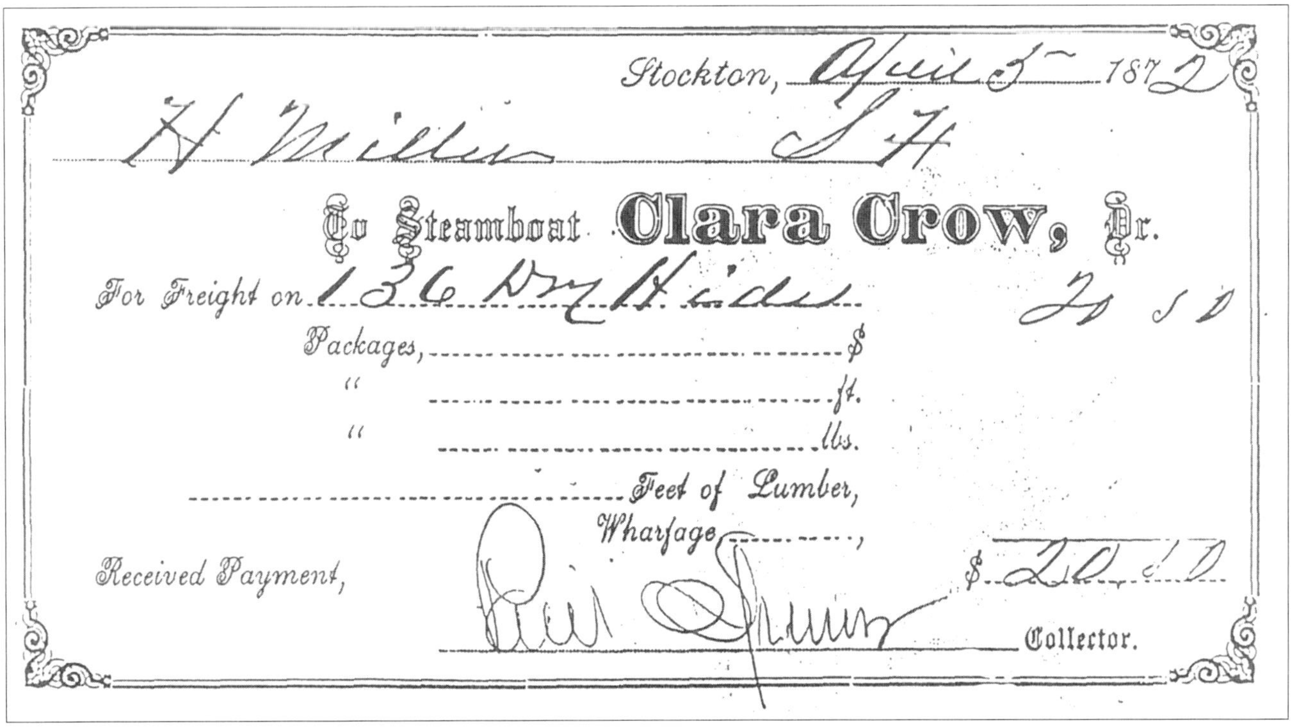
Stockton, April 5 1872
H Miller S F
To Steamboat Clara Crow, Dr.
For Freight on 136 Dry Hides 20 00
Packages, $
" ft.
" lbs.
Feet of Lumber,
Wharfage,
Received Payment, $ 20.00
Collector.

Bill of lading for hides on steamboat Clara Crow, 1872

Dear Sir

Please find out by Captain Hamilton what became of our order of fencing lumber left at Stockton from last season, which should have been 75,000 feet or thereabouts. Please send me to Merced City all letters of importance and $500.00 coin half gold half silver.

Yours very respectfully Henry Miller

Canal Farm

April 7, 1874

Charles Lux Esq.

Dear Sir

In regard to the water in the canal it is all wrong. Nothing is fixed the gates where the water is let out into the ditches are too far abouve the bottom of the canal and half of the water is wasted without any need. Everything runs slow, very slow and thing are getting badly mixed up.

Yours Henry Miller

Visalia

April 19, 1874

Charles Lux Esq.

Dear Sir

There are between 2500 head and 3000 of stock cattles which are very cheap for $10. – This will be the last change to buy them. I'll think of buying them today and tomorrow by then I'll decide. I went on Cross Creek on my way coming home to meet all the Tule River men and the general rodeo. Will meet them tomorrow at the Lakeshore.

Yours Henry Miller

Tulare City

1874

Charles Lux Esq.

Dear Sir

I arrive here the evening. From the Lakeshore seen a great many cattles, but no bveef and very few steers. Have bought no cattle yet. They are asking 12 and $15 per head for stock cattles the knowing ones. But I think I can buy for $10 dollars you will ship by steamer, two buckeyes mowers #2 with all the extras to the Home Landing with two hay rakes. Last night I slept at George Crossmore Ranch and this morning bought about 2800 half meters and half cows to be received at the North Side of the Kings River in July for $1.75 per head also 2,700 head of early young calves and 100 cows. They are to be received north of Kings River in August.

Yours Henry Miller

Los Banos

Canal Farm

April 29, 1874

Charles Lux, Esq.

Dear Sir

This evening on my return to the Canal Farm I took a walk over the ground irrigated and found we have very fair results considering everything. Were cutting alfalfa now and it is turning out well. About Schmitz I told him to get his supplies at Merced City also barley. There is plenty of grass until I can send hay from Gilroy. About the lumber for building Camp there hauling lumber from the river which will arrives if the vaqueros gets back. I shall get the ranch by Saturday. I will start a band of cattle from the Home Ranch. Have them started Monday next send to Merced City $500. Dollars in sliver for mills and $500 dollars for myself inc are of Chamberlain so that it will be waiting in Merced by next Sunday,

Yours Henry Miller

Canal Farm

April 30, 1874

Charles Lux Esq.

Dear Sir

The following article we want for this place shipped to Salt Slough. Materials to repair clipper wagons, harness headers buckeyes #2

Yours Henry Miller

Berenda Station

May 1, 1874

I arrived here this morning went to Schmitz's camp and found him getting along well. He will require 11,500 pots to put up the fence we will want 50 coils of wire, three sacks of wire staples we are putting on 3 wire and two boards the top board we are using 1 1/4x8 which we haul from the river and the other board we want from the City shipped by rail. You want to buy 6-16-24 ft long 50,000 ft or enough to fill 8 cars.

Yours Henry Miller

Santa Rita Ranch
May 1, 1874
Charles Lux, Esq.

Yesterday I went about the farm, they commenced cutting alfalfa which turns out beyond expectations. We will have a good crop of barley and some of the renters will have a very good crop of what. Yesterday bought 2-quarter sections land in the Onoffert Township from L. Hansen for $1,050 dollars. The tiles clear and his to pay out exchanged 2 quarters sections with Fessant for 2 quarter sections 22 in township 10R10 East also bought one quarter of Therman for $500 dollars paid out and from Parkin I bough half section with crop and hogs for $3,000 clear title. The land will enable us to fence our own land which has been no use to us for several years. We have to buy one more section and or quarter before we can fence. There will be no trouble the land I bought of Parkin in section 34 eastern half and is well worth the money. Land was sold last year adjoining for 20 and 30 dollars per acre we had a rain shower, yesterday and last night water commenced to rise very fast. I received a letter from R. M. Brereton stating he wished to sell a portion of his stock to us and gives his reason. I will ship the wheat. Make a price of $3 per ton. It is enough to the City. They're taking it at that price from Hills Ferry.

Yours Henry Miller

Berenda Station
May 3, 1874

The lumber and other articles for this place try to get it sent as soon as possible as soon as harvest will commence you know most of the men will find work.

Yours Henry Miller
Korn & Hirschfeldt

Charles Lux Esq.
May 4, 1874
Dear Sir

I came to the Home Ranch yesterday noon, parted out 167 head of cattle. This morning finished up band collection from Lone Willow 309 head in all and 50 calves. On my arrival yesterday David Wood was at the ranch waiting to see me to sell out his land. He wants $10 per acre he owns 7,280 acres of land and has 14 miles of fencing. I offered him $50,000 and would pay him $5,000 per month until the final payment without interest. He told me he wouldn't sell them, he would never sell but divide among his boys. I went to the rodeo and on my return Wood again made his proposition. I took his offer.

Yours Henry Miller

Santa Rita Ranch
May 4, 1874
Charles Lux, Esq.
Dear Sir

I have just arrived from Santanella Hills Ferry and Canal Farm. It rained very hard last night this morning. No I will go to Firebaugh to meet Alvin Ward and receive the Canal Co. Cattles.

Yours Henry Miller

Firebaugh Ferry
May 7, 74
Dear Sir

I came her this morning to receive the Canal Co cattles but Chandois is not at home. It rained very hard here – Last night and spoiled a good deal of dry feed.

Yours Henry Miller.

Hills Ferry
May 21, 1874
Charles Lux, Esq.
Dear Sir

I arrived here this noon to see Joseph Buth? He is getting along well with making hay. The milkier at the Home Ranch wants $180 dollars sent to the same place. Mr. Bolton sent about two weeks ago.

Yours Henry Miller

Bloomfield Farm
May 26th 1874
Charles Lux Esq.
Dear Sir
Mills has left our employment. J. L. Bills has charge for the present.
Yours Henry Miller

Firebaugh Ferry
May 28, 1874
Charles Lux Esq.
Dear Sir
At the Canal Co Farm they are haying very good wheat, you would not believe that this country would produce anything. Therefore it would be good judgment to fence more land and let it go for farming. I send you a list of articles which we ant to be left at the Home Landing.
2 sacks of coal the BEST
one piece of zine 10 foot for butcher
4 barrels of dried peaches
4 of apples
4 sacks of coffee
one box castile soap
2 boxes starch
2 boxes baking powder
compound for making soap
500# coffee
6 doz cans for preserves
sack twine and needle
10 barrels of middling grade sugar no more than 9 cents a pound
All this wants to be shipped to the Home Landing excepting Coal and sacks which will be left at San Luis Slough Landing
Signatures clipped off these letters are in Miller's handwriting

Visalia
June 10, 1874
Charles Lux, Esq.
Dear Sir
Write me to Merced in care of Chamberlain, but if my wife is going up send by her. I may buy a few more cattles I know this will be the last chance and they are cheap as dirt, if we will have ordinarily good season and if had we can keep them alive.
Yours Henry Miller

Merced City
June 13, 1874
Charles Lux Esq.
Dear Sir
I left Visalia yesterday. This is about the best lot of large cattles in the state. They are quite gentle on account of they are being moved and parted on so often. This evening I shall go to Chamberlain Ranch and cross the river in the morning received your letter by my wife. Will be at Firebaugh Ferry Monday morning. Tonite we go to Los Banos.
Yours Henry Miller

Firebaugh Ferry
June 16, 1874
I arrived yesterday. Received your letter by stage. The water is quite high and it is no small matter to get about. I want Jesse Potter to buy one good boat and rollocks and bars complete send by the first boat coming up river to be left at Home Landing.
Yours Henry Miller

Visalia
June 27, 1874
Charles Lux Esq.
Returned last evening from Tule River about the lumber to be bought buy the following for six houses the amount to build two on the upper end of the ranch. The lumber is to be left a point 6 miles below Firebaugh Ferry. For two houses and two barns the lumber is to be left at the Temple landing and the lumber for the barn to be taken to Firebaugh Ferry or above to the ranch house on the eastside of the river for our own use. For the last two houses the lumber to be left at Mears landing and lumber for one warehouse the hardware doors and windows to go to the Home Landing. We want about 5,000 posts left as follows.
1000 to be left at home landing
1000 to be left at McPike Landing
and all the balance at Temple's Landing. About shipping the above lumber and posts I thinks we will have a good River during the whole month of July, but it will not be safe after that.
Yours Henry Miller

Merced City
June 18, 1874
Charles Lux Esq
Dear Sir
I wish you will get Ackerson and find out if his brother can come up and build us a ware house. He could commence by the first of the month.
Yours Henry Miller

Visalia
July 1, 1874
Charles Lux Esq.
Dear Sir
I arrived today and yesterday the Tule River cattles from Callison 649 head, Hunsaker 456 head, Bentley 399 head and 34 gentle cows brought his brand for $1000 which are out. 60 head of Noble Blankenship 42 head of different partys and 290 head of steers of N. B. A. Mason and 21 head of our own cows and calves which came with the Mason cattles. The band will cross Kings River tomorrow morning. I received your letters and $5,000 coin.
Yours Henry Miller

Santa Rita Ranch
July 1, 1874
Charles Lux Esq.
I shipped today by Clara Crow steamer for San Francisco 5972 lbs of cheese to your address 29 salted hides and 12 dry hides. Weather very warm the goods that I wrote for have not arrived. They are needed very much.
Yours Henry Miller

Canal Farm
July 14, 1874
Charles Lux, Esq.
Dear Sir
The girl you send up wants to go right back. I didn't think you or whoever hired her took any pains to tell her what kind of place she was to go to. Then she should have had a letter from the office stating terms under which she was engaged. Schmitz coming down I gave him a check and promised him a suit of clothing for last years Christmas present. I wish you would go with him or send Jesse Potter so he can get what he wants. Also, give him all the notes against Maples & Co, Ward and others so as soon as they have threshed their grain, I shall demand a settlement. Yesterday we finished branding the Visalia cattles and turned them loss.
Yours Henry Miller

July 15, 1874
Santa Rita Ranch
Friend Jesse Potter
Dear Sir
None of the things that Mr. Miller ordered nor the things I ordered for the store. Mr. Millers catrazes have not come. So if you have not sent them please send them the first chance you have as they are needed very much, also send 1000 cigars 200 good ones and the rest cheap ones. Weather very hot. Come up when you can.
Yours truly Charley Warfield

Bloomfield Farm
July 20, 1874
John H Bolton
Dear Sir
I wish you will see about my buggy which to be made by Mr. Folsom on California st. I want it sent to this place or Gilroy as soon as finished – Now is the time I want to use it. Have Jesse buy a good lap robe and whip and sent it along.
Yours Henry Miller

Santa Rita Ranch
July 29, 1874
Charles Lux Esq.
Dear Sir
I have been looking around and find the stock doing well and the overflow has make us a great deal of grass. I wish you would get tank made for Canal Farm to hold a good quantity of water. It is for bringing water into the house, kitchen and washhouse. All of this is to be shipped to Salt Slough if practical. If not to be left at Hill's Ferry.
Yours
Henry Miller

Visalia
July 30, 1874
Charles Lux, Esq.
Dear Sir
I received a letter from J. Wiener address 429 Bush St about buying out a party at a point southeast of Korn's Store in Los Banos. Fred Caran he pays us $50 dollars a year and has the privilege to occupy 20 acres of land if needed for his business. Tell him we have no objection if buying the building but in regards to leasing the land we will want to make a change and he better wait until he sees me personally.
Yours Henry Miller

Merced City
Aug 5 1874
Charles Lux Esq.
Dear Sir
About Brereton renting the Canal it will be a failure.
Yours Henry Miller

Bloomfield Farm
Aug 20 1874
Charles Lux, Esq.
Dear Sir
I have to make a new arrangement about sending the cattles from the ranch. It don't work smoothly.
Yours Henry Miller

Visalia
Aug 26, 1874
Charles Lux Esq.
Dear Sir
You wrote in your last letter wanting to know if I meant to purchase the Carlye Please on my private account or partnership. In answer I will state that I have not been trading on my own account for a good many years passed.
Yours Henry Miller

Merced City
Aug 29, 1874
Charles Lux, Esq.
Dear Sir
I hope you don't feel poor and worry about what we owe. We are getting more than value received, but I shall not make any further purchases without consulting with you. Carlyle land let it go.
Yours Henry Miller

Visalia
Sept 9th 1874
Charles Lux Esq.
Dear Sir
Tell Jesse Potter to have my shoemaker make me a pair of boots like the last pair he made. I want him on my return as I am about bare footed.
Yours Henry Miller

Business Managers Office
S.J.K.R. R. C. and Co SF
California St upper floor Rm #3
Messer's Miller & Lux
Gentlemen
Mr. Malarin having rented the San Luis Gonzaga Ranch does not require the Telegraph office to be kept there at his expense and as he had not paid the operator there since the lst of August the office is now closed. Mr. Korn who keeps the store at Kreyenhagen's is very anxious to have the office at San Luis transferred to this store and is willing to provide two rooms for the purpose and to pay the operator as Malarin did. As this office would be three miles of the present office at the Canal terminus it will be a considerable economy to this Company to close the latter and have all the messages to be sent to Korn Store. If however you wish the present office to remain for the conveniences of your Canal Farm please say so and the operator's salary can be charged to your account. On early reply to this will oblige.
Yours very respectfully
F. L. Palmer, Secy

Bloomfield Farm
Sept 21, 1874
Charles Lux Esq.
Dear Sir
Enclosed find bill of articles wanted for the Joaquin. Let Charley Warfield buy them and sent to Gilroy. We shall have to haul the

wood over from there and load from over ere.

Yours

Henry Miller

Gilroy

Sept 23, 1874

Charles Lux Esq.

Dear Sir

I start for the Rancho tomorrow. Have sold out my place here and arranged with Mr. Miller to move my family to the Rancho. I am compelled to give my entire attention to business on the Rancho and find it too much expense and inconvenience to keep my family here. I have recovered my health so that I can go to work. Enclosed is a bill of article which I can take here as I have to send teams to depot after my furniture will write you on my arrival in Bakersfield.

Resp. yours, J. C. Crocker

PS Ship to terminus of Southern Pacific Road near Bakersfield

Wagon Timber 5 lbs black paint

1 plank 2 inches

25 lbs patent pointed HS nail

1-plank 2½ inches

1 can boiled linseed oil

1 plank 4 inches1 sack coffee

6 pounds sugar1 can boiled coal oil

500# coarse salt200# fine sugar

25 lbs red point

100# buckwheat flowers

5 boxes ball & waggoner bleaching soap

4 boxes candles

50# good chewing tabacco

51 lbs dried codfish

26 lbs good smoking sack tabacco

5 sacks good potatoes-2 sacks onions

Yours

Henry Miller

Santa Rita Ranch

Sept 24th 1874

Charles Lux Esq.

Dear Sir

I have commenced settling with renters and find we have to make them some advances, so they can go on for another season. We shall finishing clearing about Tuesday next and commence hauling wool by this time. They turned out very satisfactory.

Yours Henry Miller

Merced City

Oct 2, 1874

Charles Lux Esq.

Dear Sir

I arrived here this evening parted out at the old Dairy filed a lot of cattles and calves to be shipped Monday. We had the hardest thunderstorm here night before last that I ever witnessed in this state. It rained very hard along the river, but not enough to hurt on the plains but on the eastern side of the river and about here.

Yours Henry Miller

San Joaquin Valley Bank

Stockton, Ca.

Oct 9, 1874

Messer's Miller & Lux SF

Gentlemen:

By request of Harrold & Harrison we write you on regards to the rent of the land leased by them to you.

Yours respectfully

J. F. Higgins asst cash

Santa Rita Ranch

Oct. 20, 74

Charles Lux Esq.

Dear Sir

I have just returned from Columbia Ranch and started five teams hauling the canal lumber and from that I can judge it will cost not less than $5 per thousand to get it up over and the point it should have been left you can settle with the party at that rate so it will be settled and done with. I received letter from Mr. Brereton wanting to know if we want to rent the Canal Farm with all the improvements and land adjoining for a term of five years. I would like to get shot of the trouble which it has been to me but I don't think I'll let it go out of our control without it is to someone being responsible person which give us security for rental payments and for settlement of an agreement to keep every *thing in order and repair. I have I have not found any of that kind which wished to rent land, but should he know anyone which will measure up to our conditions I shall be willing to part company with it if you are. There is no changes that anyone will take over and or pay us anything like we paid for our improvements, implements hay and grain. I shall send you the Canal Co. water bill as soon as I have time to compare and send our meat bill and sundries. About paying for the stubble I want to settle that part myself.*

Yours Henry Miller

San Francisco
Oct 21, 1874
Henry Miller Esq.
Sir

I wrote for and obtained from the agent of Harrold & Harrison a list of their lands, and have compared with map. According to said list all the lands we have to pay more us TR125 R14E-5,635.69 acres. The lands on which Chapman has to pay sent to us are in T10SR15 in Section 38-39 640 total 4556.66 acres. Genl Beale request me to say that he has about 125 head of cattle that he wants to sell you. They are at Fort Tejon they are very fat and are now in grass up to their knees. He don'ts know what proportion of steers some are 8 to 12 years old. He requested me to ask you to send someone at once to look at them and to make an offer. I told him cattle are low and if his are on good feed, he had better not sell them now, but he says he has a special reason for wanting to sell them now. He also requests me to ask you why you don't give him $10,000 for the 100 acres of land at Gilroy belonging to his daughter. He says the interest in undivided but her deed calls for 100 acres to come out of the choice bottomland.

Yours
John Bolton

San Francisco
Oct 24, 1874
H. Miller Esq.
Sir

Miller & Lux buckaroos roping and branding, 1890's

I thought of someone to fill the clerking position up there I have been to see him and he is willing to go. He come well recommended as being very trustworthy and willing and knows how to keep accounts. He is 20 years old today. Is not in very robust health having been sick not long ago, but he is not sick now and being young I guess he will be all right. Will send him up by Saturday morning train to Gilroy there to Los Banos. Market is about the same as when Mr. Lux left were killing about beeves a day.

Yours J Bolton

PS The young mans name is I. A. Ham Jr. He's not unused to ranch life. I told him wages were $50.00 per month and board. With promise of wage raise if satisfactory.

Merced City
Nov 5, 1874
Charles Lux Esq.

If you could find me a carpenter who knows all about constructing slaughter house and hog killing establishment I wish you would send him up at once.

Yours Henry Miller

PS find out about alfalfa seed I want 5 tons

Canal Farm
Nov 5, 74
Charles Lux Esq.
Dear Sir

I find Alvin Ward started only 300 head of cattles from the outside, that is between the River and Canal above Temple. Tomorrow we will commence selecting out beef from the McPike field and I shall see to getting all there is. It commenced storming this noon rain and very heavy wind from the southeast, The prospects of plenty of rains should then be a big rain the cattles out of Soap Ranch have to be moved to comer over there. About 2,500 head and a good many of our own stock cattles have to be turned outside the Canal toward the Panoche our stock are living on green grass and are doing well.

Yours Henry Miller

Santa Rita Ranch
Nov 29, 1874
Charles Lux, Esq.
Dear Sir

I suppose you have leaned by outside party as I laid sick at Hill's Ferry. I cough fell as well as now as wish but will try and manage so we get out of debt. Then should I not feel any better than I do now quit forever. We want to get flax seed to sow 60 acres for one of our renters. You can find out what discount it will take per acre and send it to the clerk and let him forward it to Canal Farm as soon as practical.

Yours
Henry Miller

Firebaugh Ferry
Dec 1, 1874
Charles Lux Esq.
Dear Sir

I Telegraphed to you this morning about the page land which we have rented this past season to try him once more because we will take just 3 miles of fence to fence our land off from his. You can see by the map. Should your answer be that you cannot rent Schmitz will more camp tomorrow and will take down the part built last year and rebuild it on our lines. I found out why Mr. Black will not rent the land one of his connections is in partnership with White in a store they have recently opened at Watson Ferry and they want the land for shearing sheep.

I received a letter from Mr. Brereton on regard to exchanging land for his canal stock. We don't want to exchange land no buy his stock at any price tell him to wait until we can raise something on our land and have something show new settlers and there will be no trouble selling all the water they can furnish. Then we will put of price on some of our land and put it on the market providing the Co. will not except our proposition which we made to Mr. Moss the other day. Write me at Los Banos and let me know about the market and how the cows selected are butchering and about how long the cattles will last which are at hand and on the road and count about 150 head with them

out of the Ward Field. That will give a idea what time I shall start the next band from here and I will send enough to last until Christmas in the last band went over 420 head of beef and 175 head of calves.
Yours
Henry Miller

These are early letters that have had the headings cut off and some of the signatures. The body of these letters were found pasted in a notebook with a notation that they were written by Henry Miller. It was a common practice for letters of well-known persons to have the signature cut off as it was thought that the signature was more important than the actual letter. Other letters are intact and all are typed as written with all their misspellings, phrasing and grammar.

1868
Charles Lux, Esq.
Send me up by next steamer 500# of cheap sugar and 500# of San Francisco refined No 2 to be left at Chamberlin Landing
Henry Miller

May 8 1869
San Luis Gonzaga
Charles Lux, Esq.
Dear Sir:
About the renting of the land Malarin we can gid along without and for that I want to give him a cowe. I think et police.
Yours
Henry Miller

June 2, 1869
Bloomfield Farm
Charles Lux, Esq.
Dear Sir:
I want Thomas Dias to start Friday morning for San Luis Camp he wants to bring along $1000 Silver – tell him to take the Visalia stage as far as San Luis Gonzaga and ourer man will meet him there with the bugge.xxx ef you have enove silver sent me by Dias also $500.
Henry Miller

Oct. 25, 1869
Bloomfield Farm
Charles Lux, Esq.
I sent by the last load 3000 tubs of flour and 700 pounds Beans – some apples 6 kegs of nails. The other wagon loaded with Pickets.
Henry Miller

1870
Charles Lux Esq.
About the lumber, if there is no 1x6 send up one barge lot of Fenzing and Post to Chamberlin Landing
Henry Miller

1870
Charles Lux Esq.
If the steamers runs send the Articles by Steamer to Chamberlins Landing if not to Bantas Station.
Henry Miller

3/20/1871
Charles Lux, Esq.
Yesterday I went agrose the River and rented Chamberlins Land for one thousand Dollars half cash and the ballens the first of July we can put over 2000 cattles emitiadle.
Henry Miller

3/28/1871
Charles Lux, Esq,
Art Ward is taking an the Chamberlin Land 2000 head of Cattles.

1871
I have to remove about 1500 head from the Temple Ranch up and also som out of the San Luis Island to take them over agrose the River below Chamberlin Land
Henry Miller

Dec. 1871
Charles Lux, Esq.
You must make arrangements to send hay to the Joaquin by the first steamer going up the river xxx five ton to be left at the old Chamberlin landing and 10 or 15 tons to the

Santa Rita Ranch horse breaking barn

upper Ranch at I told you an the extreme end of Navigation.
Henry Miller

1872
Tell Jesse Potter the article my wife bought for the Joaquin if not send to Gilroy yet ship the same to Merced City in care of A. Chamberlin
Henry Miller

1872
Charles Lux, Esq.
Should you get this before Thomas Dias start for up here give him $500 dollars in silver and send him by way of Merced City or as far as the railroad is complet xxx there the stage to Firebaugh connect and I will send him a saddle horse—please tell Mellisa Chamberlin my Familie are well and send regards Henry Miller

1872
Charles Lux, Esq.
Mrs. Lux is here arrived safed.
The following note was added:
Dear Charles We have arrive as far as Cham have stopped to rest a few moments. Everything has been very pleasant so far wish you was here. Much Love Miranda

Bloomfield Farm
June 20, 1872
Thomas Ias arrived from the Joaquin with 350 head of. Cattles and 200 Calves.
Henry Miller

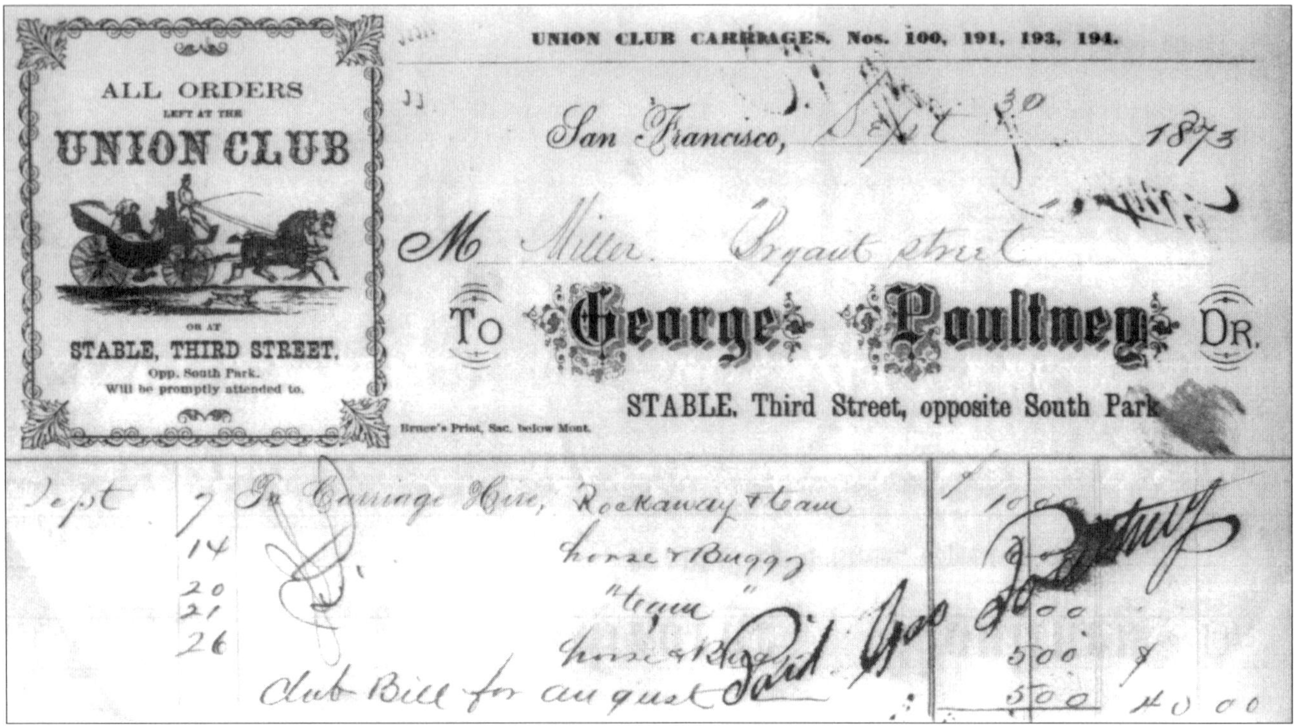
UNION CLUB CARRIAGES. Nos. 100, 191, 193, 194.

ALL ORDERS LEFT AT THE UNION CLUB OR AT STABLE, THIRD STREET. Opp. South Park. Will be promptly attended to.

San Francisco, Sept 30 1873

M Miller. Bryant street

To George Poultney Dr.

STABLE, Third Street, opposite South Park

Bruce's Print, Sac. below Mont.

Sept	7	To Carriage Hire, Rockaway & team	10 00
	14	horse & Buggy	
	20		
	21	team "	00
	26	horse & Buggy	5 00
		Club Bill for august	5 00	40 00

Paid Geo Poultney

Henry Miller's town livery bill, 1875

Gilroy
July 22, 1872
Thomas Dias started this morning with 334 head of Cattle for the city.
Henry Miller

Auzerais House, San Jose
Oct. 12. 1872
We could not gid no Cars until Tuesday morning and there for I started them by land. I send 300 head and ther will be at your Place Tuesday morning.
Henry Miller

Los Banos
Oct. 18 1872
I meet the Cattles Shafer started 200 Beef 220 Cowes and CalvesxxxxxI want the Man to return with a band of Calve Cowes after Wednesday give Shafer $500 Dollars in small coin and silver to bring over here.
Henry Miller

San Luis Camp
Oct. 25 1872
I am down here giding up a band of Cattles to send by Thomas Dias will have about 400 head and some calves xxxxx there will not be at city until a week from next Monday.
Henry Miller

1874
Charles Lux, Esq.
Send to Merced $500 dollars in silver for Mills and $500 dollars for myself in care of Chamberlin.
Henry Miller

1874
Charles Lux, Esq,
Write me to Merced in care of Chamberlin but if my wife is going up send by her.
Henry Miller

1874 - Merced
Charles Lux, Esq.
This evening I shall go out to Chamberlin and cross the river in the morning. Received your letter by my wife.
Henry Miller

1874
Santa Rita Ranch
Charles Lux, Esq.
I arrived here Thursday noon by way of Chamberlains. Find the water has gone down some.
Henry Miller

1874 – Santa Rita Ranch
Charles Lux, Esq.
I need a Girl up here to do housework no cooking tell Melisia to try and gid one such one which can do a good deal work. Send her to Gilroy and Clark can send her up.
Henry Miller

Dec. - 1874
Charles Lux, Esq.
I have today ben over the range wheir youre cattle runs. I find verey few on the riges Whair the Most Grass isxxx I road sum 2 hours and I did not see a hundred head When I came out Wher there is a slew towards the lone tree I saw sum 4 or 5 thousen head of cattle
A. Chamberlain

1875
Gilroy Jany 22d. 1875
Chas Lux, Esq.
I do not know when Mr. Miller will here as there is no communication between here and the San Joaquin.
T. B. Clark

Jan 15, 75
Salinas City
County Assessors Office, Monterey Co
Henry Miller Esq.
Dear Sir
Yours with notes of Panifilo Sabranes and Jose B Gomes Rec. J.B. Gomes is worth nothing the one on Sabranes will be outlawed tomorrow since I do not have time to present it as he lives some 30 miles away. I filed it for collection with Justice but not to issue on it. That makes it hold for one year more and in a short time I will see Sabranes and get a new note as called for.
Very Truly yours W. O. M. Garvey

Canal Farm
Aug 25, 1875
Charles Lux Esq.
Dear Sir
The Canal Co during the last few days have been acting very strange. They have waited on most of the renters with a paper for them to sign to pay $2.50 per acre for water and saying that they should not have a drop of water until they signed. The renters have been running here but I have told them the same as Miller & Lux paid. Neither young no old renters will take the land only at $1.25 per acre. This is the secret of Parkers not taking land in the Maples field.
Yours J. H. Ham

Chamberlain Ranch
Aug 28, 1875
Henry Miller
Dear Sir
I have today been over the range where your cows run. I find very few on the ridges where the most grass is. I rode some two hours without seeing or run into a head. When I came out where the slough towards Lone Tree is I saw some 4 or 5 thousand head of your cattle.
Remain respectfully A. Chamberlain

Canal Farm
Nov 3, 1875
Charles Lux Eq.
Dear Sir
There is a fair head of water in the Dos Palos Slough and also in the Wright ditch in the renter's field.
Yours Henry Miller

Santa Rita
Nov. 5 1875
Jesse Potter
Dear Sir
Are feeding cattle in Temple, Woods, Dos Palos and German boys field.
Respectfully yours
Charles Warfield

Canal Farm
Nov 8, 1875

Chas Lux
Dear Sir:
I answer to yours received last night. I would say that there is no water in Sec. 6 as yet. I am irrigaing 24 in front of the house sending the water continues as it is will have it in Badger Flat field in two days.
Yours Truly
H. G. Tanner per J. H. Ham

Kiltons
Nov 13, 75
Messers Miller & Lux
Dear Sirs
Tom Hildreth is now out of his ranch trying to sell 1,000 steers to go East at $22.50 per in currency delivered 300 miles east of here. There has been quite a lot of cattle taken east for market How is the San Francisco market?
Very respectfully yours J. C. Shirley

Mariposa Island
11.19.75
Mr. Lux
Dear Sir
I wish to know if your Dairy on the San Joaquin is to be rented the coming season. I mean to lower one below Canal Farm.
Respectfully Geo. Galbrath

Santa Rita Ranch
Charles Lux Esq.
Dear Sir
Every slough and creek is running on the Eastside of the river to get home this morning I will go out to Dutch boys camp today.
Your obedient servant
Charles Warfield

Gilroy Nov. 22, 1875
For Miller
It took the stage three days to go from here to Los Banos last week on account of high water.
Chas. Lux.

Cottonwood
Nov 24, 1875
Messer's Miller & Lux
Dear Sir
I answered it from Hill's Ferry on the 22 I will take the section 22 & 26 at this price which you state in your letter 37½ cents per acre please write me leases for same and send them to Hill's Ferry
I remain your truly
N. E. Bunker

Canal Farm
Chas Lux
Dear Sir
Mr. Bibby the blacksmith at the Canal Bridge is quite anxious to rent 520 acres of section 22 lying outside of the canal. He says he had it rented in 1874 and that Mr. Miller promised him the same for the season of 75. I leave the rent in Korn & Hirschfeldt's leave 864 acres blank. Hiderman on section 12 adjoining pays 25 cents and Korn expects this for the same.
Respectfully
I. H. Ham Jr.

March 9, 1883
Assessor
Merced County
We have to maintain a canal seventy-four miles in length which irrigates no more land that would a canal of fifteen or twenty mile more favorable situated and is absolutely of no more value than such a canal.
Very respectively yours
President
Henry Miller

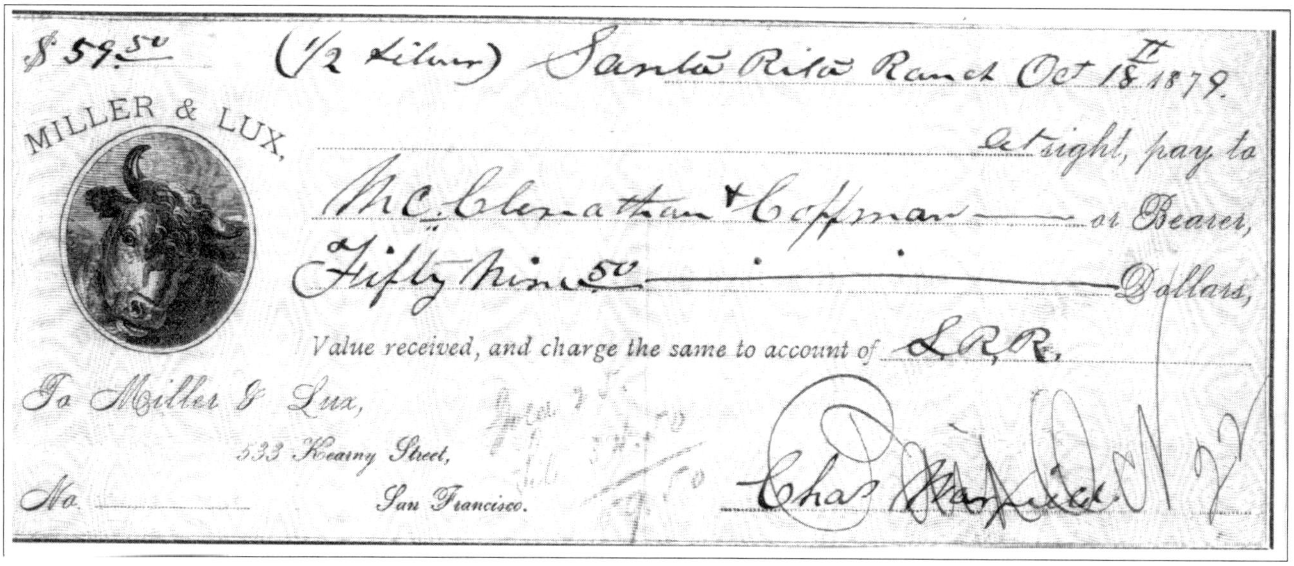
$59.50 (1/2 Silver) Santa Rita Ranch Oct 18th 1879.

MILLER & LUX,

at sight, pay to McClenathan & Coffman or Bearer,

Fifty Nine 50 Dollars,

Value received, and charge the same to account of L.R.R.

To Miller & Lux,

533 Kearny Street,

No. San Francisco.

Famous Miller & Lux Bullhead check, payment to be made in half silver.

Canal farm Sept. 27. 1875

Jesse Potter

Dr Sir

I would, like you to send me for private use,

✓ 500 Good mild cigars, a little better than the 500.

✓ ½ doz. bottles Medoc Claret,

✓ ½ " " Cognac - Gautiers Freres,

✓ ½ " " Good Whiskey,

Liquors have always been kept in the house & being out, the supply needs replenishing, before Mr Miller should return, Mr. Hand, has sent an order this date for cigars to Mr Bolton. intended for such as we have been having, but the 500 above want to be a better quality

Yours truly

H.G. Tanner

pr. J.H. Hand Jr.

Tanner's order for liquor and cigars, 1875

CHAPTER NINE

Clippings - 1860 to 1890

1/6/1877 Express
Henry Miller, Esq., the great cattle king of Merced County, was in town on Thursday looking for rain.

3/2/1878 Merced Express
New Ferry: We are informed by Mr. Henry Miller of the firm of Miller & Lux, that he will in a short time establish a ferry on the San Joaquin River near his Santa Rita Ranch. This will enable persons having business on the West Side during the high water season to visit that section without, as is now the case, of going via Hill's Ferry. This is a convenience that the community has long been in need of and one for which Mr. Miller will receive the thanks of a large number of citizens.

5/11/1878 Express
Purchasing stock – Messrs Miller & Lux have been buying hogs for the San Francisco market on a large scale in Kern County. Mr. Henry Miller has been visiting all the stock ranches in that county and inspecting the stock. They are also said to be buying cattle to drive to their magnificent San Joaquin Ranch in this county. It is further stated that they bought 1000 beef steers to be shipped to San Francisco.

6/15/1878 Merced Express
It is the intention of Miller & Lux to shortly connect their San Francisco office and the various stock stations of Santa Clara and adjoining valleys by telephone.

6/22/1878 Merced Express
Large Sale – Alexander Gody of Kern County, sold to Miller & Lux last week eleven hundred of beef cattle, said to be the finest lot of beef cattle ever sent out of the county

2/8/1879 Merced Express
On Thursday last Miller & Lux shipped several carloads of cattle from Merced to their slaughtering establishment in San Francisco.

3/1/1879 Merced Express
On Tuesday last Miller & Lux shipped from this place to their slaughter establishment in San Francisco ten carloads of beef cattle.

3/8/1879 Merced Express
Messrs Miller & Lux drove in from their ranch on the West Side of the San Joaquin river on Monday last, another large lot of beef cattle, which they shipped to San Francisco.

6/14/1879 Gilroy Advocate
Died: Miller at Bloomfield Farm near Gilroy, June 13, 1879. Sara Alice, daughter of Henry Miller, aged eight years.

9/20/1879 Express
Horses are selling in Merced at $1.00 each.

10/2/1880 Express
The well-known firm of Miller & Lux Shipped eight carloads of beef cattle from Merced Thursday last.

4/2/1881 Los Banos Enterprise
The grove at Los Banos Farm has been cleaned up in good shape and ready for the May picnic.

Miller & Lux started out ten or twelve mowing machines this week to cut their large crop of alfalfa located around Los Banos.

4/20/1881 Express

Mr. Henry Miller is very low, and little hopes for his recovery is entertained.

5/28/1881 Merced Express

Gone East – Mr. Henry Miller, of the firm of Miller & Lux passed though Merced on Last Saturday's overland, for a tour through the Eastern States. Mr. Miller has been in ill health for several months, and takes this trip through the advice of his physician, who accompanies him on his journey. The party consists of five, Mr. Miller and son, physician, secretary and waiter. Mr. Miller alighted from the train at this place, and bade his many friends goodbye and stated that his trip would very likely be extended to Europe. We understand that his wife will join him in the East should a trip to Europe be advisable.

6/4/1881 Merced Express

Charley Warfield informs us that he received a letter from Mr. Henry Miller last Tuesday, stating that he had arrived in New York, and his health was very much improved.

Mr. Miller also says he will visit Europe before returning to California.

Chas. Dickinson's property on the San Joaquin has been purchased by Miller & Lux.

Last Sunday while one of Miller & Lux's teams was crossing the railroad track on M Street, the wagon was smashed by the freight train, which was backing up to couple with another section of the train.

6/18/1881 Express

Mr. Henry Miller of the well-known firm of Miller & Lux is sojourning at Saratoga Springs. Mr. Miller's many friends will be glad to learn that his health is greatly improved.

7/9/1881 Merced Express

We learn from Mr. S. W. Sheldon, who was over from the West Side this week that his brother-in-law Mr. Henry Miller, who is now in the state of Rhode Island, is rapidly improving in health. Mrs. Miller will go east next month to join her husband.

7/14/1881 Hollister paper clipping

Henry Miller is not the greedy land man that he is sometimes pictured. Davie Wood tells that after he beat Miller to the land office and got his land, Mr. Miller came in and told Wood that if there was anymore land he wanted for him to pick it out. "I am coming to get all the land I want," Miller told Wood. "If you haven't enough money to get all the land you want now, I will let you have the money to get it."

7/30/1881 Merced Express

Charles Warfield and S. Sheldon of Miller & Lux's ranch were in town Monday on business before the Board of Equalization.

Thos Ripperdan, one of Miller & Lux's foremen passed through town last Thursday with twelve carloads of fat cattle for the San Francisco market.

8/13/1881 Express

The application of Miller & Lux for license to maintain a ferry on the San Joaquin River, at a place known as Dickinson's Ferry, for the term of ten years, is ordered granted. Bond to be given by said applicants fixed at $3,000, and amount of license tax to be paid by said applicants fixed at $3.00 per month, payable annually. The following is hereby fixed as the rates of toll which may be collected for crossing said ferry.

Footman .10
Horseman .25
Buggy\wagon 1 horse .60
“”””””””””””” 2 horse .75
“””””””””””””'4 horse1.00
“”””””””””””” 5 horse1.25
each additional span .25
each additional wagon .25
to one team
Sheep or hogs per head .01
Loose cattle or horses head .05

8/20/1881 Fresno Expositor

Clarke brought some of the best melons to town this week that ever grew. He will be in again next week with more. The Los Banos stage commenced running on its former schedule last week – leaving Merced Saturdays and returning to Merced Sundays. Sol. Waldenberg is over from Firebaugh's this week, and informed us that the shearing of Miller & Lux's sheep at that place, about 60,000 head was finished last Saturday.

8/25/1881 Gilroy Advocate

Since the transfer of 2400 acres of the Carlisle Ranch to Henry Miller, a partition fence has been erected over the southern boundary and sundry improvements made about the homestead where German Bros., Carlisle, Littleton and other pioneers lived for years. Additions have been made to the house and the back part is entirely new. It is not a commodious family residence, bearing a clean, neat and inviting appearance. A new fence surrounds it. Ingham & Sons have been very busy with the paintbrush on the house and fence and done excellent work. The ground near the house is being prepared with fertilizers for an orchard, which will be set out in the spring. Water from two perpetual springs in the hills will be conveyed in pipes to the premises. It is now the home of Mr. Miller's sister, a widow, and her two sons. The lady's name is Pfeiffer and she has but recently arrived from Germany.

9/24/1881 Express

Mrs. Tanner and Mrs. Allen were over from the West Side this week. The latter lady is a niece of Mr. Lux of the firm of Miller & Lux.

1/6/1883 Express

Miller & Lux shipped this week from this place twenty-three carloads of beef cattle.

11/18/1883 Merced Express

Mr. Charles Warfield and Henry Miller Jr. were in Merced this week

Dated 1884 Merced Express

Miller & Lux at Firebaugh's Ferry are just now finishing up their spring clip of 130,000 head of sheep. The work of shearing requires ninety men, and the average number of sheep shipped daily is about 4000, thus taking over 30 days to complete the work. The fleeces should weigh about eight pounds but some are weigh in at 15 pounds.

1/12/1884 Express

Miller & Lux shipped 216 head of fat cattle to San Francisco from this point Saturday.

6/8/1884 Merced Star

Miller & Lux, at Firebaugh's Ferry are just now finishing up their spring clip of 130,000 head of sheep. The work of shearing requires ninety men, and the average number of sheep clipped daily is about 4000, thus taking over 30 days to complete the work. The fleeces are a year's growth, and some weight as high as fifteen pounds. The superintendent thinks the entire lot will average eight pounds, or about a million pounds in all. Captain Hamilton brought down the river and shipped yesterday eight carloads for Miller & Lux, and has at least twenty-four carloads in all to ship for that firm. The wool goes by the Southern route to New Orleans, thence to New York and on to Boston.

6/28/1884 Express – A Card to the Public

A report is current in Merced County that I agreed to construct a bridge across the San Joaquin River at Chester, for the sum of five thousand dollars. The facts in the case are these: Sometime last winter, when the citizens and taxpayers of the West Side were urging upon the Board of Supervisors of Merced County the necessity of a free bridge across the San Joaquin river at Chester, and being a large taxpayer the members of Board spoke to me about the matter, saying that to construct a bridge at this point would cost the county about $15,000. Although I am not engaged in that line of business, I told the Board that I could build a bridge for five thousand dollars, merely to show that they had over estimated the cost of the structure; judging what other bridges had cost within my knowledge, I thought that it could be placed there for that money, but at the time I made this statement I did not take into consideration the draw, which would make it cost about double for the amount I said it could be placed there for. When I said a bridge could be built for $5,400. I had reference to a stationary structure. The board has shown me in the last few days plans and specifications for a drawbridge, which Mr. Burrell agrees to construct for the sum of $8,470, which I consider a good price, very reasonable, and should they accept Mr.

Burrell's plans they will in my judgment, get a bridge which will be a credit to the county and themselves, and at a price that will be satisfactory all taxpayers. Henry Miller 6/26/1884

6/28/1884 Express

The Board of Supervisors have let the contact for the construction of a drawbridge across the San Joaquin River at Chester, in this county to Mr. A. W. Burrell, for the sum of $8,470, and that gentleman is now having the timbers prepared at Oakland, and as soon as they are ready he will ship them by steamer to Chester. Mr. Burrell is the civil engineer who planned and constructed the Snelling Bridge, and the high waters in the Merced River of late have proved it to be a substantial structure, and a credit to the builder. From the plan and specifications of the Chester Bridge, and the reputation of the contractor, we have no hesitancy in saying that it will be superior to any structure of the kind in the county. We are satisfied that the Supervisors have done well in awarding the contact to Mr. Burrell at $8,700. The price is considered by engineers who are familiar with such work, to be very reasonable. The west siders now congratulate themselves that they will soon have the long needed desideratum.

6/28/1884 Express

It is reported that Miller & Lux lost about $50,000 worth of hay on their various ranches by the recent rains.

9/20/1884 Express

An Extraordinary Conspiracy!

A dispatch dated Hollister, Sept 16th says: A conspiracy to extort a large sum of money from and afterwards murder Henry Miller, of the well known firm of Miller & Lux, was first made public today, when the confession of one of the alleged conspirators resulted in the arrest of the leader of the gang, William Ritchie, who was taken into custody by Sheriff W. F. Burnett of this county. The informer is a young man about 22 years old named Thompson Davis, who has a father, mother and brother residing here. He himself has been stopping most of the time of late at Gilroy, Santa Clara County, where the plot was conceived. His statement to Luis Arques, Deputy District Attorney of Santa Clara County, Judge Montgomery of this place, and the officers of this county, is substantially as follows: Ritchie, Thompson Davis, and his brother, and David Davis his father, and an employee at Miller's home ranch the Bloomfield Farm near Sargent Station, put up a job to waylay Henry Miller and his foreman in Pacheco Pass as they were making their monthly trips between the various ranches. Miller was to be compelled to sign a check for $30,000, after which he and his foreman were to be murdered or such disposition made of them as circumstances might suggest. The check was to be drawn in favor of the old man, David Davis, who was formerly quite well off and within a few years has had considerable dealing in stock and property with Miller, and has handled quite a number of his checks. The young man, Thompson Davis, became alarmed, and communicated the matter to the Santa Clara County officers, who it seems, had information that Ritchie was a notorious and desperate character, formerly connected with criminal organizations in the Eastern States, and the leader of a similar one here. The officers were on his trail in connection with some other desperate criminality and withheld arresting him on the charge while they kept him under strict surveillance and awaited developments in the other matter. Ritchie came here a few days ago and was followed by Thompson Davis under directions of the Santa Clara County officers. Davis, it appears, again got alarmed, and last Saturday night swore out a warrant for Ritchie's arrest. Acting under instruction from the Santa Clara officers Sheriff Burnett did not serve it until this morning, when the affair leaked out so much that an arrest was absolutely necessary. Ritchie made no resistance, simply saying he knew nothing about it. His actions, however, lead the officers to believe that young Davis' statement about his being an exceedingly desperate character was true. He was armed with a large size improved Colt's revolver. He is about 28 years old, 6 feet high, fair complexioned, light hair and smooth face, except for a light straggling mustache, he has rather deep-set dark gray eyes that indicate great determination and nerve.

On him was found a letter, which seemingly implicate him in a scheme of like character with that of which he is charged. Chief of Police Harrison of Gilroy came down this afternoon and will take Ritchie to San Jose tomorrow. Harrison is of the opinion that Davis, through fear, made the complaint too quick and ruined the plans of the officers with regard to implicating Ritchie in a very serious offense. Young Davis has not been seen by the officers today, and tonight it is thought that he has left town, as he gave evidence of very great fear that Ritchie would get away and kill him. The elder Davis sought an interview with Ritchie this evening but did not secure it. Singular enough no other arrests have been made, and it is feared that through mismanagement in some quarter there will be a miscarriage of justice in this matter.

12/13/1884 Express

The Berwald Brothers built a mammoth galvanized iron tank last week for Henry Miller's Canal Farm. It is 12 feet in diameter and about ten feet deep.

9/19/1885 clipping

A Story about Henry Miller and a calf

One of the listeners recounted a story about Miller & Lux's cattle ranch. A gunner one day killed a calf belonging to the cattleman, and was proceeding to dress one of the hindquarters and wrap it in a sack. Just then a man rode up and observed. "Hello! That's a fine calf." "Oh, yes," the poacher replied "a fellow can get his pick on this ranch."

"Aint you going to take the whole calf?" "Not much! When this quarter runs out, I'll kill another. Miller & Lux can stand it."

"Well," remarked the other, "My name is Miller. If a man wants meat, all right; but he can't waste it. Now pack that calf up, skin and all, but don't you come around again. A fellow who only wants a quarter can go hungry!"

1/23/1886 Merced Express

Two thousand wild hogs were gathered on Miller's Canal Ranch the other day, and the brood sows and young pigs were left to roam. The joke is this band was a surprise, and their presence was concealed on the immense range. They hunted their own living, and were fine, fat hogs, reared without expense.

1889 Merced Express:

A letter was sent to Henry Miller asking him what he would do to aid Merced County in preparing an exhibition for the State Fair to be held at Sacramento and the following reply was received by F. M. Ostrander, Secretary of the Board of Trade.

Bloomfield Farm, August 11, 1889

Frank M. Ostrander, Esq.

Merced, California

Dear Sir:

Yours at hand and in reply will state that I will try and comply with your request sending such articles from our Merced County property as will show what the climate and soil is capable of. Mr. and Mrs. Tanner our superintendents will endeavor to do their best in sending such articles as will make a creditable showing.

Yours very truly,

Henry Miller

Wags of the day claimed he sent a box of dirt.

10/5/1889 Merced Express

Miller and Lux are shipping cattle from their Nevada Ranch to their Merced County ranch. 50 carloads of these cattle arrived at Newman this week.

Setttlers on the Plains—1890s

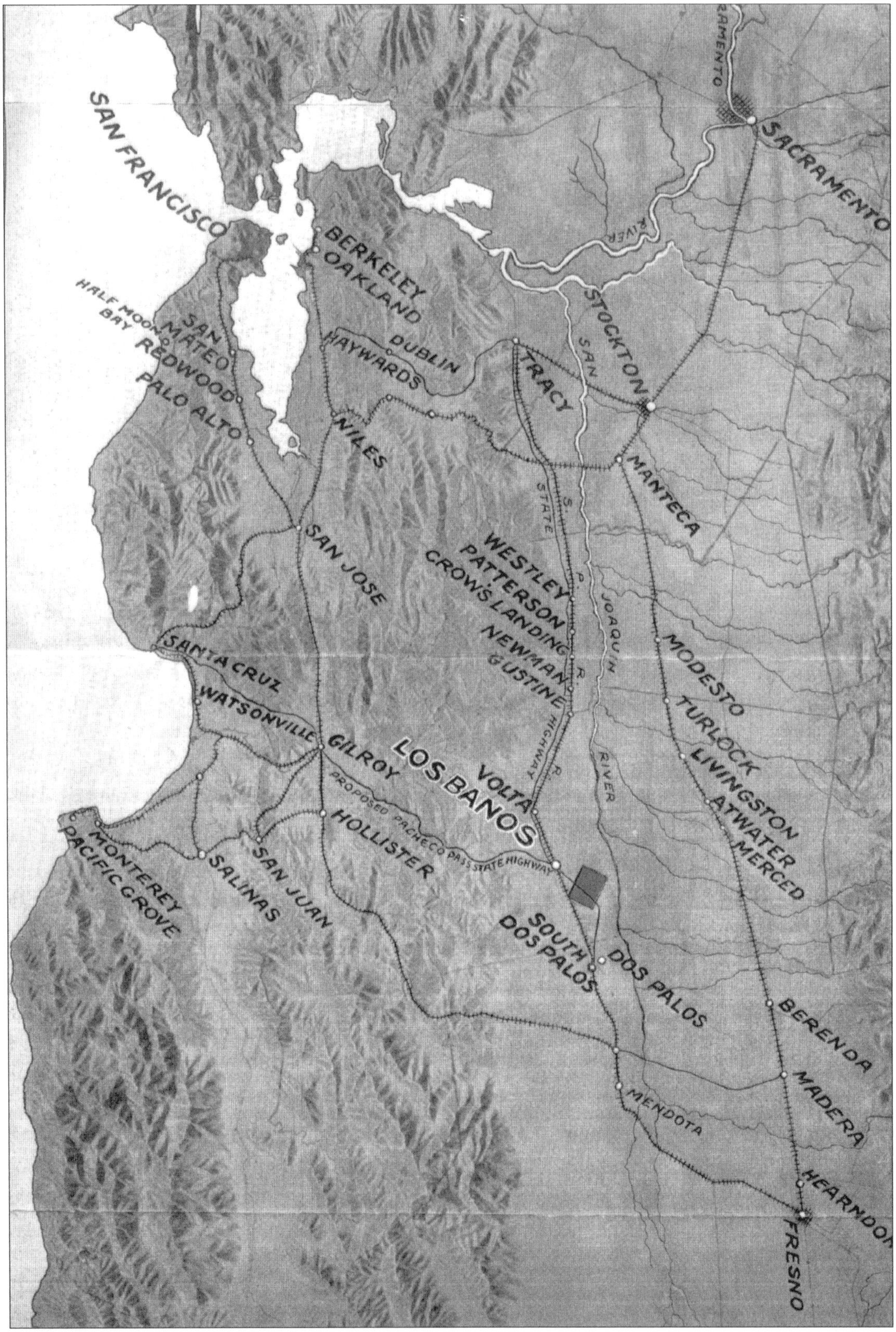

Map shows the vast area Miller & Lux covered by horseback in the early years and gives an idea of the scope of the cattle, sheep and hog drives undertaken by Miller and his men.

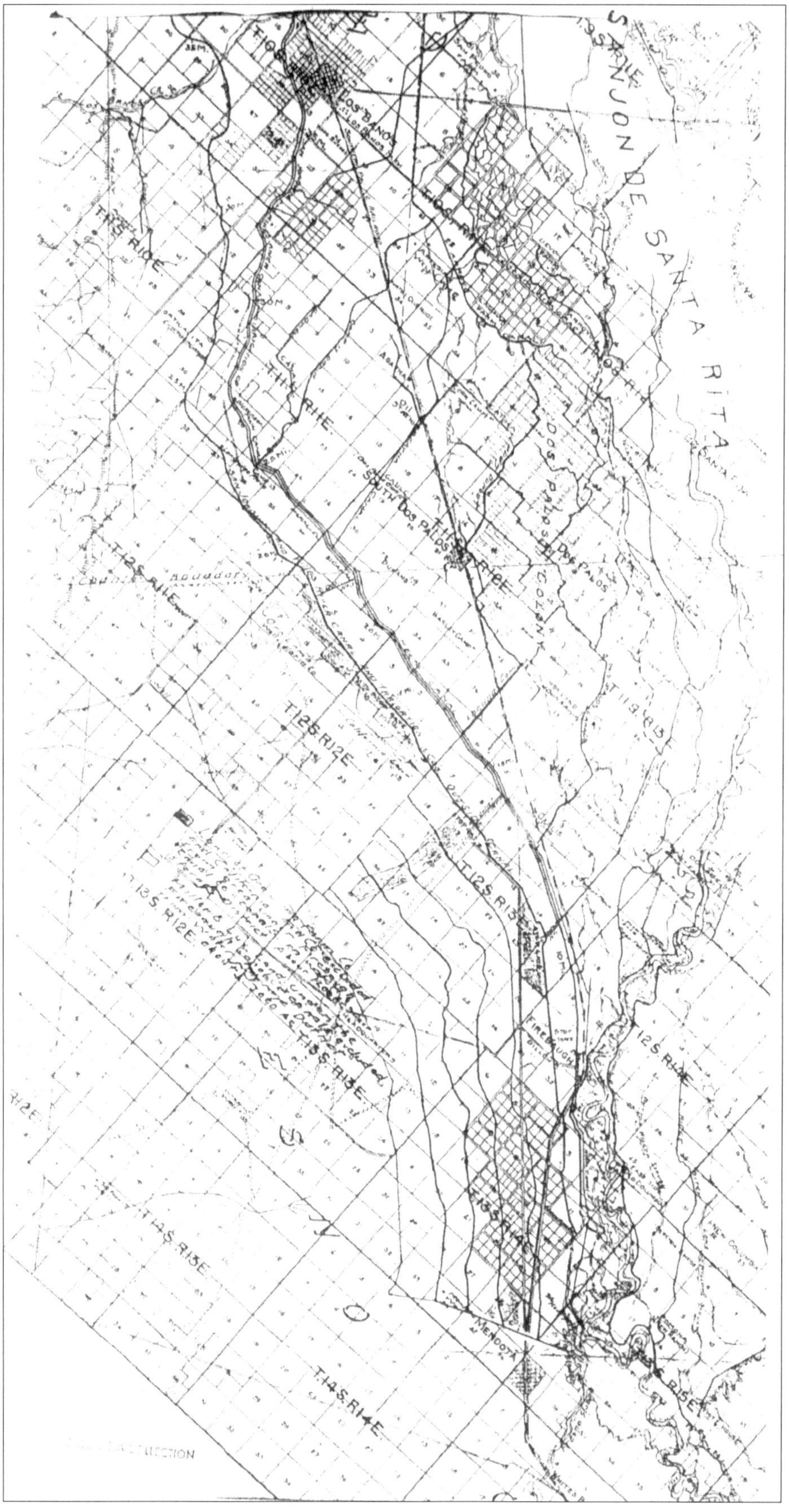

Early map of the canal system

The J.R. McDonald in 1911 on the last riverboat run from Stockton to Firebaugh. The railroads coming in the 1890's had cut down on river traffic. A group of entrepreneurs from Stockton thought to lock up the river trade with a mix of commercial and pleasure ventures. Miller & Lux had been diverting more and more water from the San Joaquin River for the canal system and the river had fallen to the point it was not feasible to go to Firebaugh. In fact on the return trip the McDonald ran aground and Miller & Lux had to divert the canal water back into the river so the boat could return to Stockton. That marked the end of major river traffic south of Stockton.

Chapter Ten

Oral and Written Interviews - 1860 to 1890

Ralph Milliken in his duties as postman to the Westside often came in contact with the old timers. He knew many of them well and often would stop to chat and hear the old stories they had to tell. He felt that if ever he could record them all, it would fill several large volumes. Many had had dealings with Miller and his superintendents and workers and had much to tell about the way Miller did business. He started recording these stories and tales in the 1920's and continued to do so until his death in 1970. Many of the stories you have read in different versions, while some remain the same. Each of these is told from the viewpoint of the speaker. These are from his notes jotted down often on scraps of paper, some dated, some not.

In reading over the many interviews that Ralph Milliken did over the years it is striking that many incidents are told and retold yet have a different twist depending on the party telling the story. Miller's being robbed on the Pacheco Pass is one; his stomping on his hat; his cattle and land dealings; how he saved his daughter's photo; they are all a part of Miller's story.

Al Cornwell, Miller & Lux employee

I first met Al in 1932. The following interview was completed on a ride from his home at the county hospital in Merced to Bell's Station and back.

At first Miller use to drive stock over the mountains to the north of Pacheco Pass. They would start in at about Cottonwood and follow the creek up into the mountains. There was an old trail there over the mountains. They would come out at Evergreen in San Felipe Valley. The drivers would take pack animals to carry their grub and blankets on. They would cross the mountains to the south of Mt. Hamilton. A man by the name of Young was the first settler in San Felipe Valley, so far as I know.

Miller had hog corrals at Canal Farm where the wild hogs were kept corralled up and fattened. This was along about 1873 or 1874. The wild hogs would be brought up from the river. At first Miller did take his stock over the mountains by way of Cottonwood. Later in the 1870's, via Pacheco Pass. When the railroad came through, by way of Berenda and Merced and after Madera was started from Madera.

At the time of the big railroad strike in 1893 Miller again drove stock to the city. The other butchers wanted to raise the price of meat, but Miller wouldn't, "No," he said, "it isn't the fault of the people. We won't raise the price." Miller had in the early days regular camping places along the route to the city. If an animal got sore footed it would be left at one of Miller's corrals and picked up afterwards and brought along. They use to have just as good tender beef before the country was fenced up and cattle were raised and fattened on grass range, as now when fed on grain.

Miller was persuaded to plant some cotton in the Cotton Field Section. They didn't know how to plant it or take care of it. "That settles it," Miller told Tanner. "I takes no advice from anyone but my foremen from now on. I hire men what's knows how to farm from now on. I don't farm any more myself."

The various superintendents at Canal Farm were: First a man by the name of Mills, then one by the names of Bills. Then came Tommy Chandler, the Middleweight champion of the Coast. Miller found Chandler drunk and fired him. Tanner at that time was baling hay. Miller put him in as foreman. This was about 1873. Mrs. Tanner was a cousin or something of Mrs. Miller.

Tanner went back East one time to see his people. Miller found something out about Tanner's way of making money for himself. Miller accused Tanner of stealing and was going to sue him. Tanner says to Miller, "You've called me that lots of times. You want to remember that every time you call me that it costs you a thousand dollars."

The old burial ground at San Luis Ranch was to the northwest of the old adobe up the creek. But few who were buried there died natural deaths. In the old adobe there used to be a bar room in the front part and cooking in the back. I have seen the place full of people. At the time I was there, there was a tame doe walking around inside the room. Grecian Bend was the name the early teamsters gave to that part of the Pacheco Pass road that is now known as the Horseshoe Bend or the Hairpin Bend. In those days fashionable women affected a certain bend in their back that was called the Grecian Bend and it was from this fad that this particular bend in the road was called the Grecian Bend. It was here that Miller was held up in the early days.

Miller was coming over the Pass with Buggy John. After Miller quit riding horseback he always had Buggy John drive him. In the buggy with them was $3,000. Miller had $200 with him. A lone Mexican held them up and took the $200 from Miller. Miller was so anxious to keep the bandit from getting the $3,000 in the buggy that he hit upon a ruse of demanding that the bandit give him twenty dollars back for expense money. "Give me back twenty dollars for expense money," he told the robber. "I needs that until I gets where I can gets some more money." Miller was so insistent that finally the bandit gave him back $20. "That's all right," Miller told him, "I's pays it back."

A couple of years or so later several teamsters wanted ivory rings for their harness and Miller took them into Briggs's store in Gilroy where Miller use to trade. One of the teamsters was Tommy Sheldon, a nephew of Mrs. Miller. Another teamster was an old man by the name of Hefner, or some such name. I was there as well. Miller took us into Briggs's store in Gilroy and said, "Give these boys what they wants." Just then a Mexican was trying to get out of sight behind some tables piled high with ready-made clothing. Miller stopped short and looked at the Mexican in surprise. He put his hand in his pocket and speaking to the Mexican said, "How do you do? Say my friend, here's that $20 what I borrowed from you over on the mountain. Take it," he insisted, "here it is."

"You didn't borrow any $20 from me," said the Mexican in very good English. "Yes, I did," continued Miller. "Yes, you're the man." Miller made the Mexican take the money. "That makes us square," continued Miller. "You needn't be afraid, there won't be any trouble." This all happened in Briggs's store in Gilroy and Cornwell was an eyewitness to it. I don't recall the Mexican's name. Jake Myer was a clerk there at that time. It was before he came over and worked for Korn & Hirschfeldt. No one had heard about Miller borrowing $20 from the Mexican bandit at the time because Miller had told Buggy John to say nothing.

Miller didn't want a town at Gilroy. Too close to Bloomfield which was only three miles away.

Miller & Lux men hauling sugar beets, 1880's

Miller would always say when he gave a man money to bind a bargain: "I binds you with your word." Miller always called Firebaugh "Firebot."

My father came to California three times. The first time before he was married. Later he came as a soldier with Fremont in 1847. In 1850 he went back East and in 1852 he brought the family out. We came in a windjammer and reached San Francisco May 16, 1852. I was born in Orleans Parish, Louisiana in 1840 and that means I was 92 last March.

My mother was a southerner and owned slaves. My father was a Yankee and never owned any slaves until he married. In 1846 when I was six my mother sold her slaves and we all went to Michigan to live where my father's home was.

When we arrived in California in 1852 we went to San Jose to live. It was only a little place. In 1853 I went with my father to mines in Mariposa. We went by pack train. We crossed the mountains north of Pacheco Pass. In 1854 I was in Arizona and became acquainted with Buffalo Bill. He scouted for a while against the Comanches. In 1857 I returned to California and drove Overland stages. In 1859 I went to work for a short time at Santa Rita for Tom Hildreth. Hildreth was not there much of the time, but had a Majordomo to run the ranch.

In 1861 I was living in San Jose. My sisters were making a large American flag to send to our father and brothers in Michigan and some Southern girls who were neighbors were over at the house and were helping them. When they got the flag done the Southern girls said in fun, "Now you must help us make some Confederate flags," so they made a couple of little Stars and Bars flags. A fellow and I came in and thought the flags would look fine on our horses' bridles. Later we rode into San Jose and tied the horses. When we came back Major Newhall of the Home Guards was waiting for us and was going to arrest us. We got away and went home. That night about 1 o'clock a friend came and told us they were going to arrest us. We made our way into the redwoods near Saratoga. We camped out for a couple of weeks and one day I shot a buck. There was no law then about killing deer. I left my gun at the camp and went down to get some water. When I got back the Home Guards were in possession of the camp. We tried to fight them off but we were arrested and taken to Mare Island where they kept prisoners. In the spring of 1862 my mother managed to get papers to free us. She went to the prison and tearing an American flag out of her bosom where she had it concealed placed the flag on me and defied anyone to hurt that flag or anyone it covered. She then produced the papers and I was freed.

I immediately got married and in two weeks was on my way to join the Confederacy. "If that's the way they are going to do, then I'll be a rebel." I and my two companions from around San Jose set out on horseback to cross the plains. We reached Virginia and joined Mosby's command. There were originally only 123 men in Mosby's command. We had a roving commission. All us three fellows came through the Civil War. I got back to California in 1867 and had a son several years old.

I went to the Santa Rita in 1867. Later I drove the Overland stages over the Pacheco Pass. Pritchard controlled the stage line over the Pass when I drove. There was lots of travel over the Pass in those days. There were so many on their way to the Southern mines. Sometimes I would have 36 passengers. A stage driver would drive to the end of his run, but if the stage had not arrived that he was suppose to meet, he would keep right on. If they met between stations the two drivers would change and the stages continue on their way. It was nothing to drive a hundred miles in those days. Stock wasn't worth anything. Mustangs they were. In driving up over the mountains from Gilroy to San Luis Ranch only stock that had been driven was used. But in coming down into the San Joaquin Valley and on the east it did matter if two or three horses of the team were broncos. Very seldom less than eight horses in a team. The stage company had men come out and cut a lot of logs at the top of the steep grade and when it was rainy and slippery the east bound stage would stop at the top of the hill and a log would be chained to the back of the stage coach to hold the stage by dragging in going down the hill to the foot of the grade. "Any of you passengers that don't want to ride down the hill, pile out," the driver would say and down the slope the stage, team and log would come. I drove the stage over the Pacheco Pass when I came back from Arizona in 1857. I don't recall any Lone Willow Stage Station and I recall passengers coming as far as Gilroy on the train. There was always a place at Los Banos where stages stopped. There was an old man by the name of Hotan who lived and farmed at Stringtown. The big stages ceased running over the Pacheco Pass when the railroad was built through on the east side. After that time only two horse stages were used over the pass.

A man by the name of Chester Dickinson who put in the Dickinson Ferry use to run a stage from Los Banos to Merced. He lived at Dickinson's Ferry. Las Juntas was about half a mile from the old Las Juntas cemetery and was on the bank of the slough.

Col. Hollister was a much older man than Tom Hildreth. Hollister came to California with Fremont. My father and Hollister were great friends. Hollister used to have lots of Mexicans skinning cattle for him. The Mexicans were so greasy that this is the reason they were called "greasers." Hollister had his hides hauled to San Francisco where they were sent by ship to the East.

McIntyre, who had a store in Firebaugh, married a sister of Jerry Firebaugh. There was a burying ground on the Helm Place and the deed calls for the fact that the burial ground is never to be removed. That is why there is a big fill necessary in the canal at a point of trees about a mile from Firebaugh. I think this burying ground was later removed to the burying ground at Las Juntas.

Chapman and Friedlander owned the Chowchilla Ranch in the early days and raised stock. What is now the Chowchilla ditch used to be called the Chapman ditch. It was built by Chapman. It was not carrying water in it when Chapman sold out. On the north of Chowchilla Ranch is the Bliss Ranch, which was built up out of government land. Miller wanted it. It was not grant land. There is an old line of picket fence running east and west that divided the two ranches and can be seen from the highway between Merced and the San Joaquin River. Friedlander was the "Wheat King" but went broke.

In the early days a person could walk from Santa Rita to Firebaugh quicker than a boat could travel up the river, the course was so full of turns which made the river trip so much longer than going a foot. Drummond was foreman for Donahue, Newhall & Co. who put the railroad in from Gilroy to Hollister. When the canal was started in 1871 Drummond came over and became foreman on the canal.

About 1875 Dustin and Conroy were partners and ran the stage from Gilroy to Berenda. Dustin drove as far as Los Banos and Conroy drove from Los Banos to Berenda. Conroy wanted to take a day off and got me to drive in his place. The Bank of California on the death of Ralston was suspended for six months. There was no pay on the canal during that time. Coffman got word that there was $38,000 coming by express from Berenda. Coffman was afraid Vasquez might try to rob the stage. Coffman gave an order to the express agent at Berenda to deliver the package to me and not to put it in the Wells Fargo box. If the stage was held up the outlaws would get the express box, but the Canal money wouldn't be in it. I was told to carry the money anywhere I thought was best. When the stage arrived at Firebaugh I delivered the money all safely to Jake Myer. I would have had a "fit" for that money. I knew Vasquez and often danced with him in Las Juntas.

The first commercial hunter of ducks and geese in the San Joaquin Valley was a man by the name of Smith who lived just north of the present bridge over the San Joaquin on the west side of the river. He came in about 1869 or 1870 before the railroad came through on the east side. He settled on the river and hunted for the market. He was the first duck and goose hunter in this country. His daughter, Lillian Smith, was born there. The son used to drive for the old man. Lillian became an expert rifle shot and used to shoot glass balls in the air when that was a novelty. A man by the name of Dr. Beaugardest was the first man to become noted for shooting glass balls with a rifle. He toured all over the world. Lillian followed him and went all over Europe and around the world shooting her glass balls. I met her years later and she was running a shooting gallery in Merced. The old Smith place was on the west side of the river not a half-mile north of the bridge. Smith made a good living but geese were cheap then. Miller has the place now.

The Anderson place was about a mile up the river south of the bridge over the San Joaquin. His wife was a Cherokee. His boys, Kee and Bill Anderson, became hog men for Miller. After the Canal came in Miller went to catching wild hogs. The San Joaquin River all up and down was full of wild hogs. Hog Camp belonged to the Canal Company originally and was so named when we drove down a band of hogs from Millerton for the Canal Company.

A man by the name of Wrightmeyer settled on the east side of the river just above the bridge toward the east. He used to catch wild hogs. Miller wanted all the wild hogs for himself and so he made a bargain with Wrightmeyer that he would give him 100 head of milk cows if Wrightmeyer would quit "hogging." Al Ward was a Texan and was Majordomo for Miller at the time. The cattle were gathered up and it was found there were more than the agreement called for. When Miller came to settle he says, "Well, Mr. Wrightmeyer, when shall we make this note due?" Wrightmeyer was surprised. "I's think," continued Miller, "we better make it due in two payments. That would be better than making it all come due at one time." "Yes, that would be better," thought Wrightmeyer. "Well, what dates had we's better set?" "Well," said Wrightmeyer carefully after much thought, "think we better make the first payment due when I die." After more careful thought he continued, "And we better make the second payment due when I come back."

Later on Wrightmeyer was coming home drunk from Firebaugh and got drowned in the San Joaquin River. The family moved away and settled at Athlone, then called Plainsburg Switch. Years later I saw Mrs. Wrightmeyer and I asked her if Miller ever presented that note for payment. "Yes," she said. "And I paid half of it." The grove of trees on the east side of the river east of the bridge is where the old Wrightmeyer place stood. Old Plainsburg was about three miles inland from the Plainsburg Switch.

I hauled steel for the bridge over the San Joaquin River. Leonard had the contract for hauling it. The Anderson place was on the south side of the river in a bend about a mile south of the bridge.

There were no cattle corrals at Santa Rita at first. The vaqueros did all the corralling. Cattle were herded at night when the vaqueros wished to hold a bunch of cattle anywhere. The Hildreth house was about where the store is now, or a little to the west about where the pepper trees are. Later Miller built a house. At one time it was feared the San Joaquin River would change its course and go down through the Santa Rita Slough and Miller had the Slough dammed off. About two years ago, I was told at the Santa Rita Service Station at the bridge over the Santa Rita Slough, that the Slough was dammed off and that no water gets into the Slough now from the river. Only the wastewater gets into the slough now.

A man by the name of Stillman, who is now probation officer at Merced, was superintendent at the Santa Rita years ago for eight or nine months, but he couldn't cut the ice with Miller and so got let out. The old orchard back of the building on the north was the first orchard in this country. It was planted in the sixties while I was back east at the war. There were lots of apple trees. It was only a small orchard. There were more apple trees than pear trees. There were no fig trees. They were planted later. Some of the present pear trees have been put in since to replace the early ones. The first fence on the Santa Rita was around the orchard. Miller wanted it to keep the cow herd out of the orchard. If I recall right Hildreth may have had a few fruit trees as well.

The Santa Rita was the last ranch broken up by Miller to plant grain. The first alfalfa was planted by the Canal Company at Oxalis along the railroad. That was on the Canal Company ranch. Brereton was trying to show that alfalfa would grow. The Santa Rita Slough was not so large in the early days as it is now. A person could cross it anywhere on a horse. It was true at one time that the San Joaquin River would break through. The Santa Rita slough has a greater fall than the river and has a more direct course.

There use to be two pecan trees, one got broken down. Only one is now left. These trees were planted in Hildreth's time. The cork elm nearby is also a very old tree. This was open country and it was the custom in the early days to stick trees in anywhere and everywhere. There were no trees at Santa Rita originally. All of them have been planted or have grown up along the slough. Where the Santa Rita buildings are located is about the highest land anywhere on the slough. It slopes away in every direction. They used to be a rowboat on the slough at Santa Rita. There use to be perhaps twice as much water before the slough was dammed off from the river. Miller's first store was at Santa Rita and was started before there was a store at Canal Farm.

The old original barn was put up by Miller probably about 1869 or 1870 and is made out of fence lumber. There was no barn for vaquero horses or other stock for years. Vaquero horses in the early days were kept out in a pasture. There was a small corral where the two-story bunkhouse is now. This was to put vaquero horses in.

The two wings of sheds on the old barn were built on later for vaqueros' horses. The vaqueros and butchers used to sleep upstairs. There was an old stairway where they used to go up. The upstairs was not floored clear across the barn and a person had to be careful and not walk off. Only the front end of the barn upstairs had a floor. Charlie Warfield took Al Ward's place.

The old slaughterhouse was put up by Miller in the fall of 1870 or in the spring of 1871 when Miller began supplying beef for the canal. The east part of the slaughterhouse has been put on in later times. The cellar was where they used to salt the hides. The demand for meat kept getting bigger and bigger until they couldn't handle it any longer at Santa Rita and the Abattoir was built on the railroad at Los Banos.

The breaking barn was built in about the 1890's. Not sure of the date. The old feed and shelter sheds were built originally for shade for stock. Miller got the idea somewhere of shade for the stock when feeding them for beef. The feed managers have been put in since. I ran an excavator for the Canal Company in the early days and Harry Thornton drove the team for me. All the trees around the feed sheds at Santa Rita were planted since Warfield's time.

The old road followed the slough from Santa Rita to Firebaugh. The old road is still to be seen on the south side of the slough at the bridge over the Santa Rita Slough at the service station. There are a number of large portable gas engines stored at Santa Rita. They were used to run feed cutters when they used to feed cattle.

The first field Miller fenced was the Temple Field. Miller fenced Temple in. The south boundary of Santa Rita Grant was the north line of Temple's ranch. Miller put the fence well over on the Grant side of the line.

There were cattle from as far always as Stockton down here at Santa Rita in the early days. Rodeoing would start in the south and proceed clean on up as far as Stockton. Enough vaqueros from every ranch would attend the rodeo to take care of the stock belonging to that ranch. As fast as stock was parted out it would be taken care of by the owners. Everyone herded their own cattle at night and kept them from getting away until they started back to their own ranch with them. Sometimes as rodeos go farther and farther away from the home ranch a couple of ranch outfits would go in together and take care of the stock as it was sorted out and then when they got back to their ranches they would sort out the band again for the two ranches.

I used to drive stage over the Pacheco Pass and was supposed to meet the oncoming stage driver at Kreyenhagen's. But if we did not meet I was to go on. One time I went as far as Watson's Ferry before I met the other stage. I drove out of Folsom to Truckee along in 1858. They used to have 36 passenger coaches and then broncos. It had to be awful stormy if there weren't at least fifteen passengers. Sometimes the stage could not haul all the passengers and it was necessary to run extras. The stages for Snelling from the West Side use to cross at Hill's Ferry and go straight across to Hopeton. "Old Nigger in the Woods Pritchell" controlled the stage line when I drove. He also had the Fairplay stage line to Santa Cruz from San Jose. He also has the fast freight line to Panamint.

The old original two-story bunkhouse at Santa Rita burned up. Part of the old original Miller house is still standing and is the one Will Wood said was the old Santa Rita Ranch House where his father sold the ranch to Miller. There was no barn at Santa Rita. There was an old horse corral where the two-story bunkhouse is now. When out on the range the "caviada" followed right along with the men. There were no corrals either for the cattle or the horses. The cattle were herded at night. When Miller used to ride horseback he had the very best for himself. They were gentle and only Miller was ever allowed to ride them.

One of Miller's fine horses raised at Santa Rita beat the train from Merced to the next town. Miller had given this horse to Sheriff Meany. A prisoner had beaten the chain gang. A freight train was just pulling out. It was a small engine and going pretty fast. "Deputize me and give me a horse," said a bystander. The sheriff did so and gives him Miller's horse. "I'll beat him to Plainsburg Switch and I'll bring him back." The prisoner was already on board. The train was starting while they were talking. The deputy beat the train to the Plainsburg Switch. The horse was unhurt. The deputy got the prisoner and telegraphed back to Merced for a team. A team was sent from the livery stable and the prisoner brought back.

At the Santa Rita Cornwell and I walked towards the willow corrals. Cornwell was showing me where the old Los Banos road used to go.

There was no ditch like there is now, you used to drive just anywhere, there were no fences and no road. Miller built the willow cattle corrals along in the 1880's. Each year a new supply of willow poles were put in the top to keep up the height and make up for the poles at the bottom of the fence that had rotted and disappeared. It was a pole corral. They kept putting on green poles every year as fast they would rot. The corral was built about the time when they commenced to get the country all fenced up and began dehorning. These corrals were put about the same time as the dipping corrals.

The mixed feed shed to the south was put up to

"I" Street in Los Banos looking west, 1891

mix feed in after Miller got to feeding barley. This here is the old breaking wagon. It was a strong affair. It had good breaks and the seat was high off the ground. We kept the front of the wagon boarded up with two-inch planks. The reason the seat was six feet off the ground and the planks so thick was that a young horse could kick as much and as high as they wanted to and do no harm. They soon learned there was no escape and quit. We also had a strong post with a couple of rings in it. We would tie a wild colt to it while hitching up to the wagon.

Mr. A. Leonard use to be superintendent at the Dos Palos Ranch at Oxalis. The Dutch Boys were two Hollanders by the name of Joseph Colp and Conrad Foster. They rented from Miller along in 1872 when Miller began renting out grain land. The two men were partners. Colp died afterwards. Foster later lived near Watts Valley northeast of Sanger in Fresno County. Leonard named the ranch where they farmed the Holland Farm.

Bill Knight lived with his father at Stringtown when the renters were farming there and the same time the Dutch Boys were at Holland Farm. Bill was a young man. Bill Knight's father didn't stay very long. Only one or two crops. Bill lived on the ranch with his father and stayed. There was a fellow by the name of Hoten among the Stringtown farmers.

The Temple Warehouse was built in 1875. It went out of use as a warehouse when the railroad came in. It was built originally for the renters to store their grain in. Before the warehouse was built the renters used to fetch down their grain and pile it up on the bank. When they discontinued using it as a warehouse the building was made over into a ranch house with a kitchen at the east end and the dining room on the south side and the sleeping quarters on the west end.

Canal Farm was originally the name of the Canal Company's ranch. The Canal Farm at Los Banos was called Farmer's Camp originally and was established in 1871. A man by the name of Davis was superintendent of the Dos Palos Ranch for the Canal Company. He was the second one and came after Chandois. He was sent east by the Canal Company to get settlers to come out. Afterwards he was superintendent for Miller and built the Central Ranch.

I hauled the first house for Jake Myer from Las Juntas when Las Juntas moved to Firebaugh. Jake Myer was the one who made Firebaugh.

Al Cornwell was 92 at the time of this interview. He is evidently mixed up on two periods in his life when he drove stage. The time he came back from Arizona in 1857 and the time he came back from the Civil War in 1867, or perhaps he did drive the stage at both times. I was unable to question him as he died shortly after this interview. As we returned to the County Hospital he said it was strange to see the barns so empty and still. "It's just not natural," he said.

August 22, 1932 RLM

Mr. and Mrs. C. W. Smith, Miller & Lux employees

The Smiths had worked for Miller for many years. They were married in 1874. In the fall of 1875 they came to the McPike ranch for Miller & Lux. In the spring of 1876 they moved to the San Luis Camp.

The San Luis Camp house at that time consisted of the adobe part with the large fireplace, which was Henry Miller's bedroom. The next room to the east in the wooden part was the room where the buckaroos slept, and then the room to the east was the Smiths'. There was a large kitchen where everybody ate and there was a porch on the north side.

During the June freshet of 1886 the water came up so high that a person could step off the north porch into a boat and ride all around the outside of the house. I was a fence rider and used to have to look after the fences on the McPike Ranch. Miller used to visit the ranches about once a month. Santa Rita was his headquarters then instead of Canal

Farm as he used later. He usually stayed one night at the San Luis Camp when on his way from Santa Rita to McPike.

Miller use to see everything when he came on his trips. He would go in the kitchen and lift up the covers on the kettles and see what was in them. He told me (Mrs. Smith) that the Spanish family that lived there before us were very wasteful. He said he would sometimes find two or three kettles sitting around at one time with moldy beans in them. Miller used to tell us not to let anything go to waste. If you have anything you can't eat feed it to the hogs or the chickens, but don't let it simply go to waste. He said that the people he had at Santa Nella used to dump even pies into the creek when they would see him coming. Such waste to Henry Miller was not to be allowed.

The first time Henry Miller came to McPike and had dinner he asked Mrs. Smith in his usual brusque way: "You like it here?" she said she didn't. "Where you from?" She replied Napa. "Well," said Miller we can't all stay up in San Francisco or Sacramento or Napa. Some of us have got to get out into the country."

When he finished his dinner he left ten dollars on the table for her.

When at the McPike place Schmitz the fence builder for Miller was there for some time and Mrs. Smith had 17 men to cook for. "Old Jimmy," an Englishman, was as big a man as Dr. Wade.

One day Henry Miller was there just after a lot of hay had been put in the barn. There were several kegs partly filled with nails in the room in the barn where tools were kept. Some of the hay had fallen down into the kegs. Old Jimmy was away at the time and so Henry Miller took me with him and said he wanted me to come and he would show me how he wanted things kept on the ranch. Miller went into the tool room and dumped every keg of nails in a heap on the floor. "Now," he says, "When the boss gets back you tells him I said for him to pick up every nail and put each size in a keg by themselves."

When Miller was gone and "Old Jimmy," who was the boss of the ranch, got back he told me to dig a hole and dump the whole bunch of nails in it. I dug a hole about two foot square and a couple of feet deep and put all the nails in it covered them up. So far as I know they are still buried there on the McPike ranch.

While at the San Luis Camp we had a daughter born that was sickly and died in a few days. Henry Miller happened along and wanted to see Mrs. Smith and the baby. He asked Mrs. Smith how the baby was doing and she told him it wasn't very well. Miller gave her ten dollars and told her to buy something for it. Mrs. Smith said she would probably have to buy a coffin for it with the money. "Oh no," Miller said, "You mustn't talk that way, everything will be all right."

One time Mr. Ham who was bookkeeper at Santa Rita drove down to San Luis Camp. Mrs. Tanner came with him for a ride. They had lunch at the San Luis Camp. Mrs. Smith had none-too many dishes and had put one kind of preserves in one dish and another in another dish. To her surprise Mr. Ham took one dish and ate all the preserves in it and Mrs. Tanner did likewise with the other.

A Spanish family had lived at the San Luis Camp before we came and Mr. Patman took my place. There was an old Spaniard by the name of Joaquin at the San Luis Camp when we were there and he said that the adobe house had been built by some relatives of Vasquez and that Vasquez used to live there before he became an outlaw. The adobe house was an old looking house even when we came.

Henry Miller in the early days used to give way to his anger without any attempt to control it. I have seen Henry Miller get so mad when things didn't go to suit him on his ranches that he would curse employees something terribly and then throw his hat on the ground and jump up and down on it. One day I head Miller abusing Tanner unmercifully and cursing him up and down. Later when I was talking with Tanner he told me that would cost Miller a hundred dollars!

About 1877 Miller became sick and he went on a trip back to Germany. His doctors told him that the trouble with him was in giving way completely to his anger. They told him that if he didn't control himself he would die. After that when things weren't going to suit him on a ranch he would tell his buggy man to drive on and he would go off and leave the ranch.

Henry Miller had some of the most peculiar traits. One time he came to San Luis Camp when I was out riding fences. He left three dollars and left word that when I returned he wanted me to take it to Canal Farm and give it to Wade White who was going to Arizona the next day to get a bunch of cattle. I got back to San Luis about ten o'clock at night and took the three dollars to Canal Farm about seven miles and gave it to Mr. White about midnight. Mr. White didn't know what the three dollars was for. He told me that the only reason he could see for Miller sending me with the three dollars was to try me out and see if I would carry out orders left by Mr. Miller.

Mrs. Smith used to be very much afraid when left alone at the San Luis Camp. One day she saw a Chinaman coming in a boat up the slough. When he got near the ranch house he came to shore and proceeded to make camp. Mrs. Smith was very much scared and seeing a white man on horseback in the distance she seized a shotgun and ran to him. She told him about the Chinaman and he came and told the Chinaman to get out of there and never come back. When she got back to the house she found that the gun she had been carrying all this time was empty.

The Chinaman got his belongings into the boat as soon as possible and rowed down the slough as fast as he could. He evidently had come up the river from Hill's Ferry where there were a number of Chinese at that time and was probably going to fish. Another time two horsemen came up to the ranch and asked when I would be home. Mrs. Smith told them that she was expecting me in a little while. When I later came up the two men rode away as fast as they could. I told Mrs. Smith later that she should never tell strangers again that she didn't know when I was coming, but should tell them that she was expecting me any minute!!!

There were great quantities of perch in the slough and Mrs. Smith could go out any afternoon and catch enough in a little while for supper. There were also shiners, chubs and suckers and of course in season there was salmon. Mr. Miller was very fond of perch and they always used to cook them for him when he was there. Mr. Miller would take a whole fish in his hand and proceed to eat it in one corner of his mouth and left a stream of bones fall out of the other corner of his mouth.

I planted some of the eucalyptus trees at the San Luis Camp you see today. Wade White was the cattleman for Miller in the early days and lived at Santa Rita. He was married twice. His son Will White is the merchant at Livingston. His two children by his second wife were born at Santa Rita. Henry Wade White was called "Wadie" when he was little. His daughter was Christiana. Henry is still a cattleman for Miller & Lux and his family live at Plow Camp.

Miller & Lux cattle in the very early days were the Mexican "long horn" variety. Some of the cattle had horns nearly as heavy as themselves. But when Miller bought the Santa Rita he sent east and bought the best breeding stock he could get and began breeding up his stock. By the time I got to the San Luis camp the Mexican long horns had pretty well disappeared and been replaced by American cattle. When we went to the San Luis Camp there was an old Mexican named Joaquin and his wife who had been in charge at San Luis for years. When he was showing me the boundaries he told me I was to look after the country as far as Palo Solo on the south and as far as McPike on the north. Palo Solo was the name for the Lone Willow.

In the early days the settlers were allowed to come down on the Santa Rita Grant and get all the wood for their own use they wanted. About the first time I saw Billy Stockton was one day when I found someone chopping wood. Stockton insisted that Mr. Miller had told him he could chop down all the trees he wanted. Stockton gave me his name and it turned out that Miller had given him permission to chop down as much as he wanted.

In the early days after a person left the Salt Slough there wasn't a tree anywhere until one came to the Kreyenhagen's tree where the Kreyenhagens had had a store. The Miller & Lux outside fence passed through the country at the Kreyenhagen's tree.

The Chamberlains lived on the East Side of the Santa Rita Slough about a mile or more north of Santa Rita. The Chamberlains were relatives of the Lux Family. The old Mexican Joaquin told me that the San Luis Ranch up in the mountains and the San Luis Camp were once a part of the same grant. The grant consisted of two parts—the part in the mountains and along Pacheco Pass and the part along the San Joaquin River.

In the early days Henry Miller used to give his men a bounty of $2.50 for every coyote they killed. But it got so that he was losing too many saddle horses by their being crippled in running pell-mell across the country after coyotes that he quit the country. It was the practice of the vaqueros to run into the ground.

When we lived at the San Luis Camp we use to have to come to Old Los Banos for our mail. When we moved to Badger Flat we had a fence in front of the house and running west past an orchard that was on the place at the time. The road from Central Point to Hills Ferry went right past the corner of the orchard and then in an almost straight line to Hills Ferry. There was another road along the foothills that ran to Hills Ferry as well. When we went from Badger Flat to Old Los Banos we rode straight across the country to Old Los Banos. The road in front of our house was a county road from the early days, but was not used until people began fencing up the country and people were compelled to travel the road proper. The old stage road from Old Los Banos went across country a mile or so south of the road in front of our home in Badger Flat. The old stage road west from Old Los Banos went almost straight for Pacheco Peak.

When the railroad went through, the railroad company wanted the town to be located where the railroad crosses Los Banos Creek. That would have suited the people of Old Los Banos and also it suited Henry Miller. But Uriah Wood didn't want any town on his land and so that made it necessary to look for another location.

The people of Old Los Banos set about to have a town where Volta is. Henry Miller came to see me and said that all he wanted was a shipping corral near Canal Farm. He wanted the town to be at Los Banos Creek and wanted the Old Los Banos people to all go in together and have a town on Wood's land. But Wood did not want the town on his land and put up a fight, and so the people of Old Los Banos were going to go ahead with their plan.

Miller went ahead with his plans for a town of his own as he would not go in with the others for a town at Volta. To the surprise and anger of all, one night the store building at Old Los Banos, containing the post office, as well, was put on wagons and the next morning was on the town site of Henry Miller near Canal Farm. The railroad company stood in with Miller and promptly called the new station Los Banos, after the post office, already sitting on Miller's land. Antone Pfitzer was fairly frothing when he heard that the people of Old Los Banos had lost their post office. At first the railroad company was not going to let them have a station at Volta. But there were so many influential people who wanted the station and they put up such a fight that finally they consented and put in a station at Volta and called it Volta Station.

Pastures of plenty in the San Joaquin in 1880

Our house was built by a man named Jordan, a "Gospel Swamp Preacher." We came to Badger Flat in 1878. At that time the principal settlers at Badger Flat were: Fowler, W. F. Clarke, Chandler, a bachelor, Worthington, who lived on the John Menjoulet place, Drummond, who lived east of the Menjoulet place, Gregg, the father-in-law of Dick Wilson. The Bibbys were living at Center Point where they had a blacksmith shop, but began moving about 1878 to Badger Flat. Joe Cirimele and Antone Cerubi were two Italian single men who raised vegetables where Cirimele now lives. The Cox family came in 1879.

Mr. Miller was the best man that this country

ever knew. When Billy Stockton was supervisor he hardly ever met Miller but what he would ask if anybody was suffering. "Don't want to see anyone suffering," he would tell Stockton. Always when Mr. Miller would stop at the San Luis Camp he would leave Mrs. Smith $5.00 and $10.00 to buy things with. But when it came to trading with Miller he wanted everything right down to the last nickel.

Interview C. W. Smith home November 29, 1925 RLM

In November of 1930 Mr. Smith and I attempted to find the location of Old Los Banos. We went to the Widman Place and to Amabiles this afternoon. We were unable to locate the site of Old Los Banos with certainty, although we found the remains of an old well in an alfalfa field. On our return we stopped in at Amabiles and he told us that the old well was the well where the livery stable used to stand and that the town was across the ditch on the north side. He says that when he plows and works his land he still finds old horseshoes and bits of coal from the blacksmith shop that C. D. Bambauer used in Old Los Banos.

RLM November 19, 1930

Zachariah Taylor Blankenship, cattleman

In the early days there was no good market for stock. Besides Jack Sutherland, there were few men buying cattle. During the dry years of the 1860's and 1870's there was little opportunity to get cash for our beef. The best market we had was Henry Miller. He was acquiring large bodies of underdeveloped land and needed stock. He used to come into the Four Creeks country and buy whole droves of cattle. They were generally delivered to Miller at Cross Creek.

I remember one band we sold to Miller in 1868. We drove the stock to Cross Creek. Miller met us there and took charge of the stock with his own crew. He had an American foreman and 20 or 30 Mexicans and Indian vaqueros. The cattle were counted by Miller. As fast as he counted them his men ran them across the bridge to the north side of the creek. When the counting was done my father went with Miller across the bridge to a light wagon where Miller carried his money. Henry Miller counted out several thousand dollars in twenty dollar gold pieces.

My father put the gold in his hat, and we rode back to Visalia. I was bareheaded, and father was carrying the hatful of money on the saddle in front of him. Unless a person has seen the hard times of the 1860's and 1870's, it is impossible to realize what a great thing it was for us to be able to get cash. There was no credit to be had.

Henry Miller was possessed of a number of remarkable characteristics. He was a shrewd businessman as ever lived, and drove a close bargain. But he never took advantage of a poor man who is in difficult circumstances. When my father sold him those cattle in 1868 we were in as hard of circumstances as people could be. Miller was more liberal with us at that time than he ever was at any other time.

Frank Latta interview 1930 given to RLM

Antonio Aguila, Miller & Lux vaquero

1860's-1890's

It used to take Miller & Lux from four to six days to drive cattle from the Bloomfield Ranch to the City. Miller used to tell the vaqueros to take it easy. "Yours time is mine," he used to tell them. He didn't want beef cattle driven fast. The stops along the way were: Bloomfield, Twenty-one Mile House, Fifteen Mile House, San Jose, Mountain View, Redwood City or San Mateo, Twelve Mile House or Lux's ranch and Butchertown.

Miller & Lux had a small ranch at the Fifteen Mile House where they could stop. Their stopping place in San Jose was a mile and a half or so past the graveyard. In the very early days the slaughter houses were right in San Francisco and many a time I have driven bands of cattle right past Mission Dolores on the way to the slaughter house.

A man by the name of Crocker owned the Temple Ranch. Miller backed up Crocker and Crocker went broke. For a while Miller had Crocker in charge of Santa Rita. Crocker died and his sons moved to Independence, on the Owens River. Most everybody that Miller staked went broke. Crocker was an example.

One time I was at Santa Rita in 1864 there were already good buildings at Santa Rita. The headquarters was a good family house. In the early days when there would be bands of cattle from the south being driven to market Miller would often go a couple of days ahead and try and buy them. When Miller would buy a band they would have to be branded.

They used to brand cattle at the old corral at Santa Rita. Six vaqueros would be working at a time in the corral. They would work in pairs on horseback. One partner would throw his rope and catch the animal by the horns and the other would catch by the hind feet. The animal would be thrown and branded lying flat on the ground. There were no branding chutes in those days. So skillful were the vaqueros that three pairs of them could be working at the same time in a big corral and not get in each other's way. Narcisio Castro was a great roper but I, Antonio Aguila, was a better rider.

One time Henry Miller called me aside and pointing to a fine black colt told me that he wanted me to break the colt to ride. He said he wanted it for Jesse Potter who was Mr. Lux's adopted son. So I proceeded to train the colt and when I had him pretty well under way Miller was at Santa Rita corral and asked me how he was getting along. The vaqueros had a big bunch of cattle in the corral and were roping and branding them. One big long horned steer possibly ten or twelve years old broke out and ran out on the plains. It evidently was a bad animal as none of the other vaqueros made any effort to go out and bring it back.

I started out on the black colt and throwing a rope over the animal's head soon had the steer back in the corral. Miller was watching and was very pleased with me and the black horse. Miller wanted to know if the colt was sufficiently trained. I told him I ought to work with the colt a little longer at roping and said I didn't want to take anyone's place in the corral, but that if Miller would let me go in the corral with the colt and work at roping with the other vaqueros for a hour I thought it would be good training for the colt. Miller told me to go right ahead and I did so. The colt turned out well and Miller was very pleased.

Thomas Dias was a Chilean and used to buy cattle for Miller. He was a little short fellow and was a very good man to handle cattle on the road. He was on the road most all the time driving cattle. Miller kept him until he died. The story is that Dias went down around Los Angeles buying cattle for Miller. He bought the cattle however, as though he was buying for himself and when he got back with them for some reason the money to pay for them was not sent and the owners came to collect their money. But all Dias would say was that he didn't have any more and the owners were never able to get their pay.

Miller was sick and went to Germany where he had an operation on his throat which was very successful and he returned to America much improved in health. Lux had been running the business while Miller was away and had the business in a terrible state. People say Lux had the money but Miller had the brains.

Miller was a master butcher and was the finest man with the knife that I ever saw. When Miller came to San Francisco he used to sell tripe and pigs feet and sausage. Eventually he had a little shop of his own and quit the sausage business. Miller could look ahead and could make money. Miller went to see Lux. Miller could see how he could make money but he needed money to do it. He told Lux, "We can make lots of money. But I need money to make money." Lux said all right and took Miller in partnership and they made a bushel of money. Miller rose to fortune very quickly.

One time Miller telegraphed to me to come over to Santa Rita. Miller had bought about 800 head of cattle in Long Valley, near Independence. I came to Santa Rita and Miller told me to take plenty of help and go over and bring the cattle over. We were to make two drives and were to start one band of cattle and then a day or so later start another band to be followed in turn by a third band. When we got the first half here we were to go back and get the rest.

That afternoon at Santa Rita they needed a beef killed for the ranch and the vaqueros were to do the work. They didn't know how to butcher very well and asked me if I would help them. I understood pretty well how to go about the job. The beef they were to butcher was a very fat cow and the vaqueros in starting to skin off the hide had made a rather botching job of it. Miller was at Santa Rita at the time and I was hoping that Miller wouldn't come around where they were at work. He knew the vaqueros weren't doing a very good job. But Miller came around when I had the beef hung up and the hide about half way off. Miller saw the poor butchering the vaqueros were doing and started in abusing them.

"Look here," said I, "we are vaqueros and not butchers, we don't know how to butcher but we are doing the best we can. You haven't any right to find fault with us because we don't know how to butcher the way trained butchers would do. I'm through. You can go ahead and finish this job yourself and let us see how it should ought to be done."

Miller was somewhat taken back, but he rose to the occasion and getting into a pair of overalls proceeded to finish up the job in true butcher fashion. After the men were through and were in the house getting washed up Miller called me into another part of the house and told me that he saw his mistake and was glad that I had stood up for my rights. He told me he didn't like to have employees who cringed from him. He said he wanted men who could stand up and talk like men. Miller said he saw his mistake in finding fault because the vaqueros weren't trained butchers. Miller always kept good whiskey on hand and Miller and I had a good drink together and everything was all right.

Miller then told me he wanted me to start for Independence the next morning after the 8,000 cattle he had bought. So the next day I started with about a dozen good vaqueros for the Owens River. Not all good vaqueros went. Some were left to take care of the ranch. We brought about half of the cattle the first time, driving them in three bands, each about a day apart. The owners and their vaqueros came with the cattle to help drive them and there were nearly forty men in the drive. They had pack animals and a man to cook and a man to drive the gentle horses. Their grub and blankets were carried on the pack animals.

"I receive the cattle at Madera," Miller had told the men selling the cattle. When they got to Madera, which was only a few miles from Miller & Lux lands, Miller was on hand to receive the cattle and pay for them. The cattle belonged to several different owners and were all mixed in together. When the cattle arrived they were all counted and the number belonging to each brand being kept separate. When everything was counted and figured up Miller took a lap robe out of his buggy and spread it on the ground.

Then he called me and told me to stand at one corner. Calling another vaquero he posted him at another corner as a guard. Then he poured out a pile of gold in one place on the robe and another pile of silver in another. Calling up one owner he would read off the number of his cattle and the amount due him. Counting out the money he would hand it to him and say, "Here. This is yours." The next man would step up and Miller would count out the amount due him and hand it to him. In this way all the men were paid off and the vaqueros sent back for the rest of the cattle. No checks were used in those days. The men selling their cattle were paid off in actual coin.

Interview with Aguila Nov. 23, 1930 RLM

W. T. White, Sheriff, Merced County

I came with my mother out to California about 1870. We came with Tom Hildreth's family; my mother was a companion to Mrs. Hildreth. Wade White was working for Hildreth and Dumphy up in Nevada and they became acquainted, and were married at Battle Mountain, Nevada in 1872. About 1874 Wade and his family came from Nevada to California and about 1875 went to work for Miller & Lux at Santa Rita. I first went to school in Merced, boarding with a family there. My mother passed on in 1879. I had been going to school in the Santa Clara Valley, and about 1880 I got on the stage and came over to the San Joaquin Valley to get a job and go to work. I stayed overnight at Old Los Banos and the next day Mr. Tanner was there. I knew Mr. Tanner and he asked me what I was doing over here. I told Tanner I wanted to go to work, so Tanner took me with him, and for a couple of weeks or so I rode with Mr. Tanner and opened gates. Then Tanner asked me how I would like to go to work in the store. I said that would be fine, so Tanner sent me to work in the Center Point store with Harry Dunbar.

Later from about 1881 to 1884 I ran the little store that Miller & Lux had at Santa Rita. Later I worked in the store in Old Los Banos and at Firebaugh. In 1890 I worked in the store at the present Los Banos. Later I and Joe Place and others had a store of our own at Los Banos. This venture went under and I left Los Banos. In 1901 I was back with Miller & Lux running the store in Los Banos until December 1909 when I left Los Banos. Miller & Lux had a store at Old Los Banos and also at Center Point. When Los Banos was started the two stores were moved to Los Banos and consolidated.

The Raisina School District was that stretch of sagebrush country under the old canal north of the Charleston District and included the families of Scanlon, Schornhorst, Wiley, Watkins and others. The people gradually froze out and the school was closed. The hills are gradually drying up. The change was very noticeable after the big earthquake. There use to be two big springs on the Idiart place and there was a big spring on the Arburua Grant. But since the earthquake, the springs have been failing and the country is drying up. The Poso Warehouse is some five or six miles north of Firebaugh on the San Joaquin River. Hoffman started his store at Center Point about 1876 and in 1879 sold out to Miller & Lux.

The only trouble that Henry Miller and Charles Lux had with each other was over land. Miller was always buying more land and would never sell any. Lux wanted Miller to sell off some land and get rid of it. Miller was very far sighted. As an old Indian explained it: "Do you see that tree in the distance? Yes, and do you see that next tree a little farther on? Yes, well do you see a little farther on another tree? Yes, and do you see next tree?" "No, I can't see any further." "Well," replied the Indian, "Henry Miller can, that next and the next that you can't see."

Just north of where the present Merced Bridge is over the San Joaquin River there used to be in the early days a ferry kept by a Mr. Robison who was a relative of Mrs. C. W. Smith. It was known as the Robison Ferry. Before that it was kept by some people by the name of Smith and was originally known as the Smith Ferry. There used to be four poplar trees there and people would guide their course by them in times of high water in order to reach the ferry. Now only one tree is standing. The Dickinson Ferry was run by this same man Robison,

Oddfellows Lodge 5th and "I" Street, Los Banos

before Dickinson took over. Ferries were located where there was high land leading up the river so that in high water people could get to the ferry.

The course of the San Joaquin River was constantly changing owing to the high floods that used to come down every year. Sandbars and new channels were constantly forming. But now that Miller is taking that water for irrigation all that is changed, and the riverbed remains practically unchanged. The Keyes Ford was the nearest way from Santa Rita to Merced, and people used to go that way in the fall of the year when the water in the river was low and people could ford the river there. At other times of the year it was impossible to ford the river, and people would have to cross over at the Ferry. A man by the name of John Keyes had a little ranch there; the ford was named after him. The road from Santa Rita through the Keyes Ford passed through the Sandy Mush country on to Merced.

The real May Day celebrations started at Canal Farm when the railroad came in. The railroad came as far as Newman first and stopped there. Then it was extended down as far as Los Banos. Los Banos was the terminal. Later it was extended further south.

In the old days Henry Miller would give away thousands of dollars worth of Christmas presents. About two or three weeks before Christmas, Henry

Miller would come to the Los Banos store and say to me, "Well what about the Christmas?" I would have a list all made out of all the foremen and their wives and families and I and Henry Miller would go over the entire list. Mr. Miller knew them all and would decide how much of a Christmas present they were to have.

"Now, that fellow is no good," Mr. Miller would say. "He is a gambler, but his wife is a good woman and they have two little girls, put them down for $25.00." Another family he would size up about as follows: "That woman is no good. She is dirty. She doesn't keep the floor swept. Her husband is a good man." And so it would go. Nearly every family that was in the employ of Henry Miller had either the husband or the wife that was no good. There was hardly a family where both the husband and the wife were good industrious people. Henry Miller knew them all and had them all sized up. But every family in his employ got a Christmas present.

At first it was the custom to make up a package of clothing, dress goods, etc. out of the store to the amount set down after each name of the family by Henry Miller. But as time went on and the children got older, it was difficult to make up a Christmas package of dress goods and other goods that would suit the families and so I would generally let foreman come in a few days before Christmas and select as much goods out of the store as Mr. Miller had indicated.

All the Christmas presents for the employees as far south as Bakersfield were selected from the Los Banos store and shipped to them. Every needy family in Los Banos was also always remembered with a big package of groceries and flannel and other needful things. Never less than a thousand dollars worth of merchandise was distributed and usually it ran up to $1,500 or $2,000 dollars.

Whenever Henry Miller came on a ranch the first thing he always did was go and look in the swill barrel. He wanted to know what was being thrown away. Little things worried him more than big things. He could not stand it to see anything wasted. He wanted everything taken care of and nothing wasted. His idea was to take care of things for maybe you would need them. Take care of the cents and the dollars will take care of themselves was his idea.

One time at Santa Rita a big fence of boards was taken down. When the nails were removed from the boards, many of them were crooked and bent. Miller had every one of them saved. There were several kegs of crooked nails, and Miller had them all saved up until winter when he put the men that had nothing else to do to work each with a block of wood straightening these nails for further use so that they could be used over again. In the early days people wanted long jobs. They didn't ask what the pay was or what the hours were. What they wanted was a job that would last a long time, a steady job.

Whenever Henry Miller would see a board off of a fence he was sure to holler. When he told a foreman to have a certain piece of work done, he was sure to go first thing when he came back to that ranch to see if it had been done. One time a board had come loose on a fence, and he told the young foreman to have it nailed back. When he came back to that ranch later, he found the board hadn't been fixed yet. The foreman tried to explain that he had been so busy that he hadn't had time to have it fixed. Miller exclaimed: "You have all the time there is. I hires you by the month and a month is thirty days and thirty nights."

Once when I was running the store at Center Point, there were a lot of packing cases and boxes piled up in a pile back of the store. Mr. Miller looked over and called me out. "You have all these boxes piled up so that you can get at them," he told me. "Pile them upside down, so the rain won't get in them. Put the big boxes on the bottom and the smaller boxes on the top. Then when anyone wants a little box you can get one, and when you want a big box you can get one. Take care of things. You may need them."

Miller taught me a lot, and I appreciated very much the early training he gave me. People always paid their store bills in the old days. There were no deadbeats. They didn't always pay in full, but a settlement was always made. Maybe a man who had a large store bill had some wheat he was holding for a higher price. Tanner would offer them maybe ten cents a hundred more than the market price at the time. The owner would usually sell and the store bill would be paid. Of course Miller & Lux didn't lose by paying a little high for the grain, for they always had use for grain on their ranches, and if a settler got in debt and wanted to move away a settlement was always made although Miller & Lux might have to allow more than a thing was worth in order to get a settlement and the land. But settlement was always made.

In 1890 the store was moved from Old Los Banos to the new town of Los Banos on the railroad. Drummond was the storekeeper for Miller & Lux at Old Los Banos, and it was Drummond and I who moved the stock of goods to a warehouse in Los Banos that was being used temporarily as a store. The Central Point store had already been moved and the Center Point store building was used for the front part of the new store in Los Banos. The Central Point store and the Old Los Banos store were consolidated at the new railroad town.

I think the post office had already been moved when I got there. Gus Smith was the assistant postmaster, and he had the post office located on Front Street. It took a long time to move the Old Los Banos store building to Los Banos. Tanner had secured an old man with a set of house moving trucks to move the building. He had two sets of wooden wheels and two long timbers under the building. There was no bottom to the roads. It was a terrible job for the old fellow. It was expected that the store building would be moved and the new store going months before the building was finally gotten into Los Banos.

The stocks of goods at both the Center Point and the Old Los Banos stores were allowed to get down awful low at both places before the stock of goods were finally moved to the new town and the stocks consolidated.

The May Day celebration at Canal Farm in 1891, which I was connected with, was a great time. Henry Miller paid $125.00 for the Merced Band to come over for the occasion. He also furnished the meat, bread and cheese.

In order to perpetuate the name of Mountain Brow Lodge the Odd Fellows of Chinese Camp consented to the moving of the lodge from Chinese Camp to Los Banos. They were mighty ice old men that belonged to the Odd Fellows at Chinese Camp. There was $1,000 in the Mountain Brow Lodge and Dr. Wade was anxious to get hold of that money to help build a lodge building in Los Banos. Dr. Wade was quite a worker in lodge affairs in those days. He was also one of the main ones in establishing a high school in Los Banos. When he went into anything it was heart and soul.

I was one of the officers in arresting Hale in the famous Hale murder. The boot tracks were unmistakably plain. The Hale house and the house where the murder was committed were perhaps a mile and a half apart. A road ran part way. From the road to the murder house, the boot prints were perfectly plain. It had rained just enough so that the boot prints stood out clearly. When the officers were satisfied that Hale was the guilty party, they rode to the Hale ranch and placed him under arrest. When the officers came in the house before starting off from the ranch, Mrs. Hale showed not the slightest emotion or affection for her husband. Hale was taken to Los Banos for a hearing before the Justice of the Peace and then taken to the jail in Merced. Hale was very nervous about being left at Los Banos and seemed to fear a lynching. He was perfectly willing to be taken to jail in Merced.

When public celebrations were held at San Luis Ranch by Hoffman, the dances were always held upstairs in the barn. There was balloon ascension at one May Day celebration at Cana Farm. The German settlers around Los Banos were a hotheaded bunch and did not fit in well with the other settlers.

The last year I was connected with Miller & Lux store in Los Banos I had every sheep account from Firebaugh to Los Banos. We did over 100,000 worth of business with the sheep men and Miller never lost a nickel. The sheep men would even deposit their money with the store. Miller was all business. He was a wonderful man. He was a worker.

Miller was particular about branding cattle. Wade White was the only man who could brand bullocks to suit him. Wade was up in Nevada for two years one time rounding up stock for Miller. Once when he was gone, someone else was branding cattle at Santa Rita. It was not done to suit Miller and he became very mad. He took off his hat and stamped on it. Miller would have these spells where he would take off his hat and jump up and down on it, but in a little while he would get over it. Miller was comical and would say his remarks about each man with a kind of chuckle.

The superintendents were sent a check from the San Francisco office. It was for the ranch foreman and straw bosses that Miller would send a check over with me. "Give them all good clothing," was Miller's orders. Mr. Miller always remembered them all. "Oh it was a job." If the poor needed something, it was up to my judgment what to give. Mrs. White would help and so did Bert Wilson's mother. Miller was mighty good and always wanted to be on the good side of his labor.

One time Mr. Miller sent word to me. "Come up, I want to see you." An old German fellow with his hat off was standing talking to Mr. Miller. "Henry," he said, "I've worked for you for thirty-five years. Won't you give me fifty cents?" Miller gave him some money and then asked impatiently, "For why you call me Henry? Did we ever sleep together?"

When Sid Shannon sold Plow Camp, Miller read Shannon the riot act. Said he never intended selling Plow Camp to anybody. Shannon declared that he had the letter from Miller telling to sell the land. He looked up the letter and showed Miller that he was wrong in saying he never told Shannon to sell Plow Camp. Sometime later the fellow who had bought Plow Camp wanted to sell it back to Miller. It seems that in the transaction the man had bought a span of mares from Miller & Lux for $500. After Miller and the man had made their deal and Miller had Plow Camp back, the man asked "What about the mares?" "I gives you $250," says Miller. The man protested that he have given Miller $500 for them. "That's the difference," chuckled Miller, "between buying and selling."

There was no such thing as an 1878 May Day Celebration given by Henry Miller. That talk of Henry Miller beginning his May Day celebrations in 1878 is pure nonsense. I came to live at Santa Rita about 1874. During flood times the water use to come almost to the present day Los Banos. Santa Rita was entirely surrounded by water at such times. Maybe two months at a time. There was no school at Santa Rita and in 1876 and 1877 I was sent to school in Merced where I was boarded.

1877 was a dry year. The climate was bad for my mother and in 1877 we moved to San Jose where it was thought the climate would be better but in 1878 she died. Along about 1880 or 1881, I went to work clerking and doing general utility for Henry Miller. I worked at Center Point and Firebaugh and went wherever I was needed. Miller paid me ten dollars a month to start with and room and board. Gradually my wages were increased until I was making $35 a month. In about 1885 or 1886 I quit Miller & Lux and took a job in Mexico working on the railroad at Guadalajara; then I came back.

One can't imagine the amount of quail there used to be up on the Romero Creek. In 1880 or so, Harry Dunbar, others and myself would go quail hunting up on the Romero. We would go on Saturday and come home Sunday. We would take a spring wagon and blankets and something to eat and have an awful good time. The quail would come down to some water hole for water, and the hunters would come home with from 50 to 100 quail.

Market hunters used to make a living hunting ducks and geese. There was Kirby and Bill Browning and Old Man Smith. Browning had two or three outfits. All he did himself was to haul the game up to the shipping point. Browning made money. Nigger Howard shot for Browning. He was a little short nigger. Nigger Howard's record was 224 ducks in two shots. With geese his record was 84 geese with two shots. These shots were with oxen. Nigger Howard's shots were the record. Old Man Smith lived at Smith's Ferry just a short distance down the river from the present San Joaquin Bridge. His daughter Lillian was a remarkable shot. There were two hunters from Gilroy who used to come over and hunt rabbits. They used a spring wagon to haul their game, and when they didn't get a hundred rabbits a day apiece, they considered it poor shooting. They used to send their game by express on the stage.

I have seen the San Joaquin River a solid sheet of water from the present Chowchilla Y clear across to Dos Palos and Kreyenhagen's Gate. About three months during 1883, you could not get off of the Santa Rita. In every direction one wanted to go it was necessary to go by boat. Every season someone would get drowned. Warfield the superintendent at Santa Rita sent a man on horseback to Canal Farm for the mail. In crossing the Lone Willow Slough, he was drowned. When searchers were sent out, they found his horse standing beside the slough. The saddle was wet and the horse was wet and shivering. The man's hat was seen floating on the water. They knew he was drowned. The next morning the body was finally located in the slough. There were treacherous places in the sloughs, and a rider had to know the country well. One time Warfield sent me for the mail. I knew every foot of the ground. I had to swim my horse seven times in going from Dickinson's Ferry to Santa Rita.

Interview at County Jail, Merced 8/12/1930 RLM

Ike Myer, early pioneer, Westside

Miller possessed decision of character in the most remarkable degree. Affairs that others would have weighed and discussed for many days he would determine in a few minutes. He could listen to a proposition involving thousands and arrive at his decision in an instant. He could in that brief speck of time determine the advantages and disadvantage of business dealings. He could see the gain as well as "the nigger in the woodpile" in a flash.

He gave freely to charity and quite freely to promote the innocent pleasure of the people. Though not a religious man, he never refused to aid churches or the lifting of debts from organizations within his domain. He never exhibited much interest in higher education, but schools of every class within the vicinity of his empire counted on him for liberal aid without ever being disappointed.

Miller's idea of a good man was one that was good for something, one who could accomplish things worth doing. Had his fortunes placed him in the army he would have been a great general. As they placed him in the cattle business he became the Cattle King in the best sense of the term. Los Banos *Enterprise*, undated.

Miller was always looking for way to make a dollar at others' expense. He often was involved with Charles Spreckles of railroad and sugar fame. There was dislike on the part of both men each for the other.

Miller got even with Spreckles twice. Once when he let Spreckles build the Santa Fe Grade down through the swamps where the whole country gets under water. When the Southern Pacific came along he let them build the road wherever they wanted to and he made no charge for the right of way. But he made Spreckles build down in the swamps and charged for the right of way besides.

Miller never paid anywhere near what a thing was worth. That was his idea of doing business. That's the way he made lots of money. He would make you an offer and if you didn't want to take it he didn't care. He never offered even half what a thing was worth. But if you went to him for anything then you paid for it.

Miller owned a riverboat and barge to bring up freight from Stockton to Salt Slough and Firebaugh. This boat was just for Miller & Lux business alone. The name of the boat was some German name and Mr. Wood thinks is was Friese. Miller later bought a couple of boats and did his own hauling. He would haul a barge behind the steamer. He could haul an awful load on the barges, especially going down stream. They would pile on woolsacks at Firebaugh until it looked as big as a house. Only rather small boats could run on the river, the river was so crooked. In high water boats could run as far as Firebaugh very nicely.

During the dry year of 1864 Miller would buy up drove after drove of cattle for what their hides were worth, about four dollars. He had barge load after barge load of hides stacked up along the San Joaquin River. When droves of cattle would start out from Los Angeles and drive their stock too fast the cattle would become footsore and when they would get to about Miller & Lux's ranches the owners would have to sell them to Miller for any price. Miller would distribute them along the river and get most of them through. Those that died were skinned and their hides saved.

Henry Miller always claimed that one good hand raised calf was worth two range calves. If a calf was not well cared for and Miller found it out, there was trouble. Davie Wood use to keep a couple of cows all the time. Miller would furnish the cows and the calves and the Woods got the use of the cows for raising the calves. They would keep calves in a pen so that the cows would not go away. When a cow would begin to wander too far off when the calves got older the Woods would take her back to Miller and get another cow and calf to take their place. There was no trouble to get all the cows they wanted. Whenever, Miller came to the Wood ranch he would invariably go look at the calves and see how they were getting along.

Miller told Will Wood that he could put seven pounds a day on ordinary range steers by feeding them on chopped alfalfa hay and rolled barley. Miller had feed yards at every ranch and would put up so many steers at the proper time and have them ready for Christmas beef. There was a great demand for fat Christmas beef and Miller was always able to supply it.

When Davie Wood sold out his ranch to Miller in 1874 he had some stock left and Miller offered him $149.50 for the lot, a ridiculous price. Davie Wood refused. He sold off most of the cows to the people here on the plains. The balance of young stock and a very fine bull he was taking over to Gilroy. When he got to the Mountain House he sold the whole bunch to Real for three hundred dollars. This was more than Miller had offered for the entire bunch.

Miller would boil when he would be driving over the country and find where someone had killed a beef and only taken part of it and left the rest—a quarter for instance. If they used all the meat that was all right with Miller but he didn't want any left to waste. He also wanted the hide left. A good deal of that kind of work was done by travelers. If the settlers butchered a beef they would use all of it and divide it up among the settlers. Miller never objected to this.

They use to make butter at Canal Farm and take it to Merced to sell. Canal Farm also furnished the ranches with butter. The butter served at Plow Camp was so rancid that no one could not eat it. One day Miller and Tanner came to Plow Camp. It was after dinner but Miller wanted something to eat. The cook served the rancid butter. Miller was hopping mad, he told them to take the stuff and bury it. Is that what you feed the men on? That night there was fine fresh butter on the table.

Pierson was the woodworker at the Canal Farm for years. After the house burned and Henry Miller's bootjack with it, Mrs. Dailey said to Pierson, "Do you think you can make a boot jack?" Pierson made one and when Henry Miller tried it he told Mrs. Dailey, "Just tell that fellow he better try and make another one. If it is not better than this one he better go off someplace to die."

Miller was having some trouble with the people living at Las Juntas. They were rustling his cattle. The county bridge between Fresno and Madera County was built about 1878, Madera and Fresno County were all one county at that time. Jake Myer lived at Madera at that time and was supervisor both when he lived at Firebaugh and when he lived at Madera and was very influential in getting the bridge across the San Joaquin River built.

Las Juntas was one of, if not the earliest settlement in the San Joaquin valley. Some of the old timers claimed it had been settled by the Spanish as early as 1800. The town was located on the west bank of the San Joaquin River about 3 miles northeast of Mendota. The name means meeting place and may have had to do with the fact that the Fresno Slough met the San Joaquin River near this point.

Las Juntas was well known as the worst town in the San Joaquin and law officers seldom followed after the tough characters that congregated there. Gambling, horseracing, murder and mayhem were everyday affairs at Las Juntas.

Since Miller had built his fence most of the people there at that time were Mexicans and half-breed Indians. Miller claimed they were stealing too many cattle and sheep and calves and ordered them to move out, as he claimed the land on his side of the fence. Miller also knew that the outlaws Murrieta and the Vasquez gang obtained their supplies there. They were safe there from American officers.

A man by the name of McIntyre started the ferry and later sold to Firebaugh who sold to Hoffman. Hoffman and Turner sold the store, hotel business and town site to Jake Myer and his partner, Kahn. Myer later bought out his partner, Kahn, and in 1883 sold out to Miller.

It so happened that a good many of the Mexican families owed Jake Myer store bills and he knew that if he let them leave the country he would never be able to collect. He owned the town site of Firebaugh and so he offered to give every Mexican family that owed him anything a free town lot in Firebaugh, merely charging them for making out the deeds. About half of the population of Las Juntas, about 12 or 15 families in all, accepted Myer's offer and moved to Firebaugh.

Miller was furious mad when he found out what had happened but was unable to help himself. There he thought he was driving these thieves out of the county and they were merely becoming more securely entrenched in Firebaugh. This happened in the fall of 1876 or 1877. Las Juntas consisted of about thirty families of Mexicans intermarried with the Indians. There were more Mexicans than Indians. It had once been an Indian reservation. Henry Miller probably had not the right to drive the Mexicans and Indians off and probably they could have stayed if they had known how to go about it to fight Miller in court.

The Village of Las Juntas, Rancho de los Californios, Rancho San Emigdio and the outlaw cemeteries at each place were the earliest white settlements in the valley. Las Juntas 1800, San Emigdio about 1810, Rancho de los Californios about

Hotel Los Banos with Opera House in background, 1900

1820. Miller solved his problem in 1883 by buying out Myer and tearing most of the buildings down.

Myer knew the Langdon family well. Mike was the best of the whole outfit. John turned out to be a kind of a lawyer. Tom was a bully and was never happy except when he was having trouble with someone. John Langdon shot Lahare who owned the Arburua Grant. Lahare was a fine man. Owing to the trouble with John Langdon he sold out to the Arburua. This occurred just at the time Miller bought out Myer.

Henry Miller was a Republican and was very much disturbed by Grover Cleveland's administration when the Secretary of the Interior ordered all fences removed that enclosed government land. Miller had been using this land as his own for years and it hit Miller and other large land owners hard for they had lots of government land fenced in with their own and were compelled to take down their fences. Ike Myer 8/7/1927 RLM

Will Wood, early pioneer, son of Davie Wood

There were two reasons why Miller always made money. In the first place he never bought anything except at a low price. He would always offer a low price for anything and a person could take it or leave it alone. He never changed his offer once it was made. If a person tried to argue with him to get him to raise his offer he would simply walk away. Another reason for his success was that he always had money to buy. If anything was for sale cheap he always had the cash to buy and take advantage of the offer.

The first time I saw Henry Miller was when I was driving an ox team over near Gilroy. Miller was riding along on horseback. He wanted to buy the ox team but only offered an insignificant price, which I refused. Tanner told me he never saw anyone who could turn everything into money like Miller did. With Miller there was money coming in everywhere. I remember seeing Henry Miller in his first buggy when he began using a buggy instead of going horseback everywhere like he did at first. The buggy had side springs and was very easy riding compared to horseback. The running gear was painted straw colored and the body dark. He had a brown team and a driver. Miller always used the same style buggy. In the early days Miller rode with his vaqueros just like he was one of them. But he got so he couldn't stand it to ride horseback so much and had to take to traveling in a buggy.

I remember well the first time he came through the country in that buggy. The Davie Wood family was surprised. They couldn't imagine who was coming. They had always seen Miller on horseback. He had a man with him to drive. This was in 1867 or early 1868.

Uriah Wood was a very tight fisted man. One time he owed Miller & Lux quite a large bill. Tanner was at the ranch and Wood was going to pay the bill. He had all the money necessary to pay the bill except for ten cents. As he gave Tanner the money Wood remarked that he was ten cents short, but that Tanner could help himself to a new grain sack for the balance. There was a pile of new grain sacks there. Tanner drove over to the pile and instead of taking only one sack worth about eight or ten cents, he deliberately took two sacks and drove off with them. Uriah Wood was furious to think that Miller & Lux had got the best of him by one sack. He told Will Wood that he couldn't see how Miller could get men that were so interested in seeing that Miller got all that was coming to him.

Old time sky farmers used always to get at least one good crop in three years and one part crop in-between. There never was more than one crop failure every three years and never two crop failures in succession. The first alfalfa planted in this country was on the Ward Field and was probably two or three hundred acres in extent. It was planted about 1872 and before Davie left his ranch in 1875. Tom Ward was the foreman at the time. The first cotton was planted on the Cotton Field Section. It was a very tall variety of cotton over one's head in height and was the wrong variety for this country. It didn't yield very much cotton and was a failure. Miller didn't plant any more cotton.

In the early days "wild game" was frequently the piece de resistance in the restaurants—the other being "lean beef" due to the fact that the prevalent black Spanish cattle then roaming the plains were long-horned, scrawny critters. Miller made the effort to bring in better stock as well as another effort being made with some success to cross native stock with buffalo, the offspring being termed "catalo." It was not until the late 1860's that white-faced cattle were imported in considerable numbers from the east by Miller and other wealthy "rancheros" that beef started to grade up to the standard we know today. By the end of the Civil War, as a result of drought, California's large herds of Spanish cattle virtually disappeared.

Dan Murphy had a big ranch near Bakersfield. Miller didn't steal cattle, but he had Mexican vaqueros who understood the business perfectly. One time Murphy found sixty head of steers belonging to him in Miller's slaughterhouse corral. Murphy brought suit against Miller for stealing cattle and Miller offered to settle by paying $60,000 for the sixty head. Murphy refused to settle. There were several trials and finally Murphy heirs settled for the $60,000. During one of the trials Murphy's lawyer tried to make Miller mad by saying, "Don't you know Mr. Miller that I used to know you when you used to peddle sausages in a basket?" Instead of getting mad Miller replied, "Yes, and if I was as big a fool as you are I would be peddling them yet."

Miller slipped up on the buckaroos one hot afternoon. Crocker and the buckaroos were snoozing away after dinner. Miller came up. He cussed them and went over to Kreyenhagen's and bought all the axes he had. Miller brought them back and told the men to go to work chopping out willows. There were lots of willows between Santa Rita and the present road. Some of the buckaroos chopped a little, but Crocker defied Miller and said he wouldn't chop willows for anybody. He sat in the shade and didn't chop a willow.

Will Wood 1927-1933 rounds RLM

Joe Webb, early pioneer and farmer

The ranchers at Cottonwood killed seventeen head of Miller's cattle in one night. They sent him word. It was a community affair. It was shortly after we came here, about 1870. Miller went to fencing right away. The fence lumber came up the river on steamers. Miller had three ox teams of eight oxen each to haul the lumber from the river out to the line of fence. The three drivers of the ox teams were Owen Hughes, John McCarthy and a man by the name of Crowley I think his name was John. They were big stout young Irishmen. Regular fence wagons were used to distribute the lumber along the line of fence as fast as it was used. It took a year or a year and a half to build the fence.

Miller put a fence all around the Webb place. It ran from the new outside fence. The new fence was Miller's outside line at that time. He bought much land outside this fence later. The fence along the Pacheco Highway was built before or about the time the town of Los Banos was started.

Interview on the street 11/1/1934 RLM

James McDermott, Miller & Lux employee

James McDermott came to California in 1867. He started a little paint shop in Oakland where he had an uncle. From there he went to San Jose where he worked in a mill under a foreman by the name of White. He kept busy for the most part glazing windows. Next he went to Gilroy where he worked as a painter for Hick and Folkerson who took large painting contacts. A friend of his wanted to get a painting contact with Miller to paint the house at the Bloomfield Ranch and Mr. McDermott went with him. While there he saw Henry Miller cleaning a gun to shoot squirrels. Picking up the rifle McDermott sighted it on a chicken some distance away and remarked to Henry Miller that he bet he could clip the chicken's head off. Miller told him to go ahead. The chicken was some distance off but its head came off as nice as you please.

Henry Miller was so pleased that they went out hunting squirrels in a two-wheeled cart. Miller used a shotgun and McDermott a rifle. Miller wasn't a very good hunter, although he got a few squirrels. But McDermott picked them off every time with the rifle. Miller was so taken with McDermott's skill that he hired him on the spot for $35 a month to shoot squirrels. That was big wages then. In those days squirrels were much worse than today. But squirrels are still a problem and seem to become worse instead of getting to be less a problem.

Henry Safstrom was at the Santa Rita Store when we were there. He and I were old time teamsters for Miller. Safstrom was 18 years old when he began teaming across the Pacheco Pass for Miller. He use to drive six horses and had two wagons. He still remembers how Miller used to furnish all his teamsters with $5.00 each for "water money." The trips use to take four days and Miller wanted his horses well watered at all the watering roughs along the road. These troughs were always in front of a saloon and roadhouses and Miller furnished $5.00 each trip to go in and buy drinks from the bartender to make up for the privilege of watering his horses. Safstrom says the bartenders got very little of it as the teamsters used it to play poker instead.

Young Henry Miller was a good boy and smart and that it was Henry Miller's own fault that he went to the dogs. He claims Miller didn't look after

Miller & Lux ranch, thought to be the Poso Ranch

him at all. He would give him all the spending money he could spend and Charles Warfield, a foreman, and Steve Sheldon, a relative of Mrs. Miller, used to get him drunk and take him places where they shouldn't have. Young Henry was very comical when he was drunk and Warfield and his friend used to like to see him drunk. Miller paid no attention to this or to young Henry and he grew up to be a drunk.

August 27, 1927 RLM

Marion Watkins, early pioneer

If ever there was man that wasn't afraid of nothing it was Brad Crow, my father told me. If they jump us we sure will let them know we was here he stated. Crow went heavily armed. The trip across the plains in 1865 was a jolly trip for us kids. We had no fear. We had lots to eat and slept fine. The trip in 1869 was a hard trip for me. I had to do a man's work, drive a team and stand guard at night.

One time I borrowed money from Miller & Lux on some land in the hills. I intended when I borrowed the money not to pay it back but to turn the land over to Miller & Lux. "We don't want it," Tanner told me. "Well neither do I," I said. Then I said, "Well can I borrow $40 more." "In fact," Tanner said, "If we hadn't already given you the first we wouldn't give you the second." In the end my wife and I proved up on the quarter section up in the hills. It was worthless so I ran up a bill at the Miller & Lux store until it was around $400.00 and made Tanner take the land for the bill. I did unload that land on Miller & Lux in the end.

Rounds 6/12/1935

Mrs. Kays Martin, Miller & Lux employee

The first people I met when I came to Los Banos were Billy Stockton, Carl Freeman and Billy White. I remember seeing Francis Freeman as a baby in a box in the printing office. There was music and singing all night in the saloons. Violins and banjos going all night. Singing and fighting. One night Ben Orognen got stabbed.

Our family came from Kentucky, near Lexington. Justice Court used to be held in the Woodman's Hall. Every morning one could see the brigade coming up for trial. Fast women, drunks, robbers, all coming up for trial committed over Sunday.

One time the canal broke up near where Enoch Bibby lived there in town. The Canal Company men were trying to fix the bank and stop the water. They were filling sacks with straw and stuffing them in the break in the bank. Morgan, the canal superintendent, was there himself. Henry Miller was also there. Some of the sacks of straw was breaking loose from the bank and floating on the water. "Jump into here," ordered Miller, "and bring those sacks back." Morgan told Miller very sternly, "I am not going to jump in the water for anybody. If you want to jump in the water, Mr. Miller, and pull those sacks out, you can do so." "All right," said Miller and into the canal he jumped. Morgan fully expected that his job was over. But to his surprise Miller promoted him into the office. Miller seemed to think more of a person who spoke his mind than one that weakly submitted.

One time when Miller was very angry he threw his hat on the floor and stamped up and down on it. "When you get tired, Mr. Miller," said Morgan, "I will help you stomp." Miller, his anger somewhat cooled, picked up his hat, brushed it off and put it back on his head.

One time Miller was held up on the Pacheco Pass and borrowed twenty dollars from his captor. Later on he was in Gilroy and saw this same thief sitting on a beer barrel in front of a saloon. Miller drove up to the man and handed him twenty dollars. "Here's the twenty that you loaned me," stated Miller.

At the time of the earthquake two girls were sleeping at the tank house where the Cothrans lived in Los Banos. Later the Dilses lived there. Now it's the Gardner house. The earthquake spilled the water out of the tank and literally washed the two girls down the stairs.

Interview 8/14/1946 RLM

Mrs. Orognen, early pioneer

I remember my father Mr. Hoffman used to keep a record book and I will try and find it for you. The Hoffmans didn't live here in 1877. That was the year they had to sacrifice their sheep. Later they came to live here.

The reason Henry Miller went to Germany for medical treatment was for a nervous breakdown after his daughter's death. When he came back from Germany he spent a couple of days at the San Luis Hotel visiting with the Hoffman family. The Hoffman family was from Hessedormstadt. Both my mother and father were from there.

My family and Henry Miller were not from the same part of Germany. But Henry Miller came and visited and told my father about things in Germany and he was pleased to hear about things in Germany. Henry Miller like to come and visit. I remember that it was Henry Miller's complaint on his return from Germany that the women in Germany at that time thought more of their dogs than they did of their children.

"Now girls," Henry Miller would say to us Hoffman girls, "keep your nickel and dimes and two bit and four bit pieces. Keep piling them up. Don't think you are millionaire when you have a four-bit piece and go and spend it. Keep a piling up. Then you will get your start. That's how I got my start."

My father raised cattle all around Mt. Diablo and the Tassahara Valley. He had a hotel right there in Alamo. It was a one-story building. It was called the "Alamo Hotel." That was where all of the girls were born.

Miller use to go right along with his chuck wagon. One time it was a dry year and Miller was gathering up his stock in the Tassahara Hills. My father was helping in getting cattle out. There is a very swift creek right back of the Alamo Hotel. Miller was trying to ride his horse across this swift stream and was unable to make it. To save Miller my father lassoed Miller's horse and pulled Miller's horse and Miller across the stream. "You saved my life," declared Miller time and time again in the years that followed. Miller showed his appreciation. It was my father lassoing the horse and pulling the animal across the stream that saved Miller's life. Miller never liked to talk much about it. He would talk to my father about it but he seemed to be chagrined that he had to have help.

It was all a fake about Mr. Hoffman saving Miller's life when a wild bull was chasing him. Nothing to that story. My father once had a clothing store in San Francisco and it was in San Francisco that my father and Mr. Miller first met and became acquainted.

When Dr. Fatjo built the large brick house at the

San Luis Ranch he told me that the reason why he did not build it right along the Pacheco Highway near the other San Luis Building was because his wife would never consent to his building it right out in the open. He said there was too much Indian in his wife. She wanted to live all by herself and family back off the road. All by herself back there hidden from sight. It was an Indian trait to live back in seclusion.

Mrs. Malarin was very refined. She wore lots of old style jewelry. She was very dark. Mr. and Mrs. Malarin always came to May Day celebrations at the San Luis Ranch. Johnny Hart was Henry Miller's coachman, as Miller always called him. Other people always called him Buggy John.

One time Henry Miller was held up and robbed at the foot of the grade. Henry Miller had a fine watch that was all engraved. The robbery occurred after Gussie was killed and Henry Miller had his daughter's picture in the watchcase. Miller told the robber, "Just let me take that picture out." The robber let Miller remove the picture. Miller told the robber that he could have the watch and his money and "I'll never bother you." Miller told the Hoffmans that the robber "didn't harm me." I am not sure but I think the watch was sent back to Miller later.

These same robbers stopped at Mrs. Hammond's house and ordered her to get them a good meal. Her husband was away at the time. The robbers assured her that they wouldn't harm her but she must get them a good meal. Mrs. Hammond had the hired man kill a chicken and she got the robbers a good meal. They paid her well. Mrs. Hammond didn't have much in the house to get a meal with in those days. It was thought that these were the same robbers who held up Henry Miller on the pass.

"It is true that Henry Miller was held up on the pass years before and borrowed money from the highwayman, on the pretext that he wanted to have enough money to get where he could get some more money."

Henry Miller was at the Hoffman May Day celebrations not more than twice. He would sit in an easy chair with a great big sack of nickels to give to the children to buy ice cream with. Miller loved children. When Henry Miller was at the Hoffman celebrations at the San Luis Gonzaga on May Day he was not putting on the celebrations, he was just a guest.

Interview 2/18/1948 RLM

Fenner Chamberlain, relative

The connection between the Lux family and the Chamberlains is as follows:

Jesse Potter was the son of Mrs. Lux by a previous marriage and was Lux's stepson. This Jesse Potter married a daughter of the Chamberlain family who owned a big ranch north of the Santa Rita.

The Chamberlain cattle and the Miller & Lux cattle use to run loose on the plains. Whenever any of the Chamberlain cattle would get over on the Miller & Lux lands the buckaroos would tell Chamberlain and he would go over and bring them back, usually one or two head. But Miller used to let his cattle come over on the Chamberlain lands by the hundreds and make no effort to keep them off. One day Miller was at the Chamberlain ranch and Chamberlain was so exasperated that he reached into Miller's buggy and yanked Miller right out of his buggy and told him what he thought of him.

Chamberlain was very quick tempered. In the scuffle Chamberlain tore Miller's coat clear off of him, ripping the coat clear in two. Mrs. Chamberlain later sewed the two pieces of the coat together again for Miller.

Fenner states he often heard the story that when Miller bought the Santa Rita he got a small tub and filled it with gold coins and paid the Mexicans and they never even took the trouble to count the money, it looked like such a large amount in that small tub.

Miller always used to ride with one foot hanging out of the buggy.

Interview in Los Banos 3/31/1955 RLM

Joe Place's store, 6th and "I" Street, Los Banos, 1905

Marion Watkins, early pioneer

Once Henry Miller was talking with old man Phillips. Miller wanted to know how Phillips was getting along and Phillips replied that he was making a living and that was about all. In those days about all ranchers could do who lived under the canal was to raise a few beef cattle. Miller hastened to correct Phillips. "You are the rich man," he told Phillips. "I am poor. You are home every night with your family, I am on the go all the time. Sometimes I do not see my family for two or three months. My children grow up like jackrabbits. I am worried from morning to night and at night I can't sleep."

The stableman at Bloomfield told me that he often times received more money in a month from Henry Miller's son than he did from his wages. He said that frequently young Miller would come to the stable where he slept and wake him up for a team. "Get me such and such a team," he would say, "and be quick". When the stableman would return, he would find a dollar. Later in the night he would have to get up and take care of the team when young Miller returned, the stableman would always find another dollar on his bed waiting for him.

Poso Schmitz told me that no matter how much of a row Henry Miller would have with the cooks he would always go away leaving the cook feeling fine. Henry Miller would come up to a ranch and inspect every part—barn, house, and kitchen. If anything didn't suit him in the kitchen he would give the Chinaman the worst calling down possible. Miller was always very particular to look into the meat safe. If he found the least tainted meat, woe to the cook. Miller was always bitter on waste. But no matter how big a row he had had with the cook he would always make peace with the cook before leaving the ranch, oftentimes giving the Chinaman a five dollar gold piece and telling him to do better.

"Why yes," Miller would always say, "the cook on a ranch can lose me fifty dollars a day if he want to."

Miller could see everything on a ranch. Nothing escaped him. You couldn't hide anything from him. When he came into a barn he was always particular to see if there were any broken harness. If there was you could depend on it that he would find it. "Send that to the harness shop and get it fixed," he would order the stableman. He would also look in the feed boxes to see if the horses were eating up their rolled barley. If they weren't he would call the stableman down proper. "Why," he would say, "if I was to give you the ranch and didn't run it for you, you would go broke."

Mr. Watkins told me about the trips to Madera. They would start out at noon. They would travel along the bank of the new canal, but they would travel to one side, as the new bank was too soft to travel. This was the summer the new canal was built. There was not water in the New Canal yet. They would pull across the New Canal over to the Old Canal for water and would camp at a camping place between the Leonard Ranch and Camp 13.

From there they would start out again the next morning. At noon they would be in Firebaugh where they would feed their teams and eat lunch. At night they would stop at the Chowchilla Canal, two-thirds of the way to Madera from Firebaugh. They would get home the next day. They would stop on the way at noon about three miles beyond Madera at the Madera Colony. They would get home in the foothills about sundown. They would get the straw unloaded in the barn the next day by noon or two o'clock. They always arranged it so they would be home Saturday night.

Sunday morning the outfits would start back to Firebaugh. Monday evening they would reach the Whiting ranch here. On Tuesday they would load up and start back for Madera. They got so tired of the work, it was so hot, week in and week out the work of hauling straw went on. Mr. Watkins told me a man needed grit to keep it up.

The second season they sowed the six hundred acres they had summer fallowed the year before and got in about 700 acres more. All together they had in about 1300 acres of land and got four sacks of wheat to the acre. The winter was good with a lot of rain. Miller would always tell the sky farmers to go ahead.

The third year they got all of two sections sowed. There was a hundred acres of barley; the rest was all in wheat. It was the agreement with Poso Schmitz that Miller & Lux was to advance the money to pay the men and to furnish groceries for the house. But there was to be no money advanced to hire additional stock to farm with. Whiting had three teams and was putting in all the land possible. When it became certain that the third season was going to be a good one Schmitz came to Whiting and asked if he couldn't get some more teams. Whiting reminded Schmitz of the understanding that Miller & Lux were not to furnish any money to hire additional teams. Schmitz replied that he would advance money for that purpose now. Go ahead and get more teams he told Whiting. Whiting did so. When they got the crop due there was $275 due for hired teams. Whiting went to Schmitz and told him he wanted to pay off the additional teams and help. Schmitz wrote out a check promptly. Whatever Schmitz said he would always back up. Miller never came around where Whiting was farming. He left it to his superintendents. Whatever they did in this regard was all right with him. Miller would always be willing to back anyone who would work. A person always knew he would be eating if he worked for Miller. He Miller would always back up one with something to eat if a person was willing to work and go-ahead and farm. Miller was always looking ahead and working to get feed for his stock. "The peoples gets a living and I gets the feed."

Undated interview RLM

Joe Place, Henry Miller's nephew

I came here in 1886. I was a bookkeeper and I had to keep three sets of books. I had the accounts

William J. "Billy" Stockton circa 1937-38

for the entire sheep department to handle. It use to take about three months to shear the sheep at Firebaugh. All of Miller & Lux sheep from Hills Ferry about to around Cantua used to be driven to Firebaugh to be sheared.

Firebaugh was a great place in those days. I used to have to bring down about $10,000 every few days to pay off the shearers and the sheep men. There was a great danger of being held up and robbed and he never had any set time of arriving at the office in Firebaugh. One time I would drive in from Los Banos. Another time I would come into Firebaugh from the direction of Cantua. I always endeavored to come without notice. As soon as it became known I was at the Firebaugh office the shearers would begin coming in for their money. The next morning there would be at least fifty drunk men lying around on the streets of Firebaugh. As fast as the sheep were shorn they would be let out into a pen and within ten minutes would be run through the dipping vat where the dip was kept good and warm. Jim Wood and Dan Shafer were the sheep bosses. Both were fine sheep men.

When I came to keep books at Canal Farm I soon found that were about 25 names of supposed employees that never showed up for their pay. Each were down on the books as receiving $30 a month and had been on the books for over a year or more. I told Tanner that I was going to take a horse and drive to the different ranches and find these fellows. Tanner took no interest. I set out and enquired everywhere but could find no trace of the men, nor had anybody ever heard of them. The fact of the case was the previous bookkeeper had been putting these fictitious names on the books and collecting their wages.

Jim Wood had over a thousand dollars coming in wages and asked me one time how his account stood. I told him, "No, there must be some mistake." Wood said that he had close to one thousand dollars due. It came out that Wood had been charged $600 by that bookkeeper and that he had taken the money. That bookkeeper had transferred to Oregon and I never did hear what happened to him. After that I insisted that some system be devised to keep the payroll in proper shape so that after I was away from Miller & Lux no one could say I had taken or stolen from anyone. Mr. Merritt from the city office came down. At that time each employee's name was entered on the books, the amount of wages for the month shown, the deductions for cash and for supplies furnished and the balances due him. Some of these accounts had been running for years and the employee could have as much as $1,500 coming to him. A new system was worked out into a system that Miller & Lux afterwards followed in making out a check each month for every employee and requiring him to sign it in two places.

The first May Day picnic we all sat at the same table with Henry Miller and ate together. About all they did was barbecue one heifer at this early celebration. Later they got big and they hired a band, a dance platform was built. Jake Gardner had the ice cream concession. Then the celebration became commercialized after the first four or five years. The first one was in 1887 or '88. Miller would turn to Tanner and say "Well, will we have the picnic again this year? Tell them to all come in." Miller served barbecued meat, coffee and bread all free. When the celebrations began to get commercialized the politicians began to come. About the second picnic we had ice cream, I think Bomeo made it. There was a big tub of salt and ice. It was the first ice cream at a May Day celebration. Bomeo would stir the milk and shake the can. As fast as the cream began to freeze on the sides of the can it would be scraped off. The cream would be stirred and flapped up against the side of the can to make it freeze. The ice cream was thick as molasses. People ate it and liked it. The milk can was set in the tub of ice and salt. This second picnic had about 100 or 150 people there. There was no dance in the warehouse when I was there. The dance came when the celebrations was moved to the park.

The ball game at Center Point was held about 1889. It was before the railroad came. It was between the Cottonwood\Newman players and Los Banos. Bill Ogden, Bill Cheatham and Soper were among the Los Banos players. I was the pitcher and Ogden was the catcher. I was a pretty good pitcher then. Used curved ball. Our opponents knew nothing of a curved ball. Just a bunch of farmers. Sturgeon and Billy Stockton ran the betting. Los Banos won. The Newman bunch was run ragged. The game was played down by the canal. The bases were just marked out as best they could and played. There was no ball ground. There was quite a crowd of people; they only played this one time.

Volta was larger than Los Banos at first. Anyone with Miller & Lux couldn't go down around Volta much. There was a wonderful racetrack out here. Old man Wanzer surveyed it. Tanner, Warfield and Schmitz tried to get Miller and Newman together on the proposition of one fine town about where Newman is now. Impossible said Miller. So Miller built his town of Los Banos in his mud hole, and Newman built his town in his mud hole. Both towns are in mud holes. Where Los Banos is now used to be an awful nice duck preserve, the ducks that use to be there! Then Sawyer butted in and built Volta. That made Miller mad again. Volta was larger and had a fine hotel and the Bambauers had a wonderful machine shop. But Sawyer didn't have any money. Miller went to work and put up fine buildings in Los Banos. A three-story brick hotel and other brick buildings. Sawyer couldn't do this. If could have he might have beat Miller out. Drummond, White and myself tried to buck Miller.

Miller asked the railroad company to name the town Los Banos. It was the railroad company that named the town Los Banos. Gus Smith was the first postmaster in Los Banos. He had a little shack on Front Street about where the Wool Growers Rest is now. They got the depot from Yuma, Arizona, or someplace down there in Arizona or New Mexico. It was second hand all the way.

There was a real cloudburst I think it was in 1894. Didn't rain a drop here. The cloudburst was up in the hills. Christian and I had three or four thousand acres out on the plains on the Little Panoche. We had the harvesters all ready. It was May. Christian came down and told me, "We are busted." The flood covered the grain about six inches deep with silt. Laid the grain all down, never cut a spear of it. The flood just took our grain. It went right up to Jim Sullivan and then stopped. We had a fine crop, lost it all; I never did have much luck farming. The Little Panoche Creek came down and overflowed, it broke our dam and everything.

The winter of 1890 was wet, too. Mrs. Tanner wanted to get over to Center Point. It took all day to get over and back to Canal farm about 2 miles. In moving the store building from Center Point they got it about as far as Orognen's and had to leave it until spring. Miller & Lux didn't have enough horses to pull it; the road was so wet and muddy. The railroad got to Newman in 1889. It stopped in Mendota the next year. There was no boom in Los Banos at all when the railroad came through. Miller had taken all the Los Banos lots that were any good. Reserved four corner lots on nearly all the corners. Brick for the store and hotel were made here in Los Banos. Chinamen made them.

It is said the Opera House cost Miller a dollar for every brick in it. There were twelve masons getting fourteen dollars a day. They would go to work in the morning and lay one thousand brick. Then they would all go over to Holm's Saloon and drink beer the rest of the day. The lather got paid by the thousand lath. When the earthquake came and the north end of the opera house fell out it was discovered that all the space in the wall between the studding was jammed full of new laths just as tight as it would hold. If people stole nowadays like they did from Henry Miller a firm wouldn't last long. What Los Banos needs is Henry Miller's old enthusiasm. People use to say, "Just wait till Henry

Miller dies then this town will boom." Will Ogden, supervisor, had the Dos Palos road to Merced built.

Miller & Lux used to drive cattle from Gilroy and Berenda. It took four days to drive cattle from here to Gilroy. All the cattle from around Mendota and south shipped from Berenda the rest from Gilroy. It use to take four hours to drive to Gilroy. One hour to San Luis, one hour to Bell Station, one hour to San Felipe and the last hour into Gilroy.

On the Dickinson Ferry road in times of high water it you got off the grade you were gone. Stakes were stuck all along for guides. Men on horseback were kept at Dickinson's Ferry and at San Luis Camp to pilot people through the water. They would pilot people across the island. There were some forty bridges. The bridges were what you had to look out for. If you missed a bridge you were gone. Old man Henry Schaffer, sheep man for Miller & Lux, got drowned on this grade. Also Mr. Theircof. It was a dangerous road.

A person working for Miller & Lux never needed to mention Simon Newman. You didn't want to cash a Simon Newman check and have it in your cash drawer when Henry Miller was around. Attempts were made by many to have the west side town where Gustine is now. But Miller and Newman wouldn't come together. They hated one another. They were at outs all the time. They sure did hate each other.

About all the alfalfa there was in the country when I came was south of Canal Farm. They were putting in quite a lot at Santa Rita when I came here. There was very little grain left at Salt Slough warehouse to wait for winter. Most all the grain got shipped out during the high water in summer. I could have bought land around Cottonwood for $70 an acre at one time. It seems the Canal Company owned a lot of land up there. Miller came down and put a price of $70 an acre on it and Drummond went to work selling it. Several pieces were sold. When Miller came down again he was very mad. He had wanted the land himself and had to give a big price to get some it back again. In the early days nobody was starving. People didn't eat much of this fancy stuff in those days but they were not starving.

The reason why the people of Los Banos incorporated the town was because they wanted a sewer system and they had to incorporate in order to get it. It took the stages about four hours to go over the Pass. They went in those days. Fresh horses at every stage station. The stages used to run to Madera. From Old Los Banos they went to Charleston and on to Madera. Later the stage line was changed to Merced and the stage traveled by way of the Dickinson Ferry Road.

Baseball was a great game when the town was young. They would charter a train and go up to Newman to play ball. Newman would charter a train and come to Los Banos to play. I was pitcher and later manager of the Los Banos team.

Interviews 1934 and 1935 RLM

Billy Stockton, early pioneer, historian, and author.

Lots of Miller & Lux cattle starved in 1877. Miller was hit pretty hard that year. He told me so himself. Much of his stock were still Spanish longhorns. It didn't pay to buy feed for them. Miller rented lots of land in the tules around Stockton and endeavored merely to keep the cattle alive. Had Miller known it he could have kept lots of this cattle right in this country on salt grass. The overflow from the river and the rain in May that I spoke of made lots of salt grass on the alkali country. The settlers who kept their stock here on salt grass got through in good shape. People in general didn't know at the time how valuable salt grass pasture really is. The few cattle and sheep that Miller left here 1877 did first rate. Miller never drove his cattle out of the country again.

The first May Day celebration I was called on to get Henry Miller on his feet and have him give a talk.

The celebration was being held at Canal farm on the occasion of the opening of the New Canal. It was a big Miller & Lux party. All of Miller's superintendents were there. On two occasions in my life I have spoken for Henry Miller. He would never get up before people and talk. I knew his aversion to speaking in public, so I simply reminded the crowd that by his presence at the gathering he showed his good will and assured them that Henry Miller was always in full sympathy with the people. I would propose three cheers for Henry Miller. The cheers were given with a will and Mr. Miller was deeply pleased.

The history of Henry Miller has never been adequately written and I fear it never will be. Henry Miller was the greatest man I ever knew. He was the biggest man that has ever been connected with the San Joaquin Valley. He accomplished so many big things. Miller was a giant among men. All the rest of us are little, insignificant when compared to him.

When the settlers out on the plains were all but starved out, following year after year of drought and poverty, they sent a committee to see Henry Miller at Canal Farm. The committee told Mr. Miller that these people couldn't pull up and move to other communities. They were too poor even to get away. What was needed, we explained, was another canal further out on the plains. Miller listened to our arguments very carefully. He himself knew the settlers' conditions. His mind was quickly made up. "All right," he snapped, "I digs you a canal!" And he did. Seven feet deep, a hundred feet wide and fifty miles long! And all this without a cent of indebtedness or bonds against the settlers' land.

Supervisor Cothran was elected in 1910 as an anti-Miller man. Of all the things he was going to do! But I had a talk with Cothran who was my cousin and I says, "Cothran if you go to fighting Henry Miller you are the biggest fool that ever lived."

During the winter months when the San Joaquin River would be high and again in the summer when the river would flood it was impossible to get to Merced by way of the Dickinson's Ferry Road. It was necessary at these times to go by train to Tracy and then around the east side back to Merced. The winter of 1910 there was an unusual amount of water pouring down the San Joaquin River and the Dickinson's Ferry Road was all but washed away. Many of the bridges were gone and the road looked hopeless. Mr. Cothran, the new supervisor, was in despair.

J. F. Clyne, the canal superintendent, together with Mr. Cothran and I, had gone to San Francisco to discuss canal rates in the Miller & Lux office. After we had finished I asked if Mr. Miller was in his office and if Mr. Cothran and I might go in and visit him. We were shown into Miller's office. It was almost the last visit I had with Mr. Miller. During the course of our visit the subject of the road to Merced happened to come up.

"Mr. Cothran," said the old man, "That Dickinson's Ferry Road was built where it never should have been built. It crosses right through the swamps. It should have been built on the high ground that runs from Los Banos eastward past the Santa Rita Ranch. I have already built a road there myself. It cost me $25,000. I am willing to make the county a present of that road. I won't charge them a cent for it. It will accommodate the Dos Palos people as well as the Los Banos country and will get rid of those swamps."

Mr. Cothran hardly knew how to express his surprise and gratitude. It would have cost the taxpayers thousands of dollars to put the Dickinson's Ferry Road back into shape. Cothran had been elected supposedly to fight Henry Miller. Without taking the least notice of Cothran as an adversary, Henry Miller offered to help the new Supervisor out of his dilemma. From that moment Cothran realized that Henry Miller was a truly public spirited man. This road given by Henry Miller to the county has since become a part of the state highway between Los Banos and the San Joaquin River. The old Dickinson's Ferry Road is now only a memory.

A few years before Henry Miller's death he came down to Los Banos to talk with all his old friends. In the pioneer days all the farmers were allowed to run store bills at Miller & Lux store. Henry Miller carried nearly the whole community during the hard times. If the people made anything they paid their accounts. If they didn't Miller carried them just the same. Many of the ranchers in this county Miller had known and helped for nearly half a century. Some owed him accounts running up into the thousands of dollars. On this visit Miller confided to his old friends: "In the natural course of events I can't last much longer. When I am gone and new hands take my place I don't want any of you people crowded for any debts you owe me. Pay me what you can. I forgive the balance."

Miller settled up with everybody. Some who owed him thousands of dollars paid him as little as only a few hundred dollars. No one in this world ever did Henry Miller a favor. He needed none. But no one ever lived in this community who was not under some obligations for favors from Henry Miller.

Mr. Stockton's booklet "The Plains Over" RLM

Terrible Dry Year of 1877

The years of 1876 and 1877 were dry years. Warfield told Dugain that he just had a letter from Miller saying for Dugain to go up to Jacks Ranch in the big Panoche and look at the cattle. Dugain told Warfield, "I want a good horse and I want a good man to go along with me that knows the country." Warfield sent Charlie Galindo, a Mexican, but a first rate fellow. Dugain and Galinda stayed at Jacks Ranch three weeks. The Jacks outfit were gathering the cattle. The cattle themselves were good grade of cattle.

Dugain was up there to buy the cattle for Miller. Jacks wanted ten dollars a head.

"This is a dry year," Dugain reminded Jacks. Then he offered Jacks $4.50 a head.

Jacks was wild! "I can't give you anymore," said Dugain. "It will take lots of feed to bring these cattle through." Jacks refused the offer and Dugain and Galindo came home. Pretty soon a letter came from Jacks saying that he accepted Dugain's offer. Warfield sent a man to Madera with the letter for Miller & Lux in San Francisco. Warfield was surprised that Dugain didn't start at once for the Pacheco.

"Let 'em sweat a while," Dugain told Warfield. Warfield was anxious to know what Dugain had offered Jacks for the cattle. "Evidently you've bought them," said Warfield. Dugain told Warfield that he had offered $4.50 a head but only those that could walk inside the gate at Santa Rita should be counted.

"$4.50!" exclaimed Warfield. "Why you are robbing the man."

Dugain set out with some men for the Panoche. The pack mules with supplies for the men followed up. Jacks had the cattle practically all gathered when Dugain arrived. They drew up the contract. Only those cattle that could reach the Santa Rita and go through the gate were to be counted. Those that died on the way or gave out were not to be counted. Anything not branded was not to be counted. There were 8,920 head that passed through the Santa Rita gate.

While the cattle were on the way Miller drove up in his buggy with Buggy John. "Can you ride a horse?" Miller asked Buggy John. He wanted Dugain to get in the buggy and ride with him. The next day when they arrived at the Santa Rita with the cattle Miller was already there waiting. Dugain asked Miler if he wanted to count the cattle as they came in the gate. "No," said Miller. "You count them yourself. This is your deal. You tend to your own business. I don't have nothing to do with your business."

Dugain picked up a lot of pebbles and every time he came to one hundred he would drop a stone in his pocket. When they were all in he would count the number of stones. That was the way they counted cattle in those days. "That's pretty good," Miller told Dugain. The next day he took Dugain to Fresno and made him a present of a suit of clothes from one end to the other. A few days later Miller sent Dugain to buy half a dozen bunches of cattle. Some of them Dugain bought, but some that were a poor grade of cattle he didn't even make an offer on.

Miller told Dugain afterwards that the next spring he sold enough beef steers out of the Jacks cattle to pay twice over for all the Jacks cattle cost. Albert Long and Dugain made millions of dollars for Miller.

Talk with Dugain, undated RLM

Ralph Milliken had many additional notes on the years of 1876-1878, those drought years that brought ruin to many a farmer and ranchers. The event was one many remembered and spoke of for the rest of their lives. Name at end is person telling the story.

The year 1877 was a terrible dry year. No rain hardly at all for two years. 1878 was a fine year. Grain at Cottonwood was as high as one's head. Headers had to raise the scythe up three or four feet high there was so much straw. — Cirimele

Miller made millions of dollars in the dry years of 1876 and 1877. He had thousands of tons of hay. Wild hay used to be three to four feet tall. Miller would put in thirty to forty mowers and put the hay up in stacks. Miller would put the mowers in the spring and make hay all summer long until the rains began. — Dugain

In 1877 was the awful dry year. In order to get through I killed all the lambs but still was unable to get through and had to sell my 4200 sheep for six bits a head, delivered to the Islands along the San Joaquin River. I used to take my sheep every summer down to the stubble fields. My camp used to be right where the town of Tracy now stands. I used to pay six cents an acre for stubble. — McCabe

Miller always looked ahead. He always had lots of straw stacks everywhere and that helped him a lot in the dry year of 1877. He also had alfalfa fields in Kern County where he sent lots of stock to tide them over. — McCabe

It was so dry in the spring of 1877 that the feed simply didn't come up. There was no spring that year. Early in the year Miller & Lux began driving their cattle out of the country. Henry Miller rented pastureland down this side of Antioch on Roberts Island. Cattle died all over the plains that year and sheep perished by the thousands. Even the antelope starved. Nothing was fat that year but the coyotes. Heminghouse reaped a fortune. He would pull the wool off the dead sheep and in the fall he had an immense quantity of wool stored up. — McDermott

During that awful dry year about 1877 Jim used to haul hay and straw for the Canal Company and for Miller & Lux from Bantas. They would make a trip a week, camping along the road enroute. Hay was thirty-five dollars a ton and baled straw from twenty five to thirty dollars a ton. — McDermott

Early in the spring they began driving the cattle out. They had corrals all along the way to stop over night on the way to Roberts Island. Al Ward would usually take the biggest bunch, followed a day or so later by Jim Lopes with another bunch. A day or so behind him would come Dan Shackleford with another drove. Charles Warfield, the head foreman, stayed down at Roberts Island most of the summer and looked after the cattle.

When the rains started in the fall Jim says he never saw feed grow like it grew then. All over the country there was nothing but the rankest filaree and clover. Then we had to bring the cattle back and that lasted for weeks. Al Ward brought back the last bunch. New Year's day the last bunch was on its way back. — McDermott

The only excessively dry year in this country since I have been here was in 1877. I took a ride that year to the Mercey Hot Springs. I went over the Big Panoche Plains and Los Banos country. More than 20,000 head of sheep perished on the plains and were lying there dead. Sheep men that year killed the lambs just as soon as they were born in order to save the ewes. There wasn't a bit of feed or water for sheep on the plains and Los Banos hills that year. Stock was too weak to be driven out of the country and there was no means of transportation to get them out. — Newton

Steam threshing machines were just coming into use in 1877. They burned straw for fuel. Some grain was raised under the canal that year with irrigation. Henry Miller would allow no straw to be burned in the steam engines as they threshed up and down the canal. Wood had to be hauled for fuel from the river. That was in the bad year of 1877. — Newton

There was a big crop in 1878. People were saying it was going to be another dry year. But on January 16, 1878 the rains came. The floods opened. Five inches of rain fell in that first storm of January. After about a month the feed question was settled and was a thing of the past. — Newton

Miller hay stack in 1880's

Part III

CHAPTER ELEVEN

Santa Rita County

Henry Miller, like most men of his time, soon became tired of the government intrusion into his business ventures. The days of doing what you wanted, when you wanted, were now under the control of bureaucrats and they were starting to hinder him in his business dealings. Regulations, red tape, taxes, and dealing with petty bureaucrats were not for the likes of a man like Miller. By the mid 1880's he was of a mind to break away from the control of Merced County and form his own County Santa Rita.

At first he put out feelers, finding many ready to break with Merced County, which had been one of five counties carved out of Mariposa County, which had been the largest in the state. Miller made his first statement in public about this issue at the 1890 May Day Picnic. While he seldom gave speeches he did take this occasion to bring up the issue. Miller's statements started to appear in the local newspapers. However, prior to his speaking out a movement was already afoot to the north.

Merced Express 12/15/1888: New County!

Newman at the West Side is full of one kind of business anyway. It is the headquarters for the proponents of a new county of which Newman will be the county seat. The southern extreme of this county will be the present southern line of Merced County and the northern will be the northern line of Stanislaus. The eastern boundary will be the San Joaquin River and the western will run to the old county line of Merced and Stanislaus. The proposed county will be about seventy miles long by an average of about eighteen miles in width.

Western Union
Sacramento 22 March 1890
Do you desire new County west of San Joaquin River?
Thomas Fowler

Express 1/24/1891: New County

It seems no prophetic fancy to foreshadow in the near future a separate, independent and prosperous county, of which Los Banos will be the county seat, on the West Side.

The first May Day picnic at Canal Farm they had a balloon. It went up a little ways and then came down in an umbrella. Miller made a speech. He said he wanted to make Los Banos the county seat, and explained the need to break from Merced. Everyone clamped his or her hands. It required 5000 people to form a new county and there weren't that many people then. Talk with Joe Cirimele 4/11/1935 RLM notes

Merced Express 2/21/1891: County Seat

Henry Miller is making a business round among his West Side ranches this week. In conversation, he made the statement that he proposes to ply his tireless energies to pushing Los Banos along the forward road, till it shall take rank among the most prosperous cities in the San Joaquin Valley. He says only a few years at the utmost will pass before Los Banos will be the county seat of a new and separate county. This expression from the concise tongued cattle king, and million-acre land proprietor, indicated that the future of Los Banos is by no means only contained in the fancies of sanguine enthusiasts.

Merced Express 1/14/1893: New County

The people of the West Side are making a fight for a new county whenever they can prove to us that they have the required population, the Express is with them in their fight. Not only that, but if we can borrow the money we will build a bridge over the San Joaquin at Hill's Ferry and donate it to the public, and grade and macadamize the streets of Newman and Los Banos. Grade and gravel the public roads in every direction for a distance of ten miles, establish a system of sewerage for both towns and give Newman a joint school district with Merced County. All this we will do provided they will vote against probhition.

Newman Banner 1/21/1893: New County…

We beg to inform the Colonian that Dos Palos is in Madera County. Therefore it need not worry about being the "Caudel Appendage" of the new county being formed by Miller.

Merced Express 1/21/1893: New County

We beg to inform the Colonian in Dos Palos that Dos Palos is in the new county of Madera. Therefore it need not worry about being an appendage of the new "county of Miller," or whatever name it may adopt.

Newman Banner 1/28/1893: Santa Rita County

Among the bills introduced in the State Senate this week was one to form the County of "Santa Rita" out of Merced and Stanislaus counties, with the County Seat at Newman. It was introduced by Senator Flint. The West Side people are determined to have a new county and some say they will get it.

Newman Banner 2/4/1893: New County…

Should our West Side friends secure the new County of Santa Rita, the Merced Express will thereafter be publishing from Huffman's Ranch, Huffman Co. instead of the City of Merced.

Merced Express 2/4/1893: Is Newman to be Tapped?

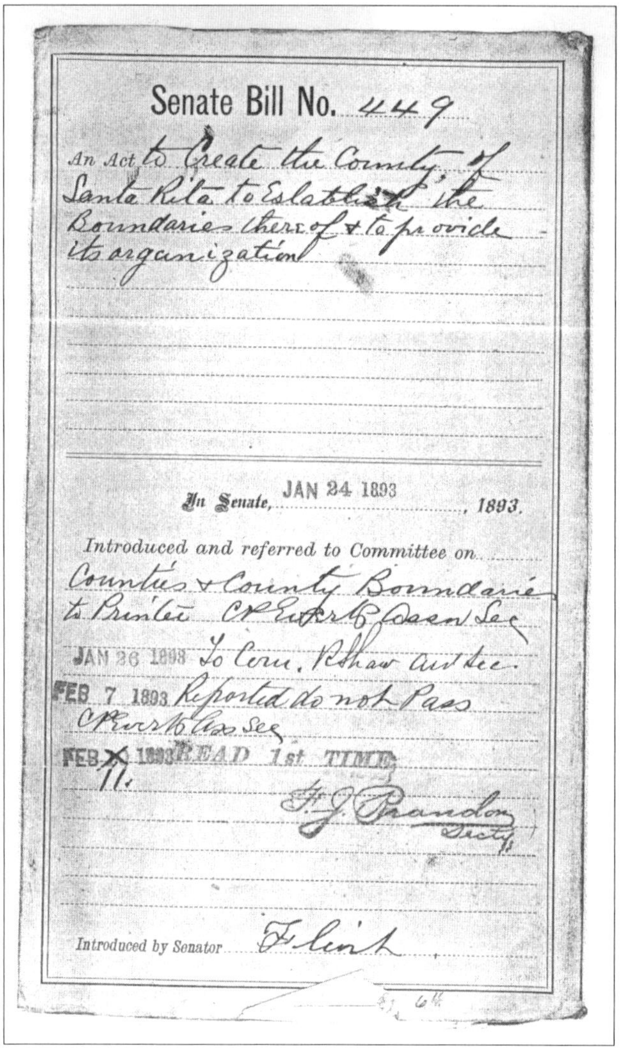
Senate Bill No. 449

An Act to Create the County of Santa Rita to Establish the Boundaries thereof & to provide its organization

In Senate, JAN 24 1893 1893.

Introduced and referred to Committee on Counties & County Boundaries

JAN 26 1893

FEB 7 1893 Reported do not Pass

READ 1st TIME

F. J. Brandon Secty

Introduced by Senator Flint

Copy of the Senate Bill No. 449 to create Santa Rita County

A correspondent of the Fresno Expositor in writing of the proposed new County of Madera and the struggle before the Senate Committee, gives out the following piece of information:

"Henry Miller, of the old firm of Miller and Lux, who had been opposing the scheme, suddenly changed front and approved that he was friendly. This is evidently a compromise and has reference to a triangular piece of territory in the southwestern corner of the proposed county, which Miller needs in his business for "Santa Rita". Of course the foregoing could only apply to probable amendment, on its passage of the present bill creating the County of Santa Rita. By taking the West Side of Fresno, Los Banos Miller's favorite town would be eligibly located for the county seat.

Merced Express 2/4/1893: New County

The West Side people are making a lively fight for the organization of the new county of Santa Rita, and if there is no opposition they may get it.

Merced Express 2/11/1893: Failure of New County

The failure of the West Side people to secure the new county of Santa Rita has caused us this week to change our future plans. We will not move to Huffman's Ranch but will remain where we are in the Hunter Building and the Express will remain an eight-column paper instead of a five-column paper.

Though the newspaper accounts are skimpy and it appears that the population of the West Side did want to break with Merced an interview done in 1932 with J. F. Muth gives additional insight into what happened.

Mr. Muth settled in Los Banos in 1889 because he had heard that there was to be a new county on the West Side and that Miller was going to make Los Banos the county seat. This new county was to be called Santa Rita and was to be formed out of land on the west side of the San Joaquin River which was the west side of Merced County, part of Stanislaus County and a good slice of Fresno County to the south. When Fresno County lost land to Madera County off its north end that put an end to any possibility of a new West Side County, for Fresno County would not consider any proposition to further decrease its size.

Miller had, at this point had the backing of the people but being the largest landowner in each of these counties he was faced with reassessments at exorbitant values for his land. Miller let the matter drop and proceeded to establish the town of Dos Palos and subdivided a lot of the land there and began building up the region. This resulted in injuring Los Banos by subdividing this land and bringing in settlers from the Midwest. As a result Los Banos did not become the big town that was expected.

THE JOURNAL OF THE SENATE DURING THE THIRTIETH SESSION OF LEGISLATURE OF THE STATE OF CALIFORNIA 1893 ON COUNTIES AND COUNTRY BOUNDARIES

Senate Chamber, Sacramento, February 7, 1893

Mr. President: Your Committee on Counties and Country Boundaries, to whom was referred Senate Bill No. 349 – An Act to change and permanently locate the boundary lines between the counties of Glen and Colusa – have had the same under consideration, and respectfully report the same back, and recommend that it do pass.

Also: Senate Bill No. 449 – An act to create the county of Santa Rita – have had the same under consideration, and respectfully report the same back, and recommend that it do not pass, as one of the counties from which territory is to be taken would not have the population required by the Constitution. 22-s

Merced County is still number one in dairy income in the state.

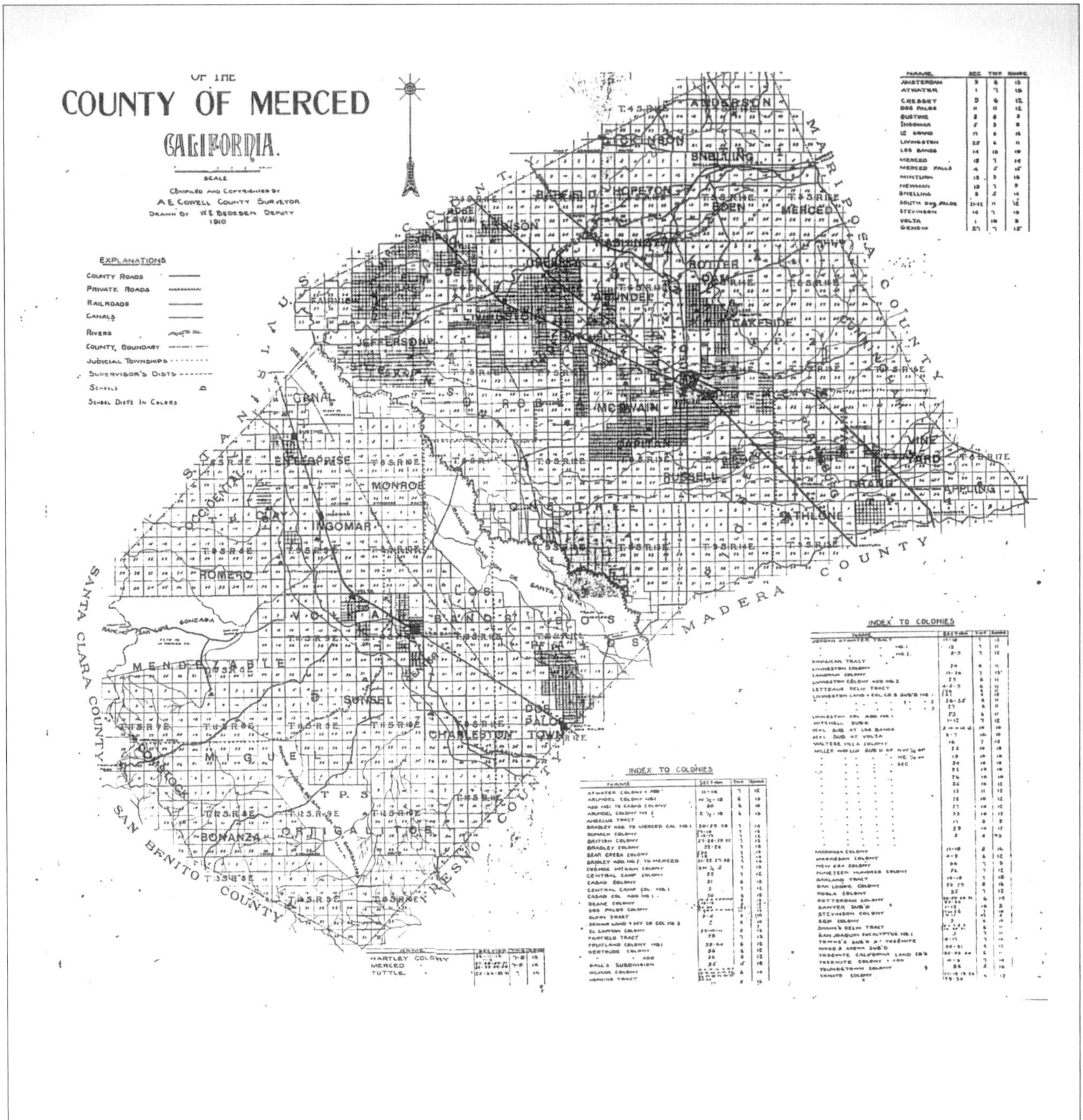

Merced County boundaries as they ended up, 1910

Chapter Twelve

Milliken's Notes - Odds and Ends

Ralph Milliken's files are full of odd facts, fiction and fancies about Henry Miller, Miller & Lux, Los Banos, the San Joaquin and the Westside as a whole. Some stories as told cannot be verified, others are events that are local gossip and lore.

Dirty Plate Route

Miller & Lux was well known for its Dirty Plate policy. There were two classes on the Route. Two distinct classes travel these whirls, distinct in speech and in appearance. One is the "burley" and the hobo of our railroad tracks who stops to recuperate for a few months on the way to or from San Francisco and Los Angeles but a class by itself, the old original order of shadow chasers, the man who never leaves the dirty plate route, who considers it home, is the "reservation stiff." The peculiar part of it all, he looked down upon the "burlies" with hard-boiled hats and coal dust down their backs.

This was once brought to my attention in a strange way. In company with Harvard Porter, a fallen mighty, I was going from North Camp to Dutch Boys stations on the "inside whirl" when a band of canvasback ducks rose out of a pound. Canvasbacks! Exclaimed he, "The first of the season. Look out now for the roughs with itch browns 19½ collars and .22 caliber eyes. It means the snows have fallen in the north and soon bands of these toughs will be floating in and they're a bad lot, take it from me. Never have anything to do with them."

Among the "burlies" one seldom if ever meets with an educated person while in the ranks of the "reservations stiffs" there are many. And it seems that the better educated, the deeper they have dipped into the fountain of knowledge, the more degraded and debased they are, especially when drunk and to be drunk is the aim and the ambition of most every rambler on the dirty plate route. I have met with many who would refuse a smoke but have yet to see one who would do the same with a drink. To satisfy this deep craving and desire they will do anything even to work and they will drink anything even a local brew known as "sheep dip" or "mule punch" so named because of its powerful kick.

The nicknames or as they have it on the dirty plate route the "ear marks" or "brands" are always apt if not very complimentary. For instance one young fellow with a large nose is called "spoon bill" another with a twisted mount "funny face." "Flipity Flop" is applied to an ex-soldier who has the sleeping sickness and who "flops" whenever the spirit moves. "Seldom Seen" Murphy avoids the other men and only appears at mealtime in their company. "Millionaire George" a youth extravagant with conversation money. "Silver Tip" a boy of 21 years whose hair is now white from a fever. There are many nicknames in the bunch. As for Swedes there are "Lone Swede," Loco Swede," Gunny Sack Swede," "one-eye Swede," "Whiskered Swede," "Crazy Swede," and the "Irish Swede." The latter speaks

English with an Irish brogue. At Seattle he deserted his ship and obtained work on a railroad with a crew of Irish navies from where he picked up his English.

Then there are the four Porters. "Porter the bum," "Porter the rake," "Harvard Porter," and "Yale Porter." The king of the dumps a 27 year loiterer on Miller's ranches," The "German Count" who every quarter day received imposing envelopes being imperial seals good for a month's steady drink, "His Nobs" a graduate of Bonn University who in a neat speech at the Dos Palos cattle corral told us his determination to return to the fold. He did not live long enough, however, to reap the benefits of his reformation because shortly after in December 1912 the newspapers recorded his death at Denver Colorado while stealing a ride on a train. The dispatches disclosed the fact that his brother at the present time holds one of the most important portfolios under the German Government.

But lastly of all the characters tramping the dirty plate route the greatest of these is the "Grand Duke" always drunk and always spouting Latin, Greek or French. My first meeting with this pearl was one morning in a saloon at Firebaugh. In the center of the room he stood with a large glass of "sheep dip" in hand leering and smirking sensually at it. Presently he sipped it. "Infamous stuff this, to soothe a thirst," he shuddered. "Excuse me," said he with a bow that displayed a flash of gentility. "Excuse me for flaunting Horace in your sodden faces but you'll never appreciate the gem even should I reduce it to words of one syllable. But hear 'Drink comrades, drink! Give loose to mirth, With joyous footsteps beat the earth and spread before the war-gods shrine, The Salinas feast, the sacrificial wine Bring Bring.'" After repeating Bring several times he slowly sank to the floor like a wet towel falling in a heap.

By your walk you are known principally on the dirty plate route and I had not traversed it an hour before I became aware of the fact. I had only started from Los Banos when under the elms that lead from that town I roused from a drunken sleep "Danny Webster" ex-lawyer and bunkhouse bum. "Hey newcomer," yelled he with a whisky hoarseness. "Take things easy. Hurry is one of the resources of a banker no one of those whose money is time. If you're going to make good here you got to wean yourself of that vice, for it's no place for he who has not the faculty or low gear."

According to the custom of the Dirty Plate Route when a wayfarer arrives at eventide to him comes a dinner, a place to sleep and a breakfast. Consequently they time themselves so as to arrive just at the meal house. Before some ranch houses are regular "roots" where tramps burn daylight between meals. When a hobo pops up if the dining room is so situated he peeks into the windows or doors to size up the table. First he looks for "one" of the boys who is working to choke off a piece of booze money. He will try to get his plate for there is always a possible that the worker pro tem has picked up an extra plate to put under his plate for the lucky hobo who gets his place. Then he reviews the regulars, the Italian laborers in order to get a second or third choice, as he is aware that there will be a rush for the chair of the man earning whiskey money. Clean looking Italians will be selected to each after, and always will a "Dago with whiskers" be avoided like a person with smallpox.

While eating it is every man for himself. Friendship ceases and modesty has its reward in the shape of an empty stomach. I found this out at the Santa Rita ranch, before helping myself to the bread I passed it to a cheek-by-jowl friend and he handed me back the empty plate. Across the table sat "The Red Lizard." Smiling. "Cheese on the pink tea stuff," laughed he. "This is the arrival of the longest and quickest reach, and if you don't put your foot in the trough like the rest of us, you'll starve."

A tombstone on the route states "Jackson" killed by a "Dago" for 40 cents March 1904. "While there's booze there's hope" is the way Yale Kelley, a scoffer and agnostic put it to us young fellows starting to follow the route.

Paul Parker Salinas RLM, undated notes

There was many a wit and poet who wrote of Miller:

Hobo's

A hobo once was lying in the
shade of Miller's stacks
His curly head was on his bed;
a roll of Miller's sacks
He turned to his partner; this is what he said;
"Can they ever run these ranches when the
old time boss is dead?"
When they go up to Heaven whatever will
they do;
They can't lay and take it easy the same
as me or you
Will they keep the same old things
as they do down here?
They'll sure play Hell in Heaven the whole
shebang will be queer
Will Buffalo Bill quit boozing when he
reaches the Golden Shore
Or a bucking bronco over Peter at the door?
Will Dan Leonard ride in an auto;
Will Wallace still drive a team?
Will Rock be irrigating out of Heaven's silvery
stream?
I wonder if Red Safstrom will still
get up at four o'clock
Will he bawl the Hobo Angels when they
hide behind the stack?
Will he leave the hump on Nick
when he enters at the gate?
Will they taboo the gunnysacks
or use the dirty plate?
—anonymous author

Santa Rita

Of all the ranches I've been on
There is one I love best
It's known as old Santa Rita
'Tis better than all the rest
Now I've been over at Belmont
And I've put in time at Lucerne
I like the ranch at San Juan
And also the one at Durham
But the place that always seems home like
And where I want to be
So take me there and I'll pay the fare
Old Santa Rita for me.
—Bob Rainly, bookkeeper and clerk Miller & Lux

Hal Safstrom next to a Miller stack, 1900

To the Late Henry Miller

A grand old man is laid to rest
And countless hearts feel sad
Because he was the truest, best
Friend the poor man had
For up and down his vast domain
Along this golden shore
No poor suppliant ever has
Gone hungry from his door
And if of all the virtues
The good book tells us of
The chiefest one is charity
We know that the God above
Will open wide the pearly gates
And 'round the great wide throne
The angel choir will welcome
Dear Uncle Henry home
—San Francisco Examiner –anonymous

Miller's Thoughts on Cats

In a letter, a copy of which is among Ralph Milliken's papers, tells of a series of letters one employee received from Miller over a period of a week or ten days about cats and rats. Miller wrote, "You should have some cats to destroy the mice in the granary." The next day he announced, "I have directed that two cats be sent you for the granary." The following day he wrote, "Have the cats arrived for the granary¿" The next day he instructed, "Do not let the cats get food around the kitchen or they will catch no mice." This was followed the next day by the inquiry, "Are the cats catching the mice in the granary¿" Followed by the reply from the foreman, that the cats had taken control and the rats\mice were no longer a problem. Miller replied, "Send one cat to San Luis Camp so they are both employed"

Miller and the Devil's Buggy

Henry Miller had long traveled via horseback in the early years and later became a familiar sight in his buggy being driven for the most part by Buggy John. Miller did not like change all that much and once when asked about automobiles he expressed the following, "I hope my daughter don't get the idea of getting one of these devil's inventions." She did and eventually he did as well.

The time came when Henry Miller finally decided the use of the automobile around the ranches was much faster than using a horse and buggy. It meant that he could drive over the Pacheco Pass with his chauffeur and save a good deal of time, something he never seemed to have enough of. He was also getting older and to ride in a horse and buggy was harder and harder. Under pressure from

Henry Agard standing next to Miller who is seated in the Devil's Buggy

his family, he decided to purchase a car. Cars were something new and he did not have that much faith in them at the time of his purchase. Miller called them "the devil's buggy."

Pierce-Arrow was the only car that Henry Miller ever owned. Henry Agard was in Los Banos and he recalled his 11 years as chauffeur for Henry Miller and his experiences in delivering the first Pierce-Arrow to his boss Henry Miller. Agard said that Miller had no use for the automobile at first, and felt it would never amount to much as long as the country had so many good horses. Shortly after 1909, the family conspired to get Henry Miller using a car so they ordered, in San Francisco, a Pierce-Arrow to be used as his own private car.

Agard was sent, by the Miller & Lux office, to the Pierce-Arrow agency to pick up the car. It was a thing of beauty and by far the finest and largest car Agard had ever driven. In his first talk to Agard, Miller made it clear that he wanted it distinctly understood that when he was riding in the car it was not to go over 25 miles per hour. Agard said, when he got used to the feel of the car and I could see that he was relaxing a little more, I would gradually increase the speed and he soon became fond of the Pierce Arrow. He would chuckle with pleasure that he could get in at Gilroy and be over at Canal Farm in two hours. He thought that was pretty nice that he could cover so much ground in such a short time.

Henry Agard

I was Henry Miller's driver and chauffeur. I was given a letter to give Mrs. Clyne directing her that whenever I was at Canal Farm I was to have my meals at the superintendent's table and that a room was to be reserved exclusively for my use at all times at the Clerks Cottage. When I was waiting for Miller at Canal Farm I of course was interested to know about him and his affairs, since I had only started to work for him in 1909. Miller & Lux was by that time a quite famous firm.

Both Mr. Wallis and Mr. Clyne confirmed the story of Miller riding around the countryside in a rowboat affixed to a wagon bed. The route as I understand it was along the south border of the San Joaquin River north of Los Banos, perhaps in the sink of the Los Banos Creek or to the east.

It was Nickel who did away with the dirty plate route. Then fires began everywhere. Granaries and haystacks. One granary was full of barley. The Miller & Lux men all pleaded with Nickel to put back the Dirty Plate Route. "No! The doing away of the Dirty Plate Route has nothing to do with these fires," was his reply. When Nickel finally put the Dirty Plate Route on again the fires absolutely stopped. I think that Miller never carried any insurance on his San Joaquin Valley properties. Feeding the hobos was his method of insurance.

Henry Miller was a wonderful man. He was thoughtful of his employees. He never forgot a favor. If Miller saw a family in need he would send up a supply of groceries, which he knew would be gratefully received without appearing to be charity. One time we stopped at Narcisio Castro's. He had worked for Miller and now was old and ill. Narcisio was in bed sleeping. Miller asked his wife how things were and how the family was making out. "Yes, we are getting by," she replied. Later Miller stopped at his store and had a load of supplies sent up. Miller knew that Narcisio's wife would never ask for anything and was too proud to accept charity. Miller knew how to treat people. Miller had given orders to his foremen to know how the neighbors struggling to make out were getting along. If there was any in need the foreman was to help them.

The Peacock Dinner

Henry Miller the great Cattle King who owned more than a million acres always felt that the rich and influential businessmen of San Francisco looked down on him as a butcher boy rather than the multimillionaire he was.

"I determined to correct their impressions," laughed the good-natured old cattle baron as he and Ralph Merritt sat on the porch of the Poso Ranch near Firebaugh. Ralph had as a schoolboy driven for Miller during his time off from school. Starting early in the mornings they would leave Canal Farm and drive from ranch to ranch, changing horses from time to time but never stopping to eat until dinner time.

Miller had a great sense of humor. He delighted to tell young Merritt stories of his life in Germany and of his later hard struggles in San Francisco. Indeed, it was Henry Miller himself who started many of the stories that have grown up about him. One of his best yarns was about the Peacock Dinner he decided to give the businessmen of San Francisco.

Miller had a twinkle in his eye and a chuckle in his voice as he reminisced. "One time I read a book. It was about a man called Alexander. By the way, Ralph," he said, "it was in this book that I learned that I already owned more land than this fellow, Alexander ever conquered. In this book was a picture of a feast this Alexander gave to the great people of the countries over which he ruled to show his greatness. In this picture all the great of the time were seated around the banquet table with Alexander at the head. A great number of husky black slaves were bearing the feast on trays on top of their heads. A whole peacock lay on each tray.

"I determined to follow Alexander's example. I decided to invite Stanford and Crocker and Ralston and all the rest of those San Francisco big bugs to a dinner at Bloomfield. I was determined to show them that although I had been a butcher boy nevertheless I was as good as Crocker and Spreckles and all the rest of them".

At Santa Rita, the home ranch, Miller had quite a number of peacocks. He told the superintendent to catch up a number of these birds and send them over to Bloomfield. The superintendent sent the buckaroos to work catching them. "But you," confided Miller, "When the cowboys tried to lasso them, these birds were so quick they would get their heads out of the noose before any of the buckaroos could pull up their ropes." The cowboys hobbling around in their high-heeled boots were having a terrible time. Finally they threw away the ropes and taking after the birds with clubs knocked the heads off.

At the Bloomfield preparation for the great feast of the peacocks was going on. Miller remembered in the picture that Alexander had black men carrying aloft the stuffed peacocks. He sent to San Francisco and had a number of colored waiters sent down. The guests arrived. Henry Miller sat at the head of the table. Each carrying aloft a stuffed peacock, the colored waiters marched in. They circled in triumph twice around the table and then deposited the fattest plumpest one in front of Henry Miller to carve.

"I attempted to drive the fork into the breast of the bird," laughed Miller. "To my consternation the peacock slid off the platter on to the tablecloth. I got it back on the platter and again made a determined stab with the fork. Once more the bird slid off, this time in another direction. Once again I got it back with no greater success than before."

Miller finally had to give up. He did not know that peacocks are so tough that no one can eat them. But the butcher boy rose to the occasion triumphantly. To his mirthful guests he explained, "Those cooks of mine know nothing. They can't even cook peacocks right. They cooked them so bad nobody can eat them!"

Interview Ralph Merritt undated RLM

Henry Miller's Infamous Boat Ride

I always doubted Henry Miller ever taking any such boat ride as many had heard of. "Utter nonsense," I declared. Just another of those colorful stories told about Henry Miller that have no foundation in fact. I remembered the whopper that Miller owned so much land that starting from the

Some of the swamp land Miller rowed around in his buggy, 1870's

Mexican border to Oregon he could camp every night on his own land. I had worked as a clerk for Miller & Lux and knew that simply was not true. I did know about the time the San Joaquin River was overflowing from five to ten miles wide throughout the west side of Merced County and the fact that Henry Miller coming from the Bloomfield by way of the Pacheco Pass, had to row some 18 miles to reach his headquarters at the Santa Rita. Since Henry Miller had passed on and I could not give him a call to ask, "Mr. Miller, is it true about your boat ride?" I did the next best thing. I made contact with his first and only chauffeur who was living in San Francisco. On my next trip to the city I made arrangements to meet with him.

"It's quite true," was the first thing he said. Government surveyors in running out the lines in the Public Domain in California shortly after the American occupation would survey down to the water's edge of any swampy lands they came to. Carefully avoiding getting their feet wet they would classify on their maps all lands beyond as Swamp and Overflow lands. Oftentimes the land would be only temporarily under water. This posed a problem to future settlers. Swamp and Overflow land could not be homesteaded nor purchased outright. Instead a settler was required to reclaim the land by putting up necessary levees or by digging drainage ditches. The land seeker when applying for such land would be more successful in obtaining such if he swore on oath that he had been all over it in a boat.

However, sometimes this land that temporarily underwater when the surveyors were mapping the country would be perfectly dry land for years after. How could a person swear that he had been all over a certain piece of land in a boat when in reality the land was perfectly dry? Henry Miller, so the story goes, took advantage of this oddity.

Miller, never one to miss an opporunity, solved the problem in his own forceful way. Putting a rowboat on the running gear of a lumber wagon he climbed up in the board and had an Italian drive him all over a tract of land along the San Joaquin River north of the present Los Banos that he was determined to possess. Miller knew that there was a market hunter by the name of Smith that used to have a rowboat fastened on the running gear of a wagon. When he was out hunting and came to a slough too deep to ford he would nevertheless plunge right in with his team and wagon. When the water got too deep his horses would swim and his rowboat mounted on wagon wheels would float. I knew of Smith as well and I thought that the story of Miller doing this had been a combination of the storytellers mixing facts.

Miller actually did ride in a boat on a wagon in order to make the affidavit that he had been all over the tract of land in question in a boat. I checked his story with the superintendent of the Los Banos Division of Miller & Lux and I heard the same story from the superintendent of the Kings River Canal and Irrigation Company. They knew all about the affair. When Miller went to the land office to sign the affidavit he truthfully could swear that he had been all over the land in a boat. Chris his Italian navigator could also swear truthfully that he was in the boat with him.

This satisfied the unsuspecting government clerk. Never once did he think to ask: "This is summer, how did you do that?" or "Was your boat on wheels?" Miller received thousands of acres this way.

Notes 1950's RLM

Rodeos in the San Joaquin

In the early days the only way to gather and sort herds of cattle was with the rodeo. The Spaniards and Mexicans have established this way of sorting what was theirs from what was not and each ranch would hold a rodeo and fiesta. With the coming of the "gringo" the tradition changed somewhat but was kept alive as the herds were large and the plains were vast.

A rodeo could be seen for ten miles or more by the spiral of dust going up into the sky where the cattle were being brought together. The cloud of dust is of a brown color and was quite a sight. Usually three or four thousand cattle were in a roundup, although one rodeo in the San Joaquin Valley had over 10,000 according to some old hands.

When it was decided where the rodeo was to be held the riders would start out as soon as it was light enough to see and ride in all directions. Usually a territory of twenty or thirty miles would be scoured for cattle. As fast as they were found they would be started for the rodeo ground. The cattle seemed to know where the rodeo was being held and would start in that direction. Soon streams of cattle would be seen coming in from all directions and headed towards the rodeo grounds. It looked like a great number of streams flowing into a large lake.

By eleven o'clock most of the riders would be back. The tired horses would be turned into the cavada and the rider would go to the cook wagon for

Branding crew, 1907

Buckaroos taking their ease on a drive, 1890's

a tortilla or flap jack. Then it was saddle up a fresh horse and work until about an hour before sundown. The beef cattle would be sorted out to be driven away while the rest of the cattle would be turned loose to wander back to their feeding grounds.

There were two kinds of roundups. One was to pick out fat cattle to sell. The other was to brand calves. In branding calves a large bunch of cows and calves are driven together. About ten calves were branded at a time. A fire is going and the irons are heated. Vaqueros ride and lasso a calf by the feet and drag it out. The mother comes rushing out of the herd and bawling for her calf. Whatever brand the mother carries is given to the calf. About ten calves are dragged out and brought over to the fire. They are branded and ear marked. The branding is done away from the herd of cattle and in the direction the cattle came from. The calves after branding are turned loose and with their mothers are headed back to the range they came from.

Cantina wagons use to follow up the rodeos. Usually about two would be at each rodeo. They were saloons on wagons where cigars, tobacco and whisky were sold. The wagons were made like flat boats. They were watertight and when they came to a stream would float across. The wagon wheels would hang down in the water under the wagon. A team of horses was used to pull the wagon from place to place.

Miller & Lux had a large rodeo ground southeast of Los Banos about a mile. You can still see the flattened ground today.

—Frank Latta interview with Z. Westmooreland, 1930

The Inspector General

In the early days Miller always had a driver and a good team. The first place he would visit at Santa Rita was the slaughterhouse. He wanted to see the butchers and what they were doing. Tapatula was the head butcher. He later died at Canal Farm.

Miller would always get up at four o'clock in the morning. "I am not better than my men," he would always say.

One time Miller caught an employee beating a mule over the head. The mule had thrown its head and struck the man. The fellow had gotten a club and had begun clubbing the mule. "Why are you hitting my mule?" Miller asked him. Miller then told the man that a mule costs money and he could get all the men he wanted for nothing.

When Miller would be on his rounds and saw a piece of sack he would stop and pick it up. That went for wood, wire, nails anything he saw. When he would get back to the warehouse he would tell the men to use the sacking to make patches and to reuse the nails. When Miller saw a horseshoe on the road he would always stop and pick them up. He would tell his blacksmiths that when a horse came in that had lost a shoe but still have three partly worn ones to use the old shoe and save a new shoe. Miller did not want new shoes used along with old shoes.

Henry Miller's son was rough. When he was here he and Jim used to play with Nellie Miller and Will White. Young Miller was the oldest. He was maybe 18 and was rough and liked to break windows. Miller used to come to my store every week to buy ten or twelve packages of candy. "I gives them a package of candy when I come to the ranches and see the foreman's children."

I knew Miller ever since I was a boy at Santa Rita. When I worked in the store about two weeks before Christmas Miller would come to me and ask me to do him a favor. He wanted me to go all over town and get the names of all the poor families that had children. They must be poor and have lots of children. "Bring me the list. I be here Wednesday." I would give Miller the list of twelve or fourteen families. Then Miller would give these families an order for $50.00 drawn on the Miller & Lux store. Miller had more confidence in me than his other employees to know those in true need.

Henry Miller did not look with favor on many religious groups in a small town. Once his office manager reported to him a request for a $500 donation for the Christian Church. Miller listened to his manager then asked, "Didn't I give $500 to the Catholics?" "Yes, Mr. Miller." "Didn't I give $500 to the Protestants?" "Yes, Mr. Miller, but you don't understand. There are some people in the community that will not go to the Catholic or the Methodist Church." "Vel den, let them go to hell," was Miller's reply.

At the Canal Farm celebrations Miller would always shake the hand of every man there. He would always ask, "Are you having plenty to eat?" He served only meat, cheese, bread and coffee. When the Mexicans started coming he added beans.

Rounds J. V. Toscano, 1945-47

Miller and the Bandits

Henry Miller had been bothered by cattle thievery for many years. These thieves were worse in the county near the New Idria mines and Miller was afraid to go into that rough country to try and track his stock and the bandits down. Miller had a man working for him named Narcissco Castro who knew the bandit Vasquez and saw him from time to time. Miller told him to tell Vasquez he wanted to see him. When Vasquez came to the Bloomfield to see Miller he was in San Francisco. His brother- in - law Sheldon sent for him. Miller told Vasquez about the theft of his cattle that was going on near the New Idria and that he wanted it stopped. Miller got Vasquez to promise that he and his men would see that no more cattle were stolen. He gave Vasquez $500 cash that day and told him whenever he needed horses or money to go to any of Miller's ranches and the foremen would have orders to help him in everyway. That was the end of the stealing.

Later Vasquez was caught and was in jail in Salinas. Miller went to see him and asked to be locked in the cell alone with him. Miller told Vasquez that if money could help him he would gladly give it to him but that the people's feelings for him were so bad that nothing in the world was going to help him. Vasquez knew this. Miller left $300 at the jail that day and a few days latter sent several thousand dollars for Vasquez to hire a lawyer. As he was taken to San Jose he knew he was going to be hung. Vasquez told the sheriff to buy him a coffin. When it was delivered Vasquez got into it to be sure it was all right and then told the sheriff to put it away till he was hung. When he was hung there was probably 10,000 people there. Henry Miller, never one to go back on a deal, had paid for the coffin.

Notes 1943 RLM

Rustlers

Miller had his own way to deal with rustlers. In the early days with 1000's of cattle and no fences, no armed cowboys and no law, and with no trouble to find a safe hideout in the hidden valleys and hills of the San Joaquin rustlers often had their own way. To Miller & Lux these rustlers were an annoyance. Miller never prosecuted a thief. The first 50 years of settlement of the area long the western side of the San Joaquin River the Miller & Lux Empire shows only two instances of what may have been summary justice for cattle thieves. Henry Miller had no part in these, and nobody on the west side ever saw who did it. Miller's men for the most part were Mexican vaqueros were unarmed and were mindful of their own business. However, in the early days two men were hanged from the cottonwood beside Los Banos Creek on Badger Flat. Nobody knew how or why they came to be there. Officers came from Snelling to cut them down and bury them. Later a couple of men were shot at Mud Slough; again nobody knew how they had gotten there. That was more or less the end of rustling on Miller & Lux lands. In fact Miller & Lux driving such vast herds to the railheads picked up more cattle along the way that they lost. He had his men out looking for unbranded strays all along the way they were quickly caught up and branded with the HH brand.

Undated notes RLM

Miller in need of job?

Down half mile out of the present Newman towards Hill's Ferry two Germans settled in the early wheat days and the place became Dutch Corners. One of them, Voight, had a blacksmith shop there. C. A. Miller tells me his father stopped there one day and there was a stranger in dirty clothes with the blacksmith. Voight introduced them: "Mr. Miller meet Mr. Miller." My father W. J. Miller needed a hand on his ranch and thought of asking the stranger if he wanted a job, but he didn't. Voight later told him that it was Henry Miller that he had thought of offering the job to.

Notes RLM

Canals

An early pioneer by the name of Bensley saw the possibilities of an irrigating dam across the San Joaquin River where the Fresno Slough and the San Joaquin River join and he filed on a quarter section on which these two rivers meet.

About 1865 Isaac Friedlander shipped a shipload of wheat from California to Liverpool. This cargo met with a good sale, being pronounced as good as the best Australian wheat. About at the time of the war between England, France and Russia over Turkey wheat became very big in Europe, which gave a great impetus to wheat growing throughout the world wherever it would grow well. California soon became a great wheat raising state.

In 1870-1871 there was a terrible drought all over California and practically no wheat was raised in the whole state. John Bensley bought out the owner of the quarter section of land at the confluence of the San Joaquin and the Fresno Slough, and started to dig a ditch to irrigate the West Side. This project was to cost about $500,000 and was way beyond the means of Mr. Bensley. But Mr. Friedlander, who had shipped the first load of California wheat to Europe and his partner Mr. Chapman, were the largest landowners in California at that time. They owned thousands of acres of grain land on both sides of the San Joaquin River. They became interested in the project. Mr. Friedlander sent to India for a certain engineer by the name of Brereton who had built extensive irrigation works in India.

Mr. Friedlander sent this engineer out to make a preliminary survey of the San Joaquin Valley and especially of their lands with the idea of irrigating all their lands on both sides of the San Joaquin River. The engineer spent six weeks on the survey and went back giving a wonderful report of the possibilities of irrigating the San Joaquin Valley and more especially the West Side. His report called for two canals. One was the San Joaquin and Kings River Canal and the other was for a much bigger canal called the Tulare Lake Canal.

Friedlander and Chapman organized a company of ten members each of whom put in $100,000. The company proceeded to buy out Mr. Bensley, giving him 100,000 shares of non-assessable stock. In the next following two years the company built the canal down to Los Banos Creek. The canal was build under the special direction of the engineer, Brereton. It was a very expensive undertaking. Hay costing $50.00 a ton was hauled all the way from Hollister by teams over the Pacheco Pass, that, too before the new grade was built. The canal extending to Los Banos Creek cost One Million Dollars!

The company proceeded for some five or six years to run the canal, but as the selling of water and irrigation was nothing like what they anticipated it was continual assessments and no dividends. Miller & Lux were in the best position to make use of the canal by irrigating grain land along the canal and by flooding vast stretches of wild pasture land. Consequently as one after another of the stockholders became dissatisfied Miller would buy them out.

While Henry Miller was gone on a trip to Europe the Canal Company got into a dispute with Miller & Lux. When Miller got back he found the canal gates were all locked and in many places his cattle were suffering for water. He took the law into his own hands and seizing an ax began chopping down the gates to let the water out on the lands. The superintendent of the canal promptly had Henry Miller arrested on three counts. Henry Miller completely turned the tables on the superintendent by rushing to John Bensley and buying up his 100,00 shares of non-assessable stock. This gave Henry Miller the controlling interest in the Canal Company. The superintendent thus found himself in the employ of Henry Miller who promptly discharged him and the case was dropped.

Henry Miller at once put John Q. Drummond in charge as superintendent, discharged all the bosses and other officers of the company, moved the offices from San Francisco to Center Point and started the Canal Company out afresh. He put the Canal Company on an economical basis, but it was several years before the canal declared a dividend. The canal headquarters were located along the canal bank a little east of Center Point.

The year 1877 was extremely dry and the settlers down towards Hills Ferry came up to talk with Henry Miller, hoping to get him to extend the canal down to their section, which was more of a grain growing country than this, owing to their usually having more rain. Mr. Miller told them that grain and hay were so scarce and high that he couldn't do anything for them that season, but that if the next year was good, and he told them he hoped it would be, that he would extend the canal down to them. The next year was a good year and he extended the canal from Los Banos to Crow Creek below Hills Ferry. Thus Mr. Miller brought water and prosperity to their very doors "without money and without price." Land worth ten dollars an acre went to one hundred dollars an acre as soon as it became known that the canal was coming.

For many years the price of all farm products was very low and many people became dissatisfied. They thought that the $2.50 an acre charged by the Canal Company was excessive and the water consumers of Stanislaus County began a series of lawsuits against Miller. In the lower courts and especially before juries they generally won but when the cases were taken to high courts Henry Miller won.

The contention was that the water rate should be computed on a six per cent rate on the physical valuation of the canals. The cost of the Old Canal and its extension to Crow's Landing was a million and a quarter dollars. Engineers went on the stand and declared that they could build the canals for $350,000. At six per cent on that valuation the water rate would be $1.50 an acre instead of the $2.50 the Canal Company was charging. Later on these same engineers went on the witness stand and declared that it would require $600,000 to replace the canal. When Henry Miller finally took the cases to the United States Supreme Court it was ruled that the Canal Company was entitled to six percent interest on the actual amount that the canal had cost to build. This was a sweeping victory for Henry Miller for by it he could have raised the water rate to six dollars an acre. But Henry Miller took no advantage of his opportunity to ruin the ranchers of Newman and Crow's Landing and kept the rate at the old charge of $2.50 an acre until the supervisors of Fresno, Merced and Stanislaus counties set the present rates of $1.50 for Fresno County, $1.75 for Merced County and $2.25 for Stanislaus County.

It was J. LeRoy Nickel who introduced the system of measuring the water and basing the

Old Canal, 1880's

charge on the amount used. This was an entirely unsatisfactory method, as the Canal Company had no accurate means of measuring the amount of water used. Sometimes a rancher would be charged as high as three dollars an acre for a single irrigation. The present rates established by the supervisors of the three counties put an end to all of this by making a fixed charge per acre per year.

The New Canal

The New Canal was built in 1897 from the head dam near Mendota to the Cottonwood district at the cost of $150,000. The locks, gates and bridges cost $100,000 more, bringing the cost of the New Canal up to $250,000. Thus the cost of the entire canal system is about one and a half million dollars, and it forms the cheapest and best irrigation system in the State of California.

During Cleveland's first administration the government lands throughout this section that had been withheld as railroad land was thrown open to settlement. Everybody rushed out on the plains and took up every available quarter section of land. For several years the settlers struggled along, planting crops only to see them dried up nearly every spring by the north winds. Finally in their land extremity the settlers made up their minds that they must have water. A water meeting was called and the people turned out in full force. Although it was a big meeting, the whole poverty stricken assembled with all their lands couldn't have raised $10,000. No bank would have loaned it to them on all their lands put together and any bank would have been foolish if it had. Yet the problem was how to get a canal out on the plains. Mr. Stockton made them a speech. "I told them," he says, "that I had seen what Miller had done for the people around Newman and thought that maybe he would do something for us. Otherwise I didn't see any hope for us, as he was the only person interested in our welfare."

Sam Smith and myself were appointed a committee to wait on Mr. Miller and see what he would do for us. He told us to keep our own council, as he did not wish any speculation in land, but that he would build the canal for us. He afterwards built the canal. Forty miles long and one hundred feet wide and filled it with water and told us to go to it. There wasn't a man below the canal that a had a piece of land as big as yours that wasn't a comparatively rich man. Miller afterwards extended the canal from Los Banos Creek to Cottonwood and added two million dollars more to the value of the country for which no land owners was charged one nickel.

The settlers offered Henry Miller a subsidy of ten dollars an acre for putting water on their lands. No one was charged a cent for the funding of the canal, Henry Miller saying that it would be too much trouble for him to collect the "subsidies." As a matter of fact Henry Miller had already made all the money he wanted and he put in the new canal in order to build up the country. He could have bought all of the Cottonwood district for $25 an acre and by putting in the canal could have made a barrel of money.

New Canal, 1890's

Miller use to stack every bit of straw for his cattle. Not one little bit was allowed to go to waste. Later as alfalfa came on and the straw was not all used up he would allow his men to burn it to get it off the fields. As a matter of fact it was afterwards proved that of his land only about 4,500 acres were fit for irrigation. The Cotton Field, the Ward Sheds and the fringe south of Los Banos was about all that was fit for irrigation. Miller began planting alfalfa but for many years it didn't pay. It was something that had to be learned how to grow. Miller & Lux got to using canal water for swamping. Thus more feed was produced for their cattle. The Canal Company didn't make enough from irrigation to pay the company's Chinese cooks.

Henry Miller never speculated in land. "I needs all the land I'd got to run my cattle on," he once told Mr. Stockton. "And I needs more yet." Not to buy land cheap in order to develop it with water and sell it high was never the plan of Henry Miller. He wanted to see the country built up. He also wanted people to help him fight for the water rights for the West Side. But in this he was mistaken for the people invariably turned against him and brought lawsuit after lawsuit against him.

If Henry Miller had been inclined to speculate in land he could at one time bought the Mitchell estate consisting of 100,000 acres around Modesto for $300,000. It is now worth 60 million dollars. But Miller was a cattleman and bought land only for his own use to run cattle on.

Eventually the canals will have to be cemented. Also the San Joaquin River itself for some 30 miles above the head dam near Mendota. One day all water will be measured.

The winter of 1889-1890 was an extremely wet one. Mr. Stockton in coming to Los Banos used to carry a board with him as he walked along the irrigation levees. When he came to an impassible spot he would lay the board down, walk across and pick up the board again and take it along with him. On April 1st he started to move his blacksmith shop from Old Center Point to the new town Los Banos. The road was so bad however he had to leave the shop setting by the road for ten days.

When the Western Meat Company came west to take over the San Francisco meat trade as they had done in Eastern cities they didn't know the kind of competition they were going to encounter in Henry Miller. Miller was actually selling one third of the meat in San Francisco. They started to freeze out competition by putting meat down as low as three cents a pound. Miller thereupon turned out 40,000 head of beef cattle back on pasture and went to buying meat from the Western Meat Company to supply his trade, killing only enough cattle in his own packing house to keep the Western Meat Company from knowing what he was doing. When the Western Meat Company saw that they had no chance of forcing Henry Miller to yield to their supremacy and came to him for a truce he told them to keep up the war as long as they liked, that he could afford to lose five million dollars a year and keep it up for 20 years! For years afterwards Henry Miller set the price of meat in San Francisco and kept the price of meat to San Franciscans at three to five cents a pound lower than it was selling in Eastern cities. The poor people of the city thus profited by Henry Miller.

In the early days dairying didn't pay, as butter was only ten cents per pound. Eggs were ten cents a dozen. Portuguese came after dairying began to pay. Portuguese expect to begin to make money right from the start. Portuguese never pioneer. As soon as things don't pay they leave immediately. Italians plant vines and trees and wait.

Undated notes and interview RLM possibly with Billy Stockton, Westside pioneer.

Sportsmanship

Henry Miller had a reputation among old timers who knew him and his peculiarities being what they termed a "good sport."

Miller was always on the lookout for fine stock and never passed an opportunity to add to his own stables if he something of value. One day while riding near the Bloomfield he saw a fine team of mules being driven along a dusty road some miles from his home place. "Fine team," he commented. "Yours¿" "Si, senor," answered the Spaniard whose knowledge of English was some what limited. "I'll give you $200 for them," offered Miller. The man acquiesced. "Take them up to my ranch and put 'em in the barn," directed Miller, counting out the mentioned sum. "Si, senor," and started towards the Bloomfield.

Some time later Miller arrived at the Bloomfield. He sauntered down to his barn to take a look at his purchase. "Pretty good buy for $200 eh¿" he queried complacently to his superintendent who stood nearby. "$200 more than you need have paid," answered the superintendent with a snort of disgust. "Them was your mules in the first place." Miller swallowed hard and replied, "Well, showed I knew a good team when I saw it," and walked away.

Another time Miller was riding around and haying was in progress. He made it his practice to keep an eye on all things and the progress of the work and the men. Half a dozen times Miller made his circuit of the big field and each time as he rode up to a certain hay stack he found one of the men loafing in the shade of the stack. Saying nothing to the man himself, he questioned the foreman about him, asked where he came from. He was told by the foreman that the man was just working until he got enough money for a saddle horse. Miller snorted then ordered the foreman to go get a saddle horse from the barn nearby. The foreman, taking it for granted that Miller wanted the horse, went to the barn and selected "the boss's" favorite riding horse who had a special made saddle and bridle kept for its use.

Miller in the meantime had fumed himself into such a state of exasperation that he paid no attention to the horse, but snatching the bridle reins from the foreman's hand strode over to the loafer, leading the horse. "Gits out of here," he barked. "Get on that horse and never show up on this ranch again. I won't have such a loafer on my place. It's worth the price of a horse to get you off here before you spoil all my workers. Git," he shouted. The man got with alacrity. Only as the man disappeared on the galloping horse in a cloud of dust did realization sweep over Miller. He suddenly recognized his favorite horse. Chagrin overspread his features, but game as he was he choked down his exasperation and with a shrug of his shoulders let the man go.

Mystery

Gilroy Advocate - 1916

Miller's First Backer Dies

Mrs. Dorothy Otto 90 years old, the first financial backer of Henry Miller, the late cattle and land baron, died in Chico on Christmas day. Mrs. Otto lent Miller the money with which he opened a meat market in San Francisco, which market became the foundation of his fortune. A few weeks before Miller's death, he sent her a check for $15,000 with a letter of thanks for her help of fifty years before. Mrs. Otto was 90.

There is some mystery here. According to her death notice, Mrs. Otto was from Miller's hometown and happened to be in San Francisco in the 1850's. It appears he received a loan or some kind of backing from her to get started in San Francisco, yet he never acknowledged it or her in any public manner. In the months prior to his death he did pay her a large sum and also remembered her in his will.

December 30, 1916 Gilroy Advocate

Gilroy's most charitable Woman crosses Silent River

Mrs. Dorothy Otto passed away at the home of her grand-nephew early on Christmas day. She left Gilroy the week before and planned to return after Christmas. She was in excellent health and spirts when she left our fair town and contracted a cold, she was in excellent spirts up till a few days before her death. The church was filled with old friends of the deceased. Mrs. Otto had lived in Gilroy for about 35 years. Her charity knew no bounds. She was ever ready with an open purse to help the poor or those in trouble. Many a soul was cheered by her generosity.

Her husband John Otto preceded her to the tomb many, many years ago. He was a bandmaster in the early days of our city and at one time mayor. Left a widow she sent back to her native town Wurtenburg in Germany for her nieces and nephews. She was a childhood friend of the late Henry Miller the great land and cattle baron and owner of the Bloomfield Farm. Mrs. Otto was among those remembered in Mr. Miller's will. Her friendship was treasured by Mr. Miller and he never forgot her many kindness to him upon his arrival in California. She lent him funds to get a start in his own butcher shop and from that his great fortune grew. Miller never failed to stop and visit coming or going from the Bloomfield Farm.

It is a strange coincidence that these two, so different, born in the same town, and friends since an early age, should live to such a great age and die within three months of each other. The Services at St. Stephens Church were conducted by Rev. Wm Wiggs, The service of the Masonic & Odd fellows cemetery were in charge of Magnolia Chapter O.E.S. in which she was one of the oldest members.

Among the relatives left are: Mrs. Richard Brem, Mrs. Louise Dietz, Miss Kate Fischer, and Mrs. Blackmore of Sacramento, George, Edward and John Matthias and Irving Otto all of whom were present at the funeral.

Miller & Lux Inc.
Los Banos Division
Los Banos, California
Dear Sir:

One of the best duck shooting seasons ever know in the western San Joaquin Valley has just come to a close. The great flights of birds in the Los Banos –Gustine area all through the season furnished important information on two subjects much discussed by the hunter.

In the first place it was shown that the quantity of ducks and geese is not diminishing. Secondly, it was demonstrated that the reputation of the western San Joaquin swamplands for being the best shooting district in the state is richly deserved.

Limits were easily brought down throughout this district on Opening Day, and just as easily on Closing Day. Every weekend during the season the shooting was excellent. Geese were more plentiful than for many years. Limit strings of spring, teal, widgeon and spoonbills were brought in by medium and even poorer shots. Expert hunters took many mallard and canvasback as well as gadwall ducks.

This wonderful shooting took place on lands owned by Miller & Lux Inc or recently sold by them to various gun clubs. Over 20,000 acres have been sold by Miller & Lux for shooting purposes. There is still for sale many thousands of acres of the best shooting lands in California.

The company desires to call the attention of sportsmen to these lands. Most of them like in Game District No. 3 in western Merced County. They are for sale at $25 per acre. The terms are one quarter down and ten years for the balance interest at 6%. Remember you are buying land, not merely a shooting privilege. It is grazing and a good investment without the shooting.

Now is the time to secure some of the choicest locations. Organize your membership and have a Club that will insure good sport for years to come. Reports of this year's great shooting around Los Banos is spreading fast. These splendid waterfowl lands will soon be taken up by sportsmen, thus preserving it for shooting purposes. Delay may mean lost opportunity.

Come to Los Banos where representatives of Miller & Lux will be glad to show you our properties. You will realize then that nothing in California can approach these lands for ducks and geese shooting purposes.

Do not wait until next spring. Even if there is still suitable land available, it will be too late to make the best use of it. BUY NOW! Fix up your levees and dams, so you can handle all available water to the best advantage; build your blinds and bait the ponds you are going to shoot over. In this way you will be assured of first-class sport from opening day until the close of the season.

Yours very truly,
MILLER & LUX
LAND DEPARTMENT

ROMANCE IN SELLING M & L CATTLE HERDS

Publicity furnished by the Miller Livestock Co., purchasers of Miller & Lux cattle, puts a new angle on the sale in the historic and romantic side lights:

Beef cattle production in California's great central valley has entered a new phase with the purchase by the Miller Livestock Company of the remaining cattle holdings of Miller & Lux, pioneer and for a long time dominant livestock producers of California. The Miller Livestock Company is owned and controlled by Louis F. Swift of Chicago, prisident of Swift & Co., and A. B. Miller of Fontana, president of the Fontana Farms Co., which in addition to other large agricultural interests operates the largest hog ranch in the world. The purchasers of the Miller & Lux breeding herd also control the B. B. Company, producer of baby beef, which operates two large feed yards, one at Fontana and the other one in the delta near Collinsville.

The transaction through which the Miller Livestock Co takes over the Miller & Lux cattle business involves the sale of the breeding herd, of the well known Double H Brand and of the 3,000 acres known as the Harrison field, together with the lease of 80,000 acres on the west side of the valley. The total consideration exceeds a million dollars.

The Miller & Lux cattle business was founded in the early fifties by Henry Miller, the penniless German immigrant who through his love and knowledge of cattle, through his daring, sagacity and foresight acquired a principality of 1,600 square miles of land on which he pastured as many as 90,000 head of cattle and twice as many sheep, for half a century dominating the livestock industry of the Pacific Coast.

Henry Miller's one absorbing interest in life was cattle. He thought in terms of cattle. For the benefit of his ever increasing herds he acquired over a million acres, most of them in California, where part of his holdings extended along the San Joaquin river for a hundred miles.

It was said of Henry Miller that he could drive a herd of cattle from the Tehachapi mountains in the southern San Joaquin valley to the Oregon line and bed them down each night on land of his own. In order to have good feed for his cattle he constructed enormous irrigation systems, a single one having over 200 miles of main canals. He kept in constant personal touch with all his far-flung enterprises in California, Oregon, Nevada and Utah, and the pride of his heart was the reputation he built up for the Double H brand.

Though he started with native California cattle and with a herd driven overland from Texas in the fifties, he soon developed a breed characteristically his own. Using the Durham Shorthorn as a base, he crossed his cows with Hereford bulls, producing a variety that had the stamina, the rustling ability and the vigor of the Herefords with the size and quick-growing trait of the Shorthorn. He bred so carefully, managed his herds with such skill that the Double H brand on a steer became the mark of acknowledged high quality.

Henry Miller, a giant who in his field ranked with the galaxy of railroad builders and mining kings that made Western history, died fourteen years ago. Now the livestock business he started, the brand he loved, will be carried on by another Miller, by the capable hands of L. F. Swift who so skillfully guided the vast enterprise bearing his name through the economic upheaval following the war that it assumed first place as the largest and strongest food-purveying institution in the world.

A. B. Miller has been farming on a large scale in Southern California for over thirty years. After dry-farming thousands of acres at the beginning of the century, Miller acquired the 18,000 acre Fontana ranch near San Bernardino, developed water, subdivided 15,000 acres and built the most successful colony of small ranches in the state, a community trible 6,000 inhabitants.

At Fontana Miller also developed the largest hog ranch in the world with a population of 54,000 swine and an annual output of 8,000,000 pounds of finished hogs. Miller's hog ranch has an international reputation.

The active entry of L. F. Swift marks a new era in the cattle business of California. The Miller Livestock Co. will specialize in large-scale production of baby beef. Hitherto the bulk of the California cattle output has been marketed as two or three-year old steers with a weight of 1,000 to 1,100 pounds. The new owners of the Miller & Lux herd will endeavor to turn out a young, perfectly finished steer weighing 800 pounds in less than half the time required under the old range methods of beef production. The old Miller & Lux method of keeping the breeding cow in the pink of condition will be carried on, but immediately after weaning the range pasturage of the cafe will be supplemented by feeding in order to stimulate rapid growth. As soon as the young stock has reached a weight of 400 to 500 pounds, it will be shipped to the huge feed yards already established and in full operation at Collinsville, on the Sacramento river, and at Fontana. There concentrates will be fed for four or five months until the young steer has reached a weight of 800 pounds. Uuder this method it is possible to produce a better grade of tender, juicy beef in a much shorter time than under the old-style range-and-rustle process. It is expected that, once a steady, all-year supply of baby beef is assured, the exceptionally high quality of this beef will give it the same popularity it has long enjoyed in the Middle West and the East.

The new owners expect to take over the present livestock organization of Miller & Lux and to expand it as the business is expanded.

Los Banos Enterprise *Feb. 28. 1930*

This sale of the Miller & Lux herd in 1926-1927 resulted in the massive land sales that took place starting in 1928 and continuing on through the 1940's.

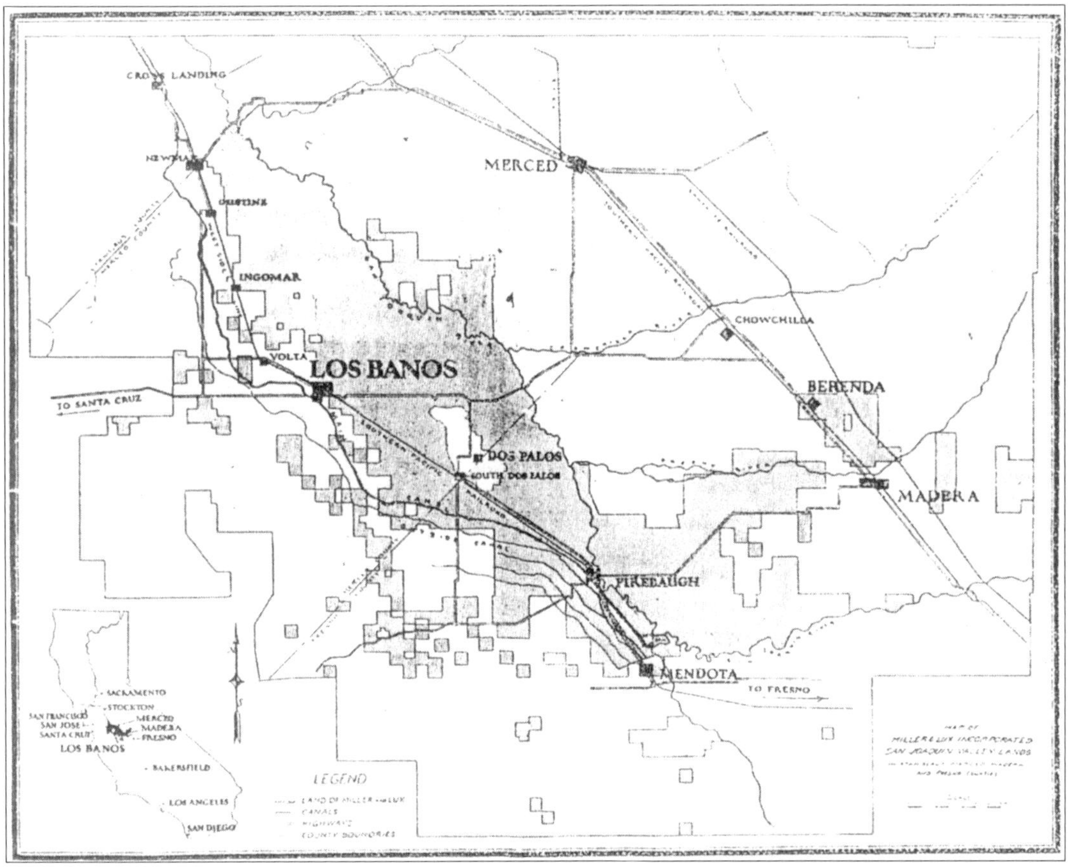

Top map is land holdings and parcels for sale in Merced, Madera and Fresno counties, 1926

Bottom map is land holding and parcels for sale in Kings, Tulare, Kern and San Luis Obispo counties, 1926

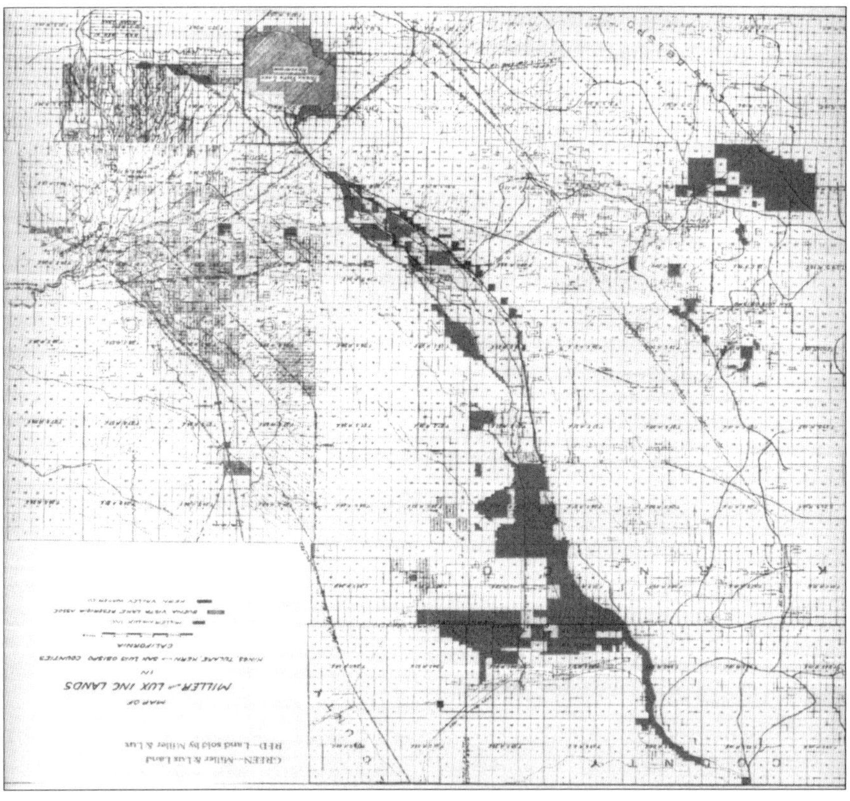

Part IV

Chapter Thirteen

Letters 1890-1916

San Francisco
May 27, 1902
Mr. D. O. Leonard
Firebaugh, Cal.
Dear Sir:

I have written to Safstrom that I want him and Vandiveer to find out all the horses that have been loaned out, to whom and under what condition and all about it. What horses have been sold, we want vented. I want a clean job make of this and when any horses are loaned out, the people should give a receipt and the receipts should be filed away. We want to do business like other people in business like manner.

Yours truly
Henry Miller

Miller & Lux Incorporated
Merchants Exchange Building
San Francisco, April 2. 1909
Mr. P. H. Turner
Hyde Ranch, Gilroy, Cal
Dear Sir:

I received yours and in answer will state that if you take the position, you need not go there before the last day of this month or two or three days before, as the man who has charge of the ranch now wants to leave the first of May. You can store your furniture and bring nothing but your clothing, excepting the crib for the child, which you will ship. The house is furnished, and your wife is expected to look after the poultry, the house, the provisions and the milk, and you will attend to the cattle and everything pertaining to them. There is a man there, - a Mexican, who has been with us for the last twenty-five years. He knows every portion of the ranch, and will show you all the details connected with the handling of our cattle. Your wages will be $50.00 per month. The property in its details is different from any you have ever attended to. There are more cattle and a larger scope of country to work over, and our way of working cattle is different from the way most people do, so you must not expect an easy job. The work is constant. The climate is hot, but there is no malaria nor other sickness.

All that is necessary for you to do is to write me that you accept and you will be there before the first of the month, as stated above. Mr. Dickenson, who is there now, will be notified of your coming. You take the train from Gilroy to Firebaugh. By writing a few lines to Mr. E. F. Ogle, Superintendent at Firebaugh, as to your coming, he will have a team at the depot waiting for you.

Yours truly
Henry Miller

Miller & Lux Incorporated
Merchants Exchange Building
San Francisco, May 13, 1909
Mr. P. H. Turner
Firebaugh, Cal.
Dear Sir:

As I have not been able to see you, and you are now left to act entirely on your own judgment in the way of doing the work, cleaning up the branding and locating the cattle as is required, I wish to state as follows: When you work the stock, you will select such cattle as are blemished and in which there is not further outcome, and which should come off the ranch. All such cattle want to be sent to Santa Rita for slaughter.

From time to time you will get up the colts and count them, and see if any of them need taking care of, as sometimes they get cut by wire or otherwise injured. You want to look through them carefully, and take out such as need care and attend to them, and no doubt Mr. Dickenson gave you the correct count.

You will kill a beef once a week, such as is most profitable, that is blemished cattle or such as are not in condition to get any better. Have some sheep there and kill one once in a while for the camp, especially in hot weather, or at times kill two small animals, but one should be enough. A little bunch of sheep must always be kept in the fields in front of the house, and at night be brought into the house yard, or put them in the corral. In the morning open the gates and allow them to go out of their own accord. In the evening, let them come up as late as possible. At times they will come up themselves, but they do not want to be left out in the fields at night.

Jesus Higuera wants to be out with the hounds to hunt coyotes, and if some of the men have nothing to do, they can assist him. Also get some wild hogs, which the water will drive out of the brush. They want to be gathered and put in the yard to fatten.

You want to see that the men are at no time idle. The men must ride such colts as should be ridden, and not allow them to get in to high condition. You want to examine the men's saddles and blankets. We do not want our horses to get their backs scalded, which is caused by bad blankets or by having them girted too tight. A horse with a sore back, which cannot well be cured, can be broke to harness. If they are serviceable horses, I wanted them doctored, the sore cured, and the blankets fixed and the men to ride them with care. To all this you should give strict attention.

Some of the cattle will be constantly moved about to eat up the feed where a second crop can be produced. You will always be ready to comply with Mr. Ogle's request, and you will be guided by his advice. Jesus Lopez has been in our employ many years, and he knows all about the ranch, and he can give you all the information necessary, but it is expected that you have charge of the cattle, and you will be responsible for the work to be property done.

When there is any killing done, I want the hide taken off carefully and salted so the entire hide has salt. If any cattle are found on the range before the hair slips off, I want them skinned and the carcass burned up. If the meat is fresh, some it can be cooked for the hogs. If they are too far-gone, and the hair skips on the hide, I want them burned up right where they are found using cow chips. They must be burned up clean and not a particle left. Jesus and the man that rides the fences and looks after dead cattle know how that should be done, but you will see that it is done properly. This is a matter I want strictly taken care of, and when there are enough hides to make it worthwhile to send them to Santa Rita, have it done. This applies also to the sheepskins. The skin of the head of the sheep wants to be left on the pelt, and the head wants to be split from the skull down and taken to the kitchen to be cooked. Every part of the animal wants to be used.

If the bulls are not distributed uniformly, have it so they will be. You will see that the man that rides the range and looks after the fences is up early and late and attends to his work. On my last visit, I found an animal bogged down in a partly dried up pond, and it had been there for some time.

You will try to get posted, so you can manage the cattle with making any comment necessary. Look around and see if any part of the ranch is overstocked.

Land which will be overflowed, and on which another crop of grass can be produced, wants to be eaten up clean. As to land that will not be overflowed, the feed that is there now wants to be saved for fall and winter.

You will notice how many calves are fit for veal. The calves that are to be shipped are principally heifer calves. A small percentage of the heifer calves are to be left to be branded and marked, but the majority are to go for veal, and need neither branding nor marking. Only a small percentage of steer calves are to be shipped for veal, those belonging to very old or blemished cows. They can be shipped for veal so as to favor the cows.

While working the cattle, if you find cows suckling calves, which should be taken off, you will have them taken off and shut up for a while, and put some place where they will not likely get back to the cows. The cows, from which the calves are shipped, you will put in a small corral for a couple of days, then put them in the small pasture, and after that turn them out where they were before the calves were taken off. If there are any cows, from which the calves are taken, and the cows are old, poor or blemished, and we want to get rid of them, you will put them somewhere with the same class of cows on the best of feed, so they can become beef.

When you ship calves, you do not want to ship them undersized unless the cow is poor and blemished, and then the size of the calf cuts not figure, but will be shipped for the purpose of favoring the cow, otherwise the calves all want to be sizeable. You will keep posted as to the number of calves ready for veal which can be called upon.

I also want to know if, in your judgment, any portion of the ranch is overstocked, or is some cases there is not stock enough.

I would like to hear from you as to how things are progressing. As I may not be able to come there for several days, a prompt answer is requested.

Yours truly
Henry Miller

Miller & Lux Incorporated
Merchants Exchange Building
San Francisco, June 2, 1909
Mr. P. H. Turner
Firebaugh, Cal.
Dear Sir:
Mr. Bird, Superintendent of the Chowchilla Ranch, wrote me that there were two horses at Columbia, which were reported as being there by our man Dickenson. Have you seen any such horses? Ask Jesus Higuera and others if they have noticed them, as I am under the impression that Mr. Dickenson only imagined the horses were there. If they were there before he left, they should be there now. I shall write to Dickenson to learn where the horses were, and how long they have been there.

As to the work on the ranch, I hope you have everything straightened out, the calves all branded, and everything in order so as to make the men otherwise useful.

If there is much meat wanted for the camps, you will kill such cattle as are not salable, at the same time they must give satisfaction. I want everything used, even the head, if the animal is not blemished, and occasionally kill some sheep. I shall send over some goats, and occasionally one or two can be killed. I want the skin of the goats tacked up head upward, the same as the skin of a deer, its full width, and when they are dry pile them away. In that way they are more salable than letting them dry like sheepskin.

If some wild hogs are driven out of the thickets by high water, try and have them caught, and put them in the yard at home, or set a trap. There are traps all the way along the river. No doubt you know how to set them.

I want the hounds taken out to kill coyotes by Jesus Higuera, and if he is not able, let somebody else go with them.

Yours truly
Henry Miller

It was about this point in time, 1909-1910, that J. LeRoy Nickel and his sons started to take over the firm of Miller & Lux and Henry Miller became more removed from the day-to-day operations of his vast empire. Miller did not have a great deal of faith in Nickel, and was unwilling to trust his judgment. Miller continued his letter writing, just as he had done for the best part of 60 years.

Mr. J. LeRoy Nickel
Vice President Miller & Lux Incorporated
Dear Sir:

In reference to our conversation on the condition of the country sheep pelts, I will give you the causes of the losses on slats.

First, in drying skins, I find that some of them are dried in the hot sun, causing a sun burnt slat. Also. These slats become broken, due to the excessive drying and tramping on them when in this condition.

To remedy this condition, I would suggestion salting the pelts with fine salt. They will be in condition to ship in about two weeks. In order to carry this out successfully, it would require a place to store them not over seventy degrees. If the temperature is very high, they should not be piled, but spread until partly cured. If reasonably cool, can be piled about twenty pelts in pile. Before salting all animal heat must be removed, otherwise wood will slip.

If temperature is too high to successfully salt, I would suggest sprinkling with salt and then start to dry the next day. This will probably give a little more moisture in the pelt than at present and lessen the chance of breaking. I would also suggest that before shipping, the pelts be put in a place containing more or less moisture, and this will soften them, and lessen the danger of breaking in transit.

Second, I note some of our pelts from the ranches are eaten by bugs. This, of course, can be remedied by proper care in the country.

To conclude, I would say that possibly none of the methods outlined may work, due to conditions that I am not familiar with. If such is the cause, I would be pleased to hear from the superintendent, then I may be able to offer some suggestions.

It is very necessary to get lined up, as from thirty to forty percent of the country pelts go into glue, and the remainder are in the condition of those you have seen.

Yours respectfully,
Miller & Lux, Incorporated,
Chas. A. Stewart,
Superintendent's Office

Miller & Lux Incorporated
Firebaugh Office
Firebaugh, Ca. March 12, 1912
Mr. P. H. Turner
Firebaugh, Cal.
Dear Sir:

Beginning with the first day of March, you salary will be raised to $90.00 per month. I don't think it is necessary to urge you to any further endeavors, as your work has always been painstaking and satisfactory to the company, but of course, you will realize that any assistance you can give in looking after the farming of the East Side Division will be much appreciated. The more I can depend upon some man to look after these matters and the more responsibility that I can throw upon you, the more time I will be able to give to outside matters.

Yours truly,
George Nickel

Miller & Lux Incorporated
Merchants Exchange Building
San Francisco, March 16, 1912
Mr. P. H. Turner
Firebaugh, Cal.
Dear Sir:

Regarding the bulls, you will gather up 30 or 35 head all serviceable two year olds or nearly so. Some may be a little over two years old. The color cuts no figure, but they must be sound and in good condition and fit for service; otherwise we will not get the good of them this year. You will send them to Santa Rita where Raven will put some more with them, and from there they will go to Los Banos for shipment. This time of year when all the balance of stock is in fine condition and bulls are not, it shows they are not thrifty, and for that reason you will not take any that do not seem thrifty and in good health. I want this done at once.

Make every one one of your men ride not less than two colts. Have a young team to drive, and get them gentle so they will be safe. Take no chances driving a team which is only partly broken. See that the mangers are cleaned every day and what is in them thrown out where the stock can pick it over.

You will keep enough hogs to eat the offal clean, and if there is not enough offal, make it up with grain. We bought some grain at a fire. You will find out where it is and send for a load, and when it comes have it scattered out where it will dry and not likely get moldy. Hogs do not do good on grain which they do not relish.

You will kill sheep and goats, goats principally, if they are in condition. Once in a while kill a big fat hog for lard. The Chinaman should get lard enough from the skimmage of the meat that is cooked.

I hope that your wife sees that there is no waste of meat or other provisions in the kitchen. Somebody should wait on the table, or have the meat carved in small pieces and the bread cut in small pieces. That makes quite a saving. There should be plenty of poultry, and on Sunday you can have some chicken, also some eggs and bacon or ham. The victuals should be well cooked and well served, and I want plenty of dried fruit and vegetables used.

The woodhouse should be constantly filled with dry wood so the Chinaman will always have plenty of it.

Yours truly,
Henry Miller

Miller & Lux Incorporated
Merchants Exchange Building
San Francisco, June 23, 1913
Mr. P. H. Turner
Firebaugh, California
Dear Sir:

Yours of the 18th inst. is received.

It seems to me you lay too much stress on irresponsible talk and take too seriously small matters. You should be able to discriminate between the talk emanating from a man like Schmiedte, and men of more character and standing. Moreover there was no occasion to write me again concerning the subject of Schmiedte's discharging you as we discussed that when I saw you and I told you there was nothing in Schmiedte's talk. I have known all about Schmiedte's character and habits for some time but it is not always practicable to make changes as promptly as desirable. Mr. Coe will be in charge of the East Side Division hereafter under Mr. Ogle's general supervisions.

I do not wish a family kept at the Monte. The work of looking after the pumps, building fences, etc. can be done by a man stopping at Bonita Farm house, and thus avoid the necessity of keeping up and establishment at the Monte. I trust you will work in harmony with Mr. Coe and not pay to much attention to idle and irresponsible talk should there be any more of it from any source.

Yours truly,
Henry Miller

Miller & Lux Incorporated
Kern County Office
Buttonwillow, Ca. June 27, 1913
Mr. P. H. Turner
Firebaugh, Cal.
Dear Sir:

Mr. Schmeidte has severed his connection with the company and Mr. Coe, now at Bonita Farm, will be in charge of the work from Ford Farm to the East Side, inclusive, with the exception of Bonita Farm, covering, in fact, all of the work covered by Mr. Schmiedte with the exception of Bonita Farm.

You will therefore, kindly co-operate with Mr. Coe and give him all assistance within your power, in order to protect and further Miller & Lux, Inc., interests to the greatest possible extent.

No family is to be kept at the Monte from now on, but the caretaker now there is to sleep in the buildings and protect them from theft or fire. All such work as fence building, pumping, or similar work not covered by this caretaker will be done from Bonita Farm. Mr. Fuller, the new foreman there, has instructions to carry out your wishes along these lines, in the way in which you suggest that they should be done.

Yours very truly
R. Adams –b.g.

The last letter:

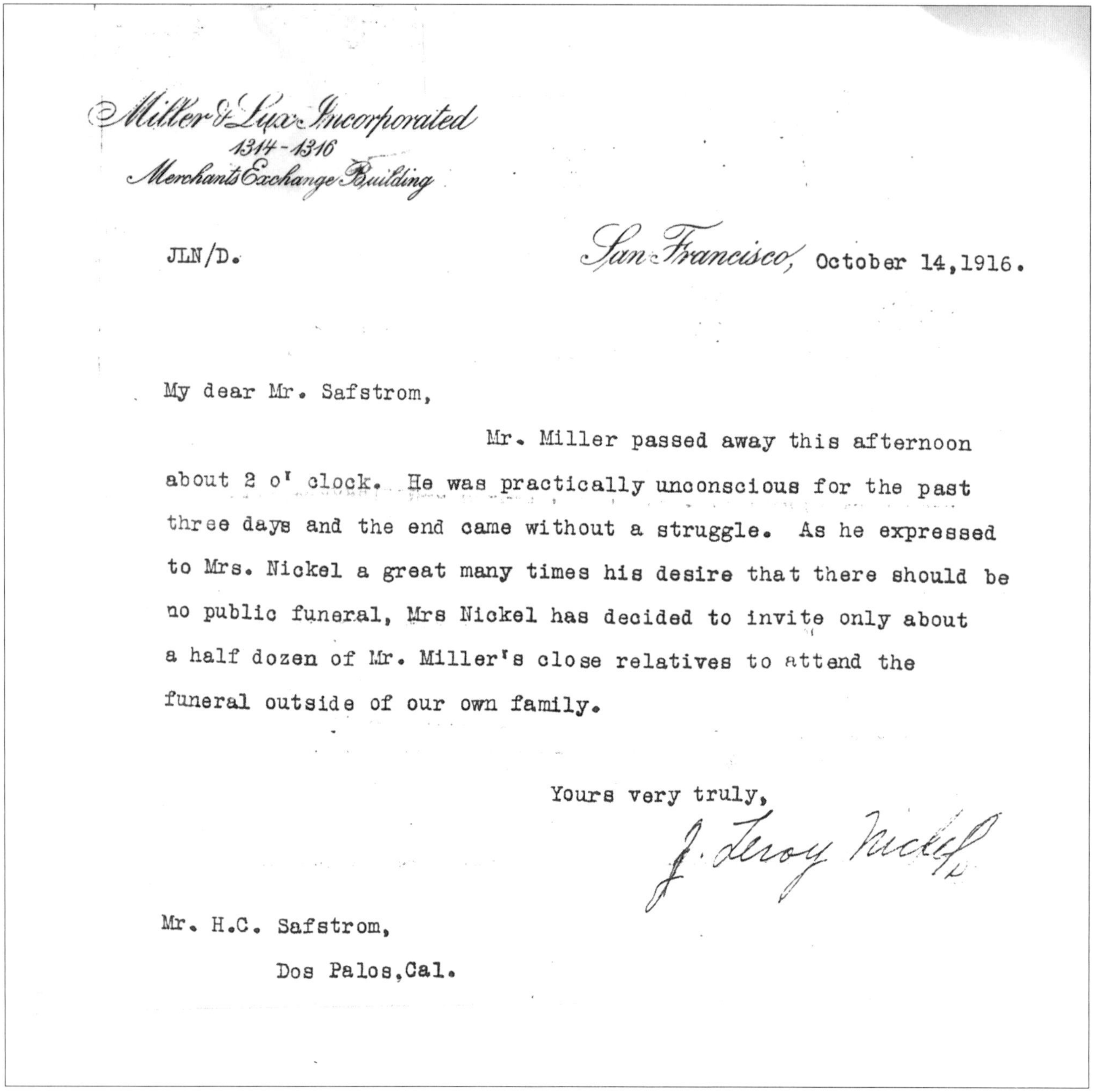
Miller & Lux Incorporated
1314-1316
Merchants Exchange Building

JLN/D.

San Francisco, October 14,1916.

My dear Mr. Safstrom,

Mr. Miller passed away this afternoon about 2 o' clock. He was practically unconscious for the past three days and the end came without a struggle. As he expressed to Mrs. Nickel a great many times his desire that there should be no public funeral, Mrs Nickel has decided to invite only about a half dozen of Mr. Miller's close relatives to attend the funeral outside of our own family.

Yours very truly,

J. Leroy Nickel

Mr. H.C. Safstrom,

Dos Palos, Cal.

Chapter Fourteen

Newspaper Articles & Clippings - 1890-1916

3/5/1892 Merced Express

Our streets are being graveled in a fine shape. Mr. Henry Miller has put quite a number of his teams at the work and it will be soon completed.

3/19/1892 Merced Express

Mr. H. G. Tanner has been superintending the planting of trees through and round the cemetery this week at Mr. Miller's order.

4/9/1892 Merced Express

Henry Miller has had the road up to the warehouse repaired and it will be in good shape this summer.

4/9/1892 Merced Express

Henry Miller is grading the streets of Los Banos. He has engaged 125 carloads of gravel and has quite a force of men at work. H. G. Tanner has had a force of men graveling H Street in front of the depot this week. A dozen carloads of gravel have been used.

4/16/1892 Merced Express

The force of men which Mr. Miller has employed grading streets are doing their work well. The streets are left in fine condition.

4/23/1892 Merced Express

H. G. Tanner has taken steps to protect the shade trees in town, by having boxes erected around them.

1/14/1893 Merced Express

It is reported that Miller & Lux sold, this week, two hundred thousand acres of reclaimed swampland in Kern County to British capitalists, for colony purposes. Eventually the San Joaquin Valley will be settled as thick as in Aztec times.

2/4/1893 Merced Express

The West Side is making a lively fight for the organization of the new county of Santa Rita, and if there is no opposition, they may get it.

2/11/1893 Merced Express

Henry Miller is very anxious that the citizens of this place organize a fire company. Mr. Miller knows that fire is liable to break out here at any time and the destruction of the property that would follow without a well-equipped fire company. Mr. Miller will help the project liberally, and it is certain that a strong fire company will be organized. Delays are dangerous and we hope for the sake of our beautiful town that immediate action will be taken.

2/30/1893 Merced Express

Henry Miller brought suit last Wednesday against Tom and John Tucker for hunting and trespassing upon his lands. Court opened at 10a.m., Justice S. A. Smith presiding. The people were represented by Jas. Peck, and Tom and John Tucker pleaded their own case. The following jurymen were sworn: H. P. Greenlaw, George Johnson, W. A. Pemberton, George Soper, J. J. Sullivan, C. Reutter, Ed Gaynor, Ed. Hoffman, T. Schottky, F. Wilcox, F. Wilkins. The case was given to the jury at 12 o'clock. At 3:30 they sent for the Judge and informed him that they had taken seven ballots and could not agree. The Judge informed them that it was very desirable to get an agreement, and gave them one hour more. The jury was again locked up, but could not come to an agreement.

3/18/1893 Hollister Alliance

We are informed that Henry Miller has offered the $50,000 bonus and the right-of-way if the Monterey-Fresno Railroad Company will run their road through Pacheco Pass. Commenting on this the Gilroy Advocate says: "Gilroy and the whole of this rich southern part of the county of Santa Clara has reason to hope so. The future advantage to land owners interested would be incalculable. Can they allow such a chance to slip though their fingers?"

3/18/1893 Merced Express

We have lately been informed that Los Banos is to have a park given by the ever-liberal Henry Miller. The site is to be on the north side of the track, and preparations for planting have already begun.

3/25/1893 Merced Express

Regardless of the rain this last week, men had been planting the trees in our new park, north of the depot and have made remarkable headway under the circumstances.

4/8/1893 Merced Express

Los Banos will soon boast one of the finest parks in the county. Henry Miller has laid out a tract of twenty acres a short distance northeast of the town which has been set apart for the park. Joseph Hogarth, of San Jose, an experienced nurseryman has the work in hand and the grove is now one of the most attractive places on the West Side. There are two avenues, the one running north and south, is planted to poplars, and the one running east and west, to black walnuts. There is a spacious walk running through that is shaded with cork elms. The park contains a great variety of trees, such as English Ewe, European laurel, Magnolia, Norwegian Spruce, Date Palms, Redwood, Orange trees etc.

H.G. Tanner keeps twenty men and ten teams busy cultivating the grounds. Pure water will be piped through the grounds so that the success of the grove is assured.

5/6/1893 Merced Express

The petition presented to the Board of Supervisors by Miller & Lux and generally signed by the businessmen of this county, asking that the Board change the ordinance in relation to the selling of liquor was denied yesterday afternoon by a vote of 3 to 2. This vote is a good criterion by which to judge the future actions of a majority of this body.

5/27/1893 clipping no paper noted

Henry Miller of the firm of Miller & Lux is going to build another big canal in the West Side country along the foothills of the Coast Range Mountains that will irrigate thousands and thousands of acres of fine land. The survey will be made next September and the canal will be ready for conducting water sometime in 1895. The people of the West Side country will now have a canal for irrigating a section of the country that has heretofore been unproductive, expect in a wet season and real-estate values will materially increase under a good system of irrigation. Mr. Miller is like the Southern Pacific, when he undertakes an enterprise he puts it though.

6/10/1893 clipping

Henry Miller, the wealthy landowner, lately gave another $200. to the public library of Gilroy.

6/25/1893 Merced Express

It is said the heirs of the late Charles Lux have asked that the co-partnership which has existed between Miller & Lux be dissolved, as they are dissatisfied with the management of the firm's business. We are of the opinion that the management of the large estate could not be in better or more competent hands than that of Henry Miller.

7/29/1893 clipping

About $400 has been subscribed for the organization of a fire department. Henry Miller is very much interested in the project and has generously offered to put in fire hydrants wherever they may be considered to be most needed.

9/2/1893 San Francisco Chronicle

Henry Miller, the millionaire partner of the late Charles Lux, has been made the chief defendant in a sensational suit in equity, which was filed yesterday in the Circuit Court. The allegations made against him are of the most serious character.

Miller & Lux 15 mule team harvester, 1890's

He is charged with a deliberate and carefully planned attempt to cheat his partner's widow, and in that attempt to defraud it is said to have played upon the credulity of others, greatly to their financial loss.

The plaintiff in the suit charged that the wealthy cattle raiser and rancher has played a shrewd game by double-dealing, and they want the Untied States Court to catch him in his own trap. In Brief, the plaintiffs ask that Miller be compelled to deed them 150,000 acres of land in Kern, Tulare and San Luis Obispo Counties for $3,000,000., to be paid under conditions which are given in detail. They charge that every arrangement, up to the last act of signing the necessary papers, has been made. They claim that they have fulfilled every obligation, and they have given way on every point, and have spent thousands of dollars to buy the property. They now ask that Miller be compelled to do his share and sell

12/10/1893 Merced Express

Mr. McBride, General Superintendent for Miller & Lux accompanied by Superintendent Torchiana of Los Banos was in town last Saturday.

1/15/1897 Merced Express

The property and business interests of the estate of Miller & Lux are to be put into corporate form with a capital stock of $12,000,000. Of the seven directors Henry Miller will choose four, the German heirs one, and interests of Mrs. Lux one. The seventh director will be chosen by the last named two. Henry Miller is to be president and general manager.

1/15/1897 Merced Express

Mrs. Eliza Chamberlain, a relict of the late A. Chamberlain died at her home, the Chamberlain ranch, on the San Joaquin early Wednesday morning, full of years and virtues. Mrs. Chamberlain came up from the city where she had spent the holidays with her daughter Mrs. Jesse Potter, only a week or so ago, since which time she has not been well, but was not considered seriously ill until within a few days of her death, which resulted from la grippe. Mr. J. F. Chamberlain of this city and Mrs. Jesse Potter of San Francisco, her only children.

1/22/1897 Merced Express

A rabbit drive is to be held on the Miller & Lux property west of Berenda, Sunday, January 31: Jim Ogle came over from Firebaugh Tuesday. He has been absent about six months engaged in superintending the construction of the big West Side canal from White's Bridge north to Newman.

2/19/1897 Merced Express

After a long period of litigation the Miller and Lux estate is to be settled, an arrangement having been made a few days ago between the attorneys for the various parties in interest. The estate is an extensive one, including land in this valley and in other parts of the State.

3/5/1897 Merced Express

A Friend of Henry Miller's came to him recently with a rueful face to say that he had been sued for breach of promise. "O" said Henry Miller, "you'll get out the difficulty without any trouble." "I'm afraid not," his friend replied. "You see she has a lot of letters from me." "My boy," answered Miller, "you should never kiss a lady through an ink bottle."

4/2/1897 Merced Express

There is a rumor going the rounds that on the first of May Henry Miller will give a reunion picnic at the Canal Farm. In the evening of the same day a ball will be given in town. Mr. Miller intends to make this the grandest event that has taken place on the West Side.

4/23/1897 Merced Express

The litigations between Henry Miller and the Lux heirs has been settled by a compromise and a corporation has been formed with a capital of $12,000,000. Henry Miller is manager of the concern and his salary fixed at $25,000.

4/23/1897 Merced Express

Over 90,000 sheep will be sheared at White's Bridge this season. It is estimated that nearly half a million pounds of wood will be shipped from there this year.

4/30/1897 Express

Miller & Lux have been moving three thousand head of cattle this week from the Shippee Ranch to the West Side range on the San Joaquin River. They have been driving them in herds of 1000 each.

5/14/1897 Merced Express

Henry Miller has announced that his new canal, which parallels the present canal but is nearer the hills, will be carried to Tracy at an early date. This will bring much of the land between Newman and Coast Range foothills water.

Undated Dos Palos Star Clipping 1897

Miller & Lux had 500,000 fence posts shipped to Dos Palos the last few days. After hauling them away with several teams each day, there were several large piles to be seen along the railroad yet.

8/13/1897 Merced Express

The German heirs of the late Charles Lux have sued the Miller & Lux Corporation, claiming that Henry Miller, who held their power of attorney, had no authority to enter into the corporation arrangements. The court is asked to declare all past transaction void, and to order the estate be divided according to the will of Mr. Lux.

9/17/1897 Hanford Democrat

Miller & Lux purchased six thousand lambs in Nevada last week, to be driven or shipped to California and kept on their ranches until there is a market for them.

10/1/1897 Merced Express

Los Banos has the finest twenty-acre public park in the State. It was laid out and planted to choice shade and ornamental trees by the order of Henry Miller seven years ago, during which time it has had much care and attention.

11/26/1897 Merced Express

A short time ago Miller & Lux gave Mrs. Ella Theircof her home place on the colony adjoining Los Banos. All will remember that Joseph A. Theircof was drowned a few years ago, while employed by Miller & Lux. He made a trip to Merced for the firm in a buggy and as the country was badly flooded, he lost his road and drove into a slough near Dickinson's Ferry. He left a wife and several children, and all his property consisted of a house and eight-acre colony lot, on which Miller & Lux held the mortgage for $350.00. At the time Henry Miller informed Mrs. Theircof that she would not lose her home, And time passed on until Oct. 6th this year when the mortgage was foreclosed and the place sold. Miller & Lux bought it and deeded it to Mrs. Theircof and now she holds her home free from all indebtedness. The place is a good one and is a comfortable home. Mr. Miller is a man of but few words, but when he says he will do a thing he never fails to keep his agreement, and our readers will be glad to hear of his generous act.

12/10/1897 Merced Express

Henry Miller has contracted with the sugar beet factory at Watsonville for 125 tons of pulp daily for cattle fattening purposes. This is about half the amount of pulp handled from the factory. Pulp is a good flesh maker.

12/31/1897 Express

The people of Los Banos will have a Christmas tree on the evening of the 24th as usual. The fact that Mr. Miller, and other who have taken part in the Christmas tree in the past, are interested insures a success, and a good time for all who attend.

12/31/1897 Dos Palos Star

Henry Miller has remembered the people of this vicinity, this year as usual, with lots of good substantial present for Christmas.

4/1/1898 Los Banos Enterprise

Word from Gilroy says that Henry Miller, Jr., son of the famous cattle king, is seriously ill. Henry Miller, Jr., will inherit more land than any other boy in the state if he survives his father. Young Miller as a student was as plain and democratic as the son of any day laborer.

5/20/1898 Merced Express

Mr. Henry Miller the cattle king was in town last Saturday as the guest of Mr. J. S. Potter at the residence of Mr. & Mrs. J. F. Chamberlain.

8/26/1898 Los Banos Enterprise

During these hot days the dining room and bar room of Hotel Los Banos are the coolest places in town. S. A. Smith went to the city Monday night to confer with Henry Miller about the advisability of starting the creamery.

9/23/1898 Merced Express

Miller & Lux expect to bale 15,000 tons of hay at Los Banos this year, which will be shipped to their various ranches. Miller & Lux have purchased $75,000 worth of tule forage at Stockton for the fall consumption of their stock from the various portions of the state

9/23/1898 Los Banos Enterprise

Henry Miller arrived in Los Banos on the train Thursday evening to look after his vast estate.

12/30/1898 Merced Express

During the past week Henry Miller and Mr. and Mrs. H. A. W. Torchiana have had several wagons out delivering Christmas presents to various families in this vicinity. Presents consisted of flour, potatoes, tea, coffee, bacon, beans, wood and all kinds of provisions. They also distributed a large wagonload of turkeys among the people. In addition to this they will place presents on the tree at Miller & Lux hall. We should judge that their presents would aggregate in value several hundred dollars. Their kindness is appreciated by all concerned.

2/10/1899 Merced Express

Henry Miller the large land and cattle owner was in town on Saturday last. While here he executed a lease to John Freer for his ranch north of Lemoore for a term of two years. One of our prominent citizens asked him how many cattle he had, which question elicited the following reply: "Don't know, can't count'em.

3/24/1899 Fresno Democrat

A horse deal was consummated at Sanger a few days ago the like of which has never before been know to taken place in this county. Last Sunday Frank Enau Brothers merchants and stock raisers, delivered to Henry Miller of Miller & Lux, 160 horses the reported price the incredibly small sum of $3.00 per head. When communicated with, one of the messrs. Frank Enau declared that the horses were of good breed and in fair condition. The reason he gave for selling them was the scarcity of feed, which render such animals too costly to keep. And what can Miller & Lux do with the herd? Comes the question. They can butcher the plugs, boil them in a large cauldron kept for that purpose, and feed them to the omnivorous hogs. That, according to report is just what the cattlemen are going to do with the 160 horses they have purchased for $3.00 apiece. Hogs are not particular what portion of the animal they are given so they will be fed. Not only will they devour the choice steaks, roast, etc. of the noble animal, but they will consume his entire carcass, except, perhaps, a few bones that cannot be digested.

4/21/1899 Express

A good story is told on Henry Miller, the millionaire cattleman of California. He recently chanced to be on one of his numerous cattle ranches and found a coyote bearing the Brand H-H, which is Mr. Miller's well-known brand. Ever since that time he has been trying to ascertain who the guilty party is that had the audacity to use one of his branding irons for that purpose. In case he should find out there will be some man out of a job, as Mr. Miller does not appreciate the joke in the least.

4/28/1899 Express

Henry Miller was on the West Side on Wednesday and Thursday visiting many of his ranches and looking after business.

7/21/1899 Merced Express

Simon & Harrell are preparing an abstract of title to all the lands belong to Miller & Lux in Merced County. That corporation owns something over 230,00 acres in this county: in addition to town lots, and the abstract will make quite a library in itself. It will take four persons four months to complete the work as many of the records are in Spanish.

12/15/1899 Los Banos Enterprise

Miller & Lux have shipped thousands of cattle into the West Side ranches during the past few weeks to take advantage of the plentiful feed in that section of the county.

1/13/1900 Merced Sun

Monday morning January 8th Martin Miles Smith died at the residence of his son C. W. Smith, at Badger Flat near Los Banos.

1/26/1900 Merced Evening Sun

Henry Miller the cattle king was thrown from his buggy yesterday at Gilroy sustaining concussions of the brain. He was being driven to the depot to catch a train for San Francisco when the team frightened at a passing train and ran about two blocks when Mr. Miller and the driver were thrown out. The millionaire was unconscious for several hours and then revived sufficiently to be taken to his San Francisco home for treatment. Owing to his advanced age his injury may prove very serious. LATER: This afternoon Sheriff Warfield received a telegram from the city stating that Mr. Miller's injuries are not serious and he will be out again in a few days.

3/9/1900 Merced Sun

M. P. Knudsen, of Los Banos late of Newman has erected a building at Los Banos on Miller & Lux lot for his own occupancy for saloon purposes and will put up a two story building adjoining.

3/16/1900 Merced Sun

Henry Miller is having his share of accidents. Recently he was thrown from a buggy at Gilroy and last week while visiting one of his ranches near Los Banos he was precipitated into a ditch by the breaking of a plank he was attempting to cross. Two his ribs were broken by the fall.

3/30/1900 Los Banos Enterprise

Henry Miller has donated to the West Side Union High school a strip 150 feet deep on K Street, extending from Sixth to Seventh, a distance of 575 feet. The piece of property is worth $1500. Mr. Miller had previously sold the property to the High School at half its value, but when outside parties insisted on Los Banos giving a free site, he refunded the money paid to him.

3/30/1900 Merced Express

Miller & Lux's pay roll will average about $5,000 a month and that of the Los Banos Creamery from $2,000-$3,000 a month. This puts a great deal of money in circulation.

Miller and Lux have a number of men at work placing water mains from the water works to their warehouses to protect them against fire.

6/1/1900 Merced Express

Sunday morning Henry Miller informed us that he wanted the local band boys to play at the May Day picnic and that he would give them $100, towards paying for new uniforms. Mr. Miller always stands ready to put up his money for anything that will be a benefit to Los Banos. He spends several thousands dollars a year to advance the interest of the town.

Mr. Henry Miller will give his annual picnic and barbeque at Los Banos Farm grove on Tuesday, May lst. The brass Band will be in attendance all day and furnish fine music. There will be as usual plenty of free meat, bread and coffee. Prizes will be offered for races, games etc., for young and old, fat and lean, the greased pig will make its appearance, as usual and everybody will have a good time

11/23/1900 Los Banos Enterprise

Bachle and Gaynor have a contract for building a neat two-story cottage for Miller and Lux at Canal Farm. Work has been commenced.

2/01/1902 Los Banos Enterprise

Henry Miller wishes, through the columns of this Enterprise, to express his sincere gratitude to the people of Dos Palos and vicinity, who on New Years presented him with a handsome silver cup, as a token of their esteem. Mr. Miller is deeply impressed with their kindness, and says that the gift is more highly appreciated by him, because it was presented by so many people in the various parts of the community.

Henry Miller has decided to place the Plow Camp and Cotton field land on the market in twenty and forty acre lots at the price of $65.00 per acre. The first part will be for sale in a week or so. The land for sale is along the road running from the main county road toward the Sam Smith ranch; and new road will be run parallel with that a quarter of a mile further south. There have been many applications for colony lots in the plow camp section; it will be sold as soon as the Miller & Lux can get it surveyed off. As soon as this is disposed of the balance will be mapped off and sold.

May 10, 1902 Merced Express

The attendance at Henry Miller's picnic and barbeque Thursday was very good, something like three thousand people being present. Most of the local people turned out, and large numbers came from the surrounding towns. Over 100 came down from Dos Palos and fully as many from Cottonwood and Volta, while Merced, Newman, Fresno, Firebaugh, San Francisco and other towns were well represented.

July 23, 1902 Los Banos Enterprise

Henry Miller is planning to put in an electric plant at the falls where the New Canal enters the Los Banos Creek to supply the town with electricity, and lumber is now being hauled for a flouring mill to be built at the same point.

The Miller & Lux monthly payroll calls for $8,000. The creamery pays out to the farmers about $3,000 and the railroad about $1,500. This neat sum is distributed and spent in this little community of 1500 people. Why shouldn't Los Banos boom?

9/27/1902 Merced Express

Died: Ward, In San Francisco Sept. 20, 1902. Thomas E. Ward Superintendent of the Santa Rita Ranch, Merced County, age 38 years.

The sad news of the death of Thomas E. Ward, Superintendent of the Santa Rita Ranch, was received by his Merced friends with regret. Mr. Ward died in San Francisco from typhoid fever, which he had contracted at the ranch several weeks ago. He had been superindent for Miller & Lux for a number of years and was held in high esteem by that firm, especially by Mr. Miller, as he had raised him on the Santa Rita ranch from a boy.

Mr. Ward was a young man 38 and a courteous and affable gentlemen and his death was an untimely one. His word in business transactions was as good as his bond, and everyone who had any business with him recognized as being as good as Henry Miller himself on all business lines. Mr. Miller endorsed all of Mr. Ward's business transactions and honored his checks and in his death he loses a faithful and valuable man, one whose place it will be hard to fill. Mr. Ward was a native of California, and leaves a wife and four children and several brothers and sisters.

February 14, 1936
Los Banos Enterprise
Joe Place

Joseph Elmer Place passed away at six o'clock last Saturday morning at the home of his son in San Francisco. He had been bedfast since last May with an incurable ailment. The remains were brought to Los Banos that day by J. A. Whitehurst. Mr. Place has been identified with the building of this community for 51 years. Native of Rhode Island he was 73 years, 4 months and 20 days. He came to this section when he was 22 years old and for several years was a bookkeeper for his Uncle Henry Miller of Miller & Lux. His mother was Mrs. Miller's sister. For many years following he was engaged in the general merchandise business in South Dos Palos and Los Banos. He married Miss Zena Clifton at Old Gilroy and is survived by her and four children.

April 1936 Los Banos Enterprise

Beloved Pioneer Passes on: Robert R Dalzell 68 a Los Banos rancher deputy peace officer and story teller passed on to his reward Friday night April 17th. He succumbed to a hemorrhage of the brain, which came upon him April 1, at his home. Bob together with his wife came to Los Banos in 1906, where he had charge of the Miller & Lux racehorses and breeding stock. He continued in this capacity for twelve years, and then purchased a stock ranch in the foothills. Mrs. Dalzell passed on in 1922. For the past twenty years he has served almost continuously as a deputy peace officer, being deputy sheriff, deputy constable and deputy city marshal. He was always interested in the Odd Fellows Lodge and Encampment of which he was Past Chief Worthy Patriarch. He is survived by 2 children, a brother and sister living in New York, and three grandchildren

8/8/1940 Merced Sun Star

H. A. W. Torchiana In Sunday's San Francisco Chronicle appeared the following obituary. Last rights for Henry Albert van Coenen Torchiana Netherland Consul General at San Francisco for 26 years will be held at 11 a.m. Monday at N. Gray & Co.

Torchiana, 72 died Friday at Stanford hospital after an illness of several weeks. He was a native of Java and became an American Citizen in 1895. He was a Republican, a member of the Dutch Reformed Church, Family and Olympic clubs as well as the Pastiempo Golf Club of Santa Cruz. His widow Catherine survives. His home was in San Francisco; the summer place was Villa Fairview in Santa Cruz. Following cremation the ashes will rest in San Francisco until they can be returned to Holland. Torchiana left Merced to take the management of the Miller & Lux ranch with headquarters at Canal Farm near Los Banos. His experiences of several years in Miller's employ were woven into a book, "California Gringos," published in 1930. In later life he became a well-known clubman and his home in Santa Cruz is one of the showplaces of the beach city.

1/15/1950 Fresno Bee

Canals of Henry Miller May become Water System of New Irrigation Unit

Los Banos: The Central Valley Project of which Henry Miller never dreamed may result in the early day San Joaquin Valley cattle baron's pioneer irrigation system now a 2 million dollar enterprise becoming the valley's newest irrigation district. Data is being assembled by officials of the Central California Mutual Water District as a preliminary to bring to a vote the question of organizing the territory served by the Miller System now the San Joaquin Canal Company as an irrigation district. If organized a board of directors would be e elected which could negotiate for the purchase of the canal company or file suit to condemn it. The canal company was constructed entirely during the lifetime of the fabulous Miller. The completion of the first canal was the occasion for the first May Day in Los Banos celebration given by Miller at Canal Farm here in 1878. Miller's custom is being continued by the present annual Los Banos celebrations famous throughout California. It was not Miller however, but others who in 1871 formed the San Joaquin and Kings River Irrigation Company to connect the San Joaquin River and Tulare Lake. Miller, never one to let slip an opportunity to turn a fast dollar, let it be known he had riparian rights and granted the canal company rights of way and subsidy for water at half rates. When mounting demand for public ownership of the canal and the railroad opposition finally forced the original promoters to seek an out, Miller took the entire system off their hands for a third of its original cost, reportedly $1,250,000.

That was just a beginning for a man whose empire at one time consisted of 22,217 square miles of land in California, Nevada, Oregon and Arizona, more than 100,000 head of cattle, $30,000,000 in hotels, shops, banks and slaughterhouses. "Wealth became his because he visualized the value of irrigation water and its sources, the value of uninterrupted feed supplies for his stock and the growing demand for meat," stated one contemporary. The first dam across the San Joaquin was brush and interlaced willow tree affairs at Mendota, and was part of the original San Joaquin River canal system. Some 78 miles were dug in that early hectic project. Today the system has 71 miles on the main canal, 56 miles on the outside canal completed in 1896, and innumerable laterals. It supplies water to 152,000 acres. The San Joaquin Canal Company as it now stands was completed before Miller's death and the main portion was completed prior to 1880.

Los Banos Opera House, 1905

Bank of Los Banos owned by Miller & Lux, 1905

Los Banos before the earthquake of 1906. The far building is the Los Banos Hotel; the one next to it is the Opera House.

Opera House destroyed by the great quake of 1906

Chapter Fifteen

Interviews Covering 1890-1956

Wallace Smith, professor and author

Wallace Smith worked for the Miller and Lux outfit as a young man. He is best known today for his history of the Great Central Valley, *The Garden of the Sun*. He also later wrote *Prodigal Sons*, dealing with the outlaws Evans and Sontag. Smith was raised in the valley and later became a college professor at Fresno State College.

Interview at Milliken Home August 1st and 2nd, 1940.

Once or twice a month Miller & Lux used to bring in cattle on riverboats and land them at the Lombard Street Wharf. From this wharf the cattle would be driven to Butchertown. Butchertown was about three miles south from Third and Market. It was necessary that these cattle be driven across Market Street by six o'clock in the morning. Miller & Lux used to drive hogs this way also.

On the morning of the Great Earthquake in San Francisco three Miller & Lux vaqueros were driving a bunch of cattle from the Lombard Street wharf to Butchertown. The names of the vaqueros were Tommy Mora, his brother Santos Mora and Dutch Chism. Just as the cattle were going down Howard Street there was the rumble of the earthquake. The horse of one of the vaqueros began prancing. The vaquero put his hand up to his face. When he looked again all the cattle were gone. When he looked more carefully he found that a brick building had fallen on the cattle and killed them all. The dead cattle were all covered up with fallen brick. At this time the horse Tommy Mora was riding jumped into the basement where the brick building had stood. The horse was out of its head with fright. It was running around and around crazy in the basement. The vaqueros shoved a 6 by 6 timber down into the basement to go down on. The horse looked at the plank in astonishment. Then ran right up the plank and out onto the street.

At the Miller & Lux slaughterhouse in Butchertown they killed from 100 to 150 beef a day. The slaughterhouse was three stories high. The beef were killed on the top floor. Sheep killing was done on the second floor. From 200 to 300 head a day were killed. On the first floor from fifty to 100 hogs a day were butchered.

Butchertown was built on a long wharf built on piles over tidewater. It was just like a street built up on piles. The street was called First Avenue. The various slaughterhouses were on one side of the street. The stables were on the other side. The pigpens were built under the slaughterhouses where the hogs could get all the waste dumped down to them from the slaughterhouses. Besides the slaughterhouses and barns along the street there were saloons and boarding houses.

At the time of the earthquake the hogs' pens were floating in the water, the hotels and barns were also floating around. Men in rowboats were rowing around trying to round up the cattle in the water. Cattle were to be seen in every direction.

Wallace Smith used to be Henry Miller's driver and often drove him around in his buggy. Miller & Lux used to keep six buggies at the Butchertown stables. These were for the use of the head men of the company. Six boys about eleven or twelve years old including Wallace Smith were hired at one dollar a day to drive the buggies. Miller used to tell the boys to save their money. He used to give one-dollar tips. The boy who drove for Dillon would be told by Dillon, "Meet me at my house at Haight and Fillmore." Henry Miller got in his buggy at his home. Nickel would meet his buggy at Third and Townsend Depot. The collector also had a buggy and driver.

Molton was superintendent of the barn. The six buggies and at least two horses apiece were kept at the Butchertown barn. Every morning before starting out the buggies were all washed and the harness blackened. The six boys employed as drivers got one dollar per day. The best driving horses from the Miller & Lux ranches came from Los Banos and Santa Rita and were sent up for use in the city. It was an awful big barn at Butchertown. All butcher teams were kept there.

At the horse parades in San Francisco in the contests for matched teams Miller's teams took lots of prizes. His four horse butcher teams of blacks, grays, bays and sorrel, were beautiful teams. The butcher teams were divided up on the hog route, the sheep wagon and the beef wagon. These butcher wagons would hold twenty-two or twenty-three beef cattle on them. The horses had to pull these wagons on the dead trot over the cobble stone streets for four hours at a stretch. A man name Dipple had a patent on these wagons. The front wheels would pass entirely under the front of the wagon.

These butcher wagons would leave Butchertown any time from two o'clock in the morning. By three or four o'clock they would be in the city. They hauled meat to the retail butcher shops. The teamster was required to put the beef on the butcher's shoulders. But the butcher had to pack the meat from the wagon into his shop.

The horses would get cobble sore. It was the duty of the horse boss to come down to the ranches and pick out teams to be shipped up to the city. The horses would be shipped up along with cattle. A partition would be put in the cattle car to keep the cattle and horses apart. Saddle horses and teams were shipped this way. In bringing the cobble sore horses back from the city to the ranches the horse boss would drive them to Mountain View. Then to San Jose. Then to the Bloomfield. Sometimes he would bring them to the Santa Rita. The best hay and barley was sent up from the ranches to feed these horses at Butchertown. A man would come down and select the best fields of hay and have the crop baled for the Butchertown stables. Barley for the horses feed also came from Miller's own ranches for these horses.

Antone Lopes and his brother, Jesus, were Mexican vaqueros and came originally from Mexico. Miller was down buying cattle down toward the Mexican border. The Lopes brothers came up with Miller. That's how they got here. Pachuco also came up. Antone Lopes became beef man at Oxalis. Jesus became the same thing at Firebaugh. Antone Lopes could write nothing and I kept his time book for him (Wallace Smith). Rafael Quinn was the same type of man and tended to the shipping of beef cattle from Buttonwillow.

It required just the right kind of knowledge and judgment to know when a beef steer had made all the gain it would ever make and should be shipped to Butchertown along with the fat cattle. "Look at that steer," Lopes would say to me, pointing out a rough steer. "It aint going to get any fatter." A couple of weeks later the animal was still not getting any fatter. "Fat's as he will ever get," Lopes would tell me and away the steer would go along with the other cattle to Butchertown.

Oxalis Ranch was the former Hog Camp. It was only a mile from the ranch to the shipping corral on the railroad. Orders would come from the city office how many cattle and hogs to ship to Butchertown. Cattle were shipped every Tuesday and Friday. A carload of hogs would also be sent along with the cattle.

Henry Miller bought the HH iron along the San Joaquin River between the Santa Rita Bridge (weir) and the Turner Ranch Bridge, between Weir Camp and Santa Rita there is the tallest tree on the river. It stands all-alone. It is called the Palo Alto tree and it stands pretty close to the Turner Ranch. There is the remains of an old ranch there. There used to be a big ranch in there. "That's the ranch," Lopes told me where Miller bought the HH brand. He bought all the cattle branded HH. Instead of going to the bother of venting them he bought the iron, too.

When Lopes died he was buried at Los Banos. There was a big vaquero funeral for him. Vaqueros came on horseback from everywhere. There were fifty or sixty of them all on horseback. Even Jesus, his brother from Firebaugh, was there on horseback. Neither Antone nor Jesus knew how old they were. The vaqueros all on horseback took off their hats as the casket was lowered into the grave. That night all the vaqueros before returning to their ranches all got drunk to celebrate the funeral.

Henry Miller, Hans Raven and Antone Lopes always talked Spanish to each other. Pachuco came up with Miller and Antone and Jesus Lopes from the southern part of the state. Pachuco never got to be a foreman and in his old days Miller pensioned him off at New Columbia.

Miller & Lux buckaroos. Wallace Smith is first on left, 1907

Pachuco became the coyote man. He rode around on horseback followed by ten or a dozen hounds. He had these dogs all named after famous prizefighters. A big black one—his fastest dog—he called Jack Johnson. A red, rough dog was John L. Sullivan. Another was Jim Corbett. Pachuco talked to his dogs in Spanish. For weapons he carried a pick handle with a buckskin strap slung over his saddle horn like a quirt. One dog alone won't fight a coyote.

When Pachuco and his band of dogs would sight a coyote the chase was on. Pachuco on his prancing horse would follow after the fleeing dogs. The foremost dog would soon overtake the coyote. A nip on the coyote's tail and the coyote would slow up. By the time others in the pack would catch up all would tear into the coyote.

Pachuco would come riding up. If the coyote was getting the best of the dogs Pachuco would reach over with his pick handle and lay the coyote out. But if the dogs were getting the best of the coyote Pachuco would call up old John L. Sullivan who couldn't keep up with the leaders of the pack. "Here, old Sullivan," he would tell the old dog, "you finish him up." Sullivan would growl. His hair would brace up and the old dog would get his licks in. Pachuco did this so Sullivan wouldn't fail to come.

There was a bounty on coyotes. Pachuco's method of skinning coyotes was first to slit the coyote's throat and start the skinning. Then he would slip the coyote over a fence post, forcing the post right up into the coyote's neck. Then he would fasten his rope to the coyote's nose. Wrapping his rope around the saddle horn he would start his horse going and yank the skin right off the coyote.

Over the bridge along the road from Firebaugh to Madera for a distance of fourteen or fifteen miles there use to be a coyote on every post, the work of Pachuco and his prize fighter dogs.

Weaning time was when all the ten month old calves would be brought in. They would continue to suck their mothers. The calves would be locked up for nine days. The cows would be locked up in another corral. Hay would be hauled to them for feed. The calves would be starved for nine days until they learned to lick green grass. Then the cows would be switched to the Los Banos Division and the calves to the Firebaugh Division. Branding time for the calves was in the springtime. Any calves that were missed this year would be caught next year. Hans Raven was the cattleman. Red Safstrom ran the land part of Santa Rita. Did the general farming. Had nothing to do with the cattle. He had his own hay crew. Hans Raven got his start with Miller driving a vaquero wagon.

When rounding up weanlings the vaqueros would also pick up all feeder cattle. These feeders were put all off in a pasture by themselves. When order would come to bring feeders they would be brought to the feeding corrals. The feeding corrals were at Temple Ranch under Ed Turner; the River Ranch under Bader; The Holland Farm under Dan Leonard, Poso Ranch under Ogle. Awful big feed sheds were located here. On the Santa Rita range the vaqueros were kept busy branding, weaning and collecting these feeders. At the Santa Rita there were 15 or 18 vaqueros. They took care of all range work. At Oxalis there was a foreman and seven or eight men. Antone Lopes was the beef picker for all. He was an expert on fat beef cattle. This crew picked the fat cattle and shipped them. The road outfit of Antone Lopes consisted of a wagon to carry the vaqueros' blankets and to do the cooking in. As a rule the cook drove the wagon. Lopes hired Raven just to drive. Raven later became cattle superintendent.

Antone Lopes told me "anytime you come I give you a job." He put me to work and I stayed six years. Miller brought you along. He built up his employees. I was a helper at Santa Rita breaking colts. The next year I was in charge.

Miller's system was stall feed beef. The feeders were fed chopped hay all summer. Miller & Lux raised barley and rolled it. This was sprinkled on the

Miller & Lux cattle killed on Market Street by earthquake, 1906 - *San Francisco Public Library archives*

Butchertown with the Miller & Lux slaughterhouse in the foreground, destroyed by earthquake, 1906 - *San Francisco Public Library archives*

chopped hay. The feeders were kept in corrals. They were the fat cattle for winter. Miller and Lux use to have big losses from black leg until they got to vaccinating the cattle. There was a stiff wagon in every division. It was the duty of this crew to skin the dead cattle and burn the bodies.

The vaqueros could catch a calf and eat it whenever they wanted to. The wagon and cook for the vaqueros would pull into a Miller & Lux store and get the commissary filled up. This consisted of canned goods and hams. Maybe the cook would take on a three or four months supply. The cook drove the wagon for the vaqueros. All the bedding of the vaqueros packed on this wagon. The foreman would say: "Meet us at the Lone Willow." When the vaqueros would get there at night the cook would have the camp all made and the supper all ready.

Miller had the Shippee Ranch rented. There the vaqueros would put the two year olds out on the range. They would bring them back as three year olds to fatten. The branding outfits were: Firebaugh (at New Columbia), Los Banos and Santa Rita. At the Santa Rita Schmitz stated there were 66,000 calves. One spring's branding. He just kept track of the number of bulls. Every ten calves he would make a mark on the fence. 120,000 calves were branded that spring. "We knew Herminghouse was safe for 4,000 cows," states Schmitz. They would brand the calves and leave them there. In the fall they would bring in the feeder steers and weaning calves to Santa Rita. Half a day's branding would be four or five hundred calves. In the afternoon three or four hundred more. Canal Farm, Los Banos had the same sized crew. The Firebaugh crew was not so large.

Rafael Quinn, Buttonwillow, was the stock cattle boss. The vaqueros would gather a thousand two year old feeders and put them in pasture. They would make a drive through Merced to the Shippee Ranch near Hornitos. They would drive through the streets of Merced. They would go the back street on the east side of town. They had lots of trouble. The cattle would stampede and run back. The vaqueros would have to work their horses to death. Then they would have to bring up the caviada and change horses. Miller wanted the horses shod with copper shoes. Steel shoes would strike on the rocks and start fires.

The vaqueros went on the year round. There was the cook wagon and the cook and twelve to fourteen vaqueros. In the caviada every man had from three to seven horses. The wrangler would keep along and go on ahead. The horses were fed up and rested up. Each vaquero would catch up a fresh horse every day sometimes several times a day. The boss would give a new fellow a horse. He would keep a good tough horse for a tough day. Then he used a gentle old horse for moving cows and calves.

The pride of the Chowchilla Ranch was in raising mules. A man by the name of Bird was the superintendent. The owners didn't raise cattle. They use to buy Arizona long horns. The breaking barn went the year round. Workhorses and buggy stock were the ones broken. Miller thought mares were no good for saddle horses. Mares were for driving horses. Geldings were for saddle horses. Fillies, two and three year old mares, were used for driving. Miller raised his own stock, both horses and cattle.

There were breaking barns both at Canal Farm and at Santa Rita. Riding horses were broke every spring. From 30 to 60 head of saddle horses were broken each season Santa Rita. The same at Los Banos. Miller found that mares fight too much. In the caviada they caused stampedes. Fernando Polatos was in charge at Los Banos; Buffalo Bill was in charge at Santa Rita.

Butcher wagons were shipped down from the city. The butcher wagon used to leave Santa Rita at six o'clock. Buffalo Bill used to drive a four-horse team to Los Banos. Also a four-horse team to Poso. They were outlaws and were the toughest. The butcher wagon for Ingomar used to leave with two horses in front and two tied behind. Jim Haggerty used to drive the butcher wagon in the city. He was known as Society Jim.

There was a vaquero wagon at Santa Rita and another one at Los Banos. The vaqueros would camp all winter at Los Banos. They would have their own cook. The vaqueros did not like the chuck at Canal Farm.

Miller had a system for everything. In breaking colts we just took the rough off. So the rider could guide them. Broke only just enough so the vaquero could ride them to work the cattle. It took about 15 days to a month to break the colts. They would be ridden once a day. That would take the greenness off of them and then we would turn them over to the vaqueros. A horsebreaker would tie up 15 or 16 colts and ride each of them, it was a good day's work. Some horses got meaner instead of being more under control. The vaqueros would put these green horses in the caviada. Each vaquero would then select their own mounts and make their one pick. Good colts went to the good men. The first day out the colts would buck more. Colt breaking was always started in September. Horse breaking went on all the year around with the horses working day in and day out. The main breaking barns were at Santa Rita and Los Banos.

I remember Mrs. Clyne. She would park all day in her buggy and watch the horses being trained. She loved horses. Horses are a menace to cattle land. They tramp more grass down than they eat. Miller used to keep them on the poorer salt grass land. Miller kept colts in the swamp field south of the river; he used to keep several hundred horses there at a time. Hans Raven would select the geldings for saddle horses. Safstrom would select the work stock and mules.

Lots of colts were raised on the river bottom at New Columbia. There would be upwards of 1,000 colts in the corrals there. Raven and Safstrom would get in the corral with them and pick them out for their use. A break wagon was used to break workhorses. Old Bob would teach the horses to drink. A pole cart was used to break the driving horses; a good driving horse was used to break the colts. Bob always had the breaking horse on the right the colt on the left.

Mrs. Clyne was a horse lover but a poor hand. Miller had given orders that a suitable buggy horse was to be found for her. Bob Dalzell was the only horse doctor Miller & Lux used to have. Later Nickel put in a veterinary at Santa Rita. It was the only good thing that Nickel ever did. Dr. Taylor was his name. Hogs and cattle were vaccinated and that put an end to black leg.

Johnny Smith was the sheep superintendent for Miller & Lux. He was the only white man ever known to learn Basco. A ranch foreman would tell the sheep superintendent that he had a field good for pasturing some sheep or the superintendent would ask for pasture for fattening lambs. Each ranch had its own fence crew. When Miller got into orchards a spraying crew was organized.

There was a store at Santa Rita. It was customary to break colts on the sand bar at Santa Rita. It was just a big sandy knoll where the buildings are located. It was pretty high. Santa Rita was an island when the Temple Slough Weir broke.

The best Italian vegetables were grown at the Santa Rita; it was just a natural good spot. The best of Miller's vegetables, fruits and butter came from the Santa Rita. Miller knew the best land he held was at the Santa Rita, from there to the river, back through Temple Slough to the Santa Rita. The very best land. On the swamp lands around the Santa Rita thousand of cattle could be kept year round. One field the Van Dusen was over 20 miles long, a mile wide and ran along the river from Santa Rita bridge to Firebaugh.

Miller fed all hobos. Miller never turned a man away hungry. There might be forty men working on some ranch and maybe twenty hobos being fed. The cooks would put extra food on the tables and let the hobos eat off the men's dishes. The hoboes were entitled to sleep and be there for breakfast. They would walk to Midway for lunch. They would stay over night at Arroyo or at Santa Rita. A nice half-day hike was eight miles. Hobos were around all winter. Among the hobos following of the Dirty Plate Route was known as breaking or being a fireman on the Dirty Plate Route. "Where are you going this winter?" one hobo would ask another. "I think I'll go breaking on the DPR."

Nickel out and out cut feeding the hobos on the Miller & Lux ranches. Barns and stacks starting burning. All the harness at Canal Farm was cut up with a jack knife. It was all cut to ribbons. Another night the Ward Sheds burned. In Miller's time the hobos would always close gates behind them. After Nickel closed down the DPR hobos would never shut a gate, the cattle would all get mixed up. In Miller's time everything was brains, everything Miller did showed foresight.

There was no reason why the Miller and Lux business could not have been continued. Everything was running. The attitude of the firm changed when Nickel took over after Miller's death. Nickel even shot and killed the peacocks at Santa Rita.

Wallace Smith as told to RLM 1940 RLM

Pete Hansen, Miller & Lux employee

If you did right for Henry Miller he did right for you. One time I was feeding sheep at Grove Farm where the walnut grove is now. There were two Italians helping me. The sheep would not eat up all the hay they were feeding the sheep. Miller came driving up. Miller was standing up in the buggy. I knew at once that Miller was more than mad. Miller was mad because the hay that the sheep wouldn't eat was being left for the sheep to trample on. He began to abuse me and I at the same time was getting riled myself. Miller seized a hayfork and began picking up the wasted hay. Finally in his wrath he threw the pitchfork as far as he could throw it. "Now go and pick that up," he said. I promptly told Miller that if he wanted that pitchfork picked up he would have to go and do it himself. Miller went and picked it up. If a person talked back to Miller, Miller would treat him right. Miller and I talked it over what ought to be done with the hay. I said that I thought it should be picked up and hauled over to the horse barn for bedding. "No," said Miller. And he told me to keep it gathered up and place it in a pile by itself near the stacks of hay. The next year the stackers used this for stack bottoms. Miller never wanted anything to be wasted. He wanted everything used.

One time I was cultivating the orchard at Grove Farm. Sometimes a fruit tree would get accidentally barked in the cultivating. Miller was very abusive about it. A year or so later I was leaving Miller's employ, and the trees in the orchard were flourishing. "Yes," said Miller, "but what if I hadn't raised a row."

Interview Farm Center Dinner 3/26/1949 RLM

Archie Smith, Miller & Lux employee

Nickel was coined too high. Should have been a penny! He was always trying to fire Wallis. Lees was a Scotchman. He was the office manager of Miller & Lux before Mr. Heron. Whenever Lees couldn't have his own way he would tell people, "I quits." One time Miller says to me confidentially, "The next time that man quits—we fires him." And Miller did. Lees was the most surprised man imaginable.

I was the stenographer for Miller & Lux when he first came here. I use to live out at Canal Farm. Whenever Miller was on the ranch Miller would be up at five o'clock in the morning. He would have me up at that time taking letters until breakfast. Miller worked every day. He didn't know any more than a hog does about Sunday. One time someone was trying to tell Miller something. "I'se wants you to know that I'se runs this business," was Miller's reply. "When I needs any help from you I will ask you. You keep your nose out of this business."

One time Miller was driving over the ranches. He came to an old man in a cart. There was a gate there and Miller wanted to drive through. "Come and open this gate," ordered Miller. "Henry Miller isn't paying me to open gates for you," was the old man's reply. He kept right on with his work and Henry Miller had to get out and open the gate for himself. "Who is that man?" asked Miller on his return to the ranch headquarters. "He is a good man. He tend to my business for me."

Henry Miller used to like to stand on the street near the Miller & Lux office and pass out two bit pieces to the tramps who would come up and speak to him. Miller would get from ten to thirty dollars worth of quarters from the office and go out on the street. One time there was a bunch of tramps standing on the corner waiting for Miller to come out. There was a tough looking fellow who ran a barbershop in Los Banos. His name was Murphy. He wanted to talk with Miller on some business and when he saw Miller come out of the office he went up and tried to speak to him. Miller paid no attention to him. Just shoved out a quarter to him. "I don't want your quarter," roared the barber. Shannon who was standing with Miller had to lean up against the building and roar with laughter. He

Miller & Lux's first tractor unloaded off train 6th Street, Los Banos, 1907

knew this barber well. Afterwards Shannon was talking with this barber and he said, "Henry Miller thought you looked like a tramp, and I'll be dammed if you don't."

Miller liked to meet the old timers. Wright, Mrs. Idiart. "Yous gives Billy Wright anything he asks for," was Henry Miller's orders to the office. "He'll never ask for anything he can't pay for." Billy Wright sold his cattle to Miller & Lux for forty-five years. It was only after Miller's death and the new management wouldn't pay Wright as much as he could get else where that Wright had to sell his cattle elsewhere. Wright felt very badly when he had to sell to a new buyer.

Miller once told Billy Wright, "You are better off than I am, I have got so much business, you have got a little kingdom of your own. You can get over it all. I can't sleep nights." One time a railroad was surveyed over the Pacheco Pass to the oil fields. Miller had given a right of way free. Miller felt reminiscent that day. He told me, "My social side has never been developed. I guess I've been too busy. Now I'm too old. All the pleasure I gets out of life is to look back and see what I've accomplished."

In the early days of Henry Miller the hills were infested with bands of robbers. The plains were considered no good. They were covered with water. One time Henry Miller went by boat from Old Los Banos clear to the Santa Rita about 20 miles. One time there was a terrible flood on the plains and any number of Miller's cattle were drowned. After the buckaroos had rescued all that they thought they were able to, Henry Miller came over from Bloomfield and took charge and got out over a 1,000 head more huddling on the knolls and surrounded by the flood waters of the San Joaquin River.

Miller never swindled people. About a month after I came to work an old couple came in to the office. Everything they had was mortgaged to Miller & Lux. "you take what we got," they told Miller. If you can give us anything, all right. Give us 50 cents on the dollar. You take the rest." But Miller advised them to hold on. To hold on to their land. Now these people are well off.

Mrs. Moran offered their quarter section to Welch for $10,000. Welch offered them $9,000. It is now worth $25,000. Years before Mrs. Moran had tried to sell the quarter to Miller. Miller said he did not want to buy it. He told Mrs. Moran to have the land all put into alfalfa. She had no money. Miller told the office to let her have the money to plant the land into alfalfa. Mrs. Moran did so and became wealthy for taking old man Miller's advice.

I think the smartest thing Miller ever said came about this way. Someone was asking him why he was always buying so much land. "Look here, there'll never be any more land. But there will be an awful lot more of people." That was the keynote of his success. He knew that the land would be valuable.

Miller said that Nickel was the poorest judge of human nature he ever knew. Nickel said that Miller was the shrewdest judge of human nature there ever was.

One time Miller was held up on the Pacheco Pass. After the robbery Miller told the robber he didn't have a cent on him and for him to loan him $20 for expense money. The robber tossed him a $20 gold piece. About twenty years later Miller saw the man in a hotel in Stockton. Walking up to the man Miller said, "Here's the twenty I owe you." The man protested ignorance of any such debt. "I never forgets a face," Miller told the man. "Yes, I owes you that twenty, you know that I knows you."

Miller's ability to recall details was absolutely marvelous. At Los Banos, Dos Palos and Firebaugh there were dozens of old timers. Every Christmas Miller gave away a fortune to these old timers. One time Mrs. Sherburn was sick. She was away from Canal Farm for five years. He remembered her and every Christmas sent her a present. Every good turn that anyone did for Henry Miller was never forgotten. The Park in Los Banos was given by Miller. "As soon as the town is able to take care of it," declared Miller, "I'se gives it to them."

In 1895 the Bank of Los Banos was established. William Lees was the first cashier, Heron was the assistant cashier. Miller never prosecuted an employee who stole from the firm. There was a bookkeeper who was the cashier from Miller & Lux at Dos Palos. He was drawing only a small salary. His cigar bills were $40.00 a month. Besides he had a livery bill, a big bill, more than his salary. Merritt, the office manager for Miller & Lux, was checking up on him. "I'll see you in the city," the bookkeeper told Merritt, "have other sources of income." That night the bookkeeper took the Owl and skipped the country. It was a great mess, this affair at Dos Palos. Merritt checked it out and found that about $10,000 had been stolen. Merritt was at the table at Canal Farm. I used to eat at the superintendent's table. Miller was there. "I'm ashamed of you," said Miller to Merritt. "Wes swallow our losses, we say nothings, but what a good time that fellow must have had laughing over us."

The Miller & Lux Bull's Head checks were famous. They were simply checks drawn on Miller & Lux in San Francisco. Before the bank was established these checks were considered as Certificates of Deposit. They were without interest. That's the way people deposited their money before the bank was started.

Interview March 3, 1946 RLM

Charles Stafford, Miller & Lux employee

Charlie Bambauer made a lot of money by buying out several of the settlers who had got cold feet. They had fields of grain sowed but the rains held off so long that they thought it was going to be a very dry year. Bambauer bought out several fields for very little and when the rains came he made a lot of money.

In 1886 Tanner had a bumper crop on the Tanner quarter where old Center Point used to stand. There was a bumper crop in the Volta country that year also. A man by the name of Sawyer was farming that country at the time. There was a bumper crop on that poor land but there has never been another good crop since.

In 1890 the Plow Camp section was all a vacant country. Miller & Lux started a camp there and Bill Fought was foreman. They put in all the country that year. Teams from Plow Camp would go as far south as a couple of miles south of the Cheatham place. John Tully used to set out before daylight with teams from Plow Camp and often it was nine o'clock at night before they got back to Plow Camp and had their teams cared for. They had seed sowers right on the plows and harrows fastened on behind and the whole operation of plowing, seeding and harrowing was done at one operation.

Orin Knight was at one time foreman at Camp 13. There was a sort of shed there and one day Miller came in and found a keg of nails of all sizes mixed together. Miller kicked the keg over and told Orin to sort them out, each size by itself. The first place that Miller went on his next visit to Camp 13 was to this shed. The keg of nails were still scattered over the floor just as Miller had left them. Orin Knight lost his job.

Miller was always very particular about not wanting any dead animals left on the range. He had stiff pits at various places on the ranches where all dead animals were to be hauled and burned. One time two eight mule teams were busy for a long time hauling wood that had been cut down along the river. The wood was used to burn dead cattle with.

A man by the name of George Johnson was foreman at the Grove Farm. When Tanner knew that Miller was in the country he would call up the different foremen and give them a tip so that if they had any corners that needed looking after before Miller got there they would have time to get things straightened up. One time Tanner called up Johnson and told him that Miller was in the country. They were doing some butchering at Grove Farm in those days and a regular butcher was kept there. They had been killing sheep from time to time and just spreading out the pelts on a pile of wood. The sun had cracked these skins and they looked very cracked and bad.

Johnson thought the easiest way to get rid of the these pelts for the time being was to throw them in a hole nearby and then take care of them after Miller had gone. There was a hole in an alfalfa field where the water in running from a higher check to a lower one had washed a big hole. When Tanner drove up with Miller, Miller told Tanner to get out of the buggy and let Johnson get in and take him over the ranch. Everything looked fine and Johnson thought everything was going to go along fine. Johnson and Miller started and Johnson tried to keep as far away from that hole as possible. Wait a minute, said Miller. Where do those tracks go? Johnson tried to get away from them but Miller insisted that Johnson should follow those tracks. I wants to see where those tracks goes, he told Johnson, so they followed them through the alfalfa and soon came to the hole where the pile of sheep pelts had been thrown.

"Who throws these pelts in this hole?" demanded Miller. Johnson realized that the only thing he could do was to make a clean breast of the matter and so he told Miller that the pelts didn't look very good and so he had thrown them into this hole until Miller had gone and then he was going to take care of them. "Well, you's sends them up to Canal Farm," he told Johnson. Miller was perfectly satisfied after he had exposed Johnson in his trick and nothing more was said about the matter. Johnson had tried so hard to steer away from that

Farmers Camp established in 1873 changed to Canal Farm 1876

field but Miller was so insistent to go and see where those tracks led to.

Wallis put in a new road between Canal Farm and Santa Rita. The old road went to the north of the present road. I worked on the building of the road at the east end. Miller was very angry when he found out what had been done. "You's spends all my money," he told Wallis. But the next time when Miller came down, the new road had had time to settle and was really in fine shape. It happened that Miller was in a hurry to get to Santa Rita and when he found how much shorter and better the new road was he was perfectly satisfied with Wallis in building the new road. I remember that Wallis told Miller when he was trying to explain to Miller the need of a new direct road between the Canal farm and Santa Rita that the old road was all right when Miller was using ox teams, that the old road was made up when Miller first used those teams.

Miller always wanted things done the way he said. "I pays for it," Miller would say and I think he was perfectly right in wanting things done his way. Miller would go out and look through the manger to see that they were kept clean. Miller always shook hands with a foreman when he finished going over a ranch and started for another ranch. When a foreman went with Nickel the foreman would take Nickel, but when a foreman went with Miller, it was Miller who took you. It was Miller who told you where to drive.

The "Bull Pens" were an old landmark and Miller always said they were never to be torn down and removed. Frank Soper told me that in the dry year of 1877 he and Antonio Lopes herded hogs on the plains that summer and used to corral them every night in the bullpens. The country between Santa Rita wasn't improved in the early days. What is now the Weir Camp used to be called the Stacy Miner place because a man by that name was in charge there. Later it became known as the Weir Camp. The Lone Willow Ranch was just a cookhouse and barn at first. The Midway was built after the Lone Willow. Mud Slough was so called because it was so boggy. There were only a few places where a person could cross it. The year 1886 was a great grain year. The rains came very late that year. On March 17, 1885, Jameson sowed a lot of barley on government land in Section 25 and it made 25 sacks to the acre.

Miller was a wonderful man and spent millions of dollars developing the country. He was very shrewd and bought cattle and land at his price.

Interview with Charles Stafford April 4, 1932 RLM

Mace Roberts, blacksmith and farmer

I came from Tennessee to California in 1891. About 1894 I and a friend were fishing on the river and decide to come over to Charleston. I was living in Madera at the time, I knew Z. G. Jameson and I and a friend went over to visit him. Jameson wanted me to work for him in his blacksmith shop. So I went to work for Jameson and thus got located in the Los Banos region. I later worked in Jack Sullivan's shop in town. I was a great hunter and the sport is what kept me here. There were antelope on Little Panoche Creek when I first came here. I shot and killed a buck with my pistol.

I made a good friend out of Miller. One time I had a summons to serve on Miller. It was some Jim Peck litigation. I went out to Canal Farm and asked Torchiana if Miller was there. "Yes," he replied. You tell Miller I want to see him I said. Torchiana went in and told Miller that the constable wanted to see him. Tell the constable to come back tomorrow, Miller sent out word. I told Torchiana you tell Miller the constable wants to see him right now! "You's tells me that constable comes right in," replied Miller. After that when I would be running for constable Miller used to tell his foremen to do what they could for me. That I was a good man. If Miller liked you he would go the limit for you.

One time Miller called all his superintendents and big men to the Miller & Lux office in what is now the Federated Clothing store. I was going by and heard loud talking going on inside. The window curtain was up about an inch. I was the constable at the time and so I peeked under the window curtain. All were standing in a circle and Henry Miller was in the center. His hat was on the floor and he was stamping up and down and just laying it off to his men. I never did hear what it was all about.

The biggest number of beeves I ever saw Miller kill in one day was 18 big beeves. He did the job himself. I worked on the canal when they were digging the new canal. I was the blacksmith and they moved the blacksmith shop right along as the camp moved. I got $40 a month and I never worked harder in my life. It was a good-sized camp. There were lots of horses and mules employed in digging the canal. There was one excavator. It was pulled by eight or ten horses or mules. I started in to work when the camp was about where Ora Loma is now and worked until the camp was almost to Newman. Scrapers followed the excavator. These were four-horse scrapers and did the finishing.

One time I had an agreement to meet Miller in Los Banos. Miller had set a price on the water works and electric light plant. $6,000 was the price and I had got a partner and we were going to raise the money and buy the water works and the electric light plant and fix things up. Miller met me in Los Banos just as agreed. Miller was sitting in his buggy. Come here and he motioned for my partner and I to come. I changed my mind Miller said. Then he went on to say that he was going to fix things up. "I'se got more money than you got" Miller said, I will fix it up. I was awful mad at the time.

I drove Miller a few times. Miller would be talking away to himself but he would never say a word to his driver. I always liked the old fellow (Miller).

Interview Mace Roberts February 10, 1943RLM

Rudolph Bambauer, early pioneer

The Bambauers settled at Camp 13 in 1872, where they built a stopping place for travelers. Lots of people were traveling through the country in those days en route to the southern part of the San Joaquin Valley. The Bambauers sold them hay and grain and served meals.

Later they moved their building and started the Charleston stage station. The Bambauer children, there were six of them, use to drive in a spring wagon ten miles to the Center Point school. That made a trip of twenty miles a day to school. I remember a teacher at the Center Point School by the name of Taylor. He was a fine old man, but very deaf and he had to use an ear trumpet. Another teacher we had was Robert Gracey who afterwards became president of the Merced Security Savings Bank. Mr. Wiley and Mr. Bambauer finally got a school established in their neighborhood it was the Charleston School.

I remembers an Irish family by the name of Talbot who farmed on Miller & Lux land at Stringtown. Stringtown was between Camp 13 and Dos Palos. The Talbots farmed for six years. At the end of six years the family was four thousand dollars in debt to Miller & Lux. Miller furnished teams, etc. and store credit to the renters. There were the father and mother and a daughter and three boys in the family.

Miller was coming along one day and Mr. Talbot told Miller that he was going to leave—that he couldn't hold out any longer. Miller asked him where he was planning on going. Talbot told him that there was lots of fine government land open up around Walla Walla in Washington Territory. Miller asked Talbot if he had any money to get there on. Talbot told him no, but that they would get there somehow. Miller wrote out a receipt in full for the four thousand dollars and also a check for four hundred dollars in cash. He handed to Talbot and drove on.

Years afterward the Talbot boys used to come back and go hunting with us Bambauers. They told us that they had located up around Walla Walla, Washington. In those days they could take up both a homestead and preemption. All together the Talbots had taken up a thousand acres or so of land. It was fine grain land there and the family became very prosperous. They sent back to Henry Miller the four hundred dollars and later also the four thousand dollars as well.

The foreman at Grove Farm use to be George Meredith. One time Henry Miller was riding by some corrals. The gate was lying on the ground. Miller told Meredith to throw the gate in the ditch. "It is not doing any good there on the ground," stated Miller.

Miller had an orchard and some grapes on Grove Farm. The peaches were simply gigantic in size and the grapes were as large as plums. The grapes were so delicious and so large that they used to ship them by express to market. Later a dairyman got hold of the place and probably tore them out to put in alfalfa.

One time Miller was going over the pass and a highwayman held him up and took all his money. Miller asked the Mexican to leave him $20 as he didn't want to be broke. The fellow handed Miller a twenty-dollar piece. A long time later Miller was in a saloon in Gilroy. It was his custom to call up everyone for a drink. He recognized at once that a Mexican sitting over at a table was the fellow who held him up. When the fellow was up at the bar drinking Miller slid a $20 gold piece over to him and said "That's the $20.00 you loaned me."

Charlie Bambauer backed up Harry Bambauer for four years after Charlie quit sky farming. There were four dry years in succession and Charlie lost $30,000. Charlie bought a prune orchard over at Cupertino for $40,000. During the war when prices were so high he was offered $80,000 for the place. Charlie figured that if the place was worth $80,000 to the other fellow it was worth that much to him. So he refused to sell. The last four years he hasn't made a thing and now he couldn't get even the $40,000 that he paid for the place.

Cottonwood Farm April 5, 1935 RLM

Adolph Salau, farmer

Miller & Lux built a big dam in the creek about 1908. Miller told me that he wanted the people having ditches to have all the water they want. But he wanted all what was left. Even when the creek was small there would be a big ditch full of water coming out of the Miller & Lux dam. I always had a crop. I would irrigate when there was water in the creek and not wait until I needed the water, for perhaps then the creek would be dry. I would take all the water the ditch could carry. If there was any water left it ran down to the next dam and Chance Waggoner took it. If there was still more water it flowed on down the creek to Miller.

One time Torchiana spoke to Mr. Salau in German and Salau astonished Torchiana by replying to him in Javanese. "Where did you learn that?" asked Torchiana in surprise. "In Java." "How long were you there?" "Eight days." "Well, if I were as smart as you are Mr. Salau I wouldn't be the common place man that I am," replied Torchiana.

Miller & Lux had freight teams constantly on the road over the mountains hauling freight. They had splendid teams of six horses each, and two wagons, a wagon and a trailer. The teams traveled at a slow, steady gait and Miller had steady drivers. They hauled freight steady the year round. Two or three such outfits were kept going all the time. The loads consisted of five or six tons on the two wagons.

Beef cattle were taken over the mountains at all times in droves of from 500-700 head. They were always at least six buckaroos with them and usually seven. Droves of hogs from one hundred to two hundred head were driven over the mountains and several droves were lost. Something would scare them and the whole drove would break for the chemise brush and once gone the drivers could never get them again. They became wild hogs and roamed the mountains until they were killed off by the wild animals and hunters who were always glad to get them. There were bunches all over the mountains that had gotten away from Miller & Lux. Sheep were driven over the Pass in the same way the hogs were.

Undated interview RLM

Carrie Jameson, pioneer family member

It was a great treat to the early settlers to get to have a picnic at Canal Farm under Miller's shade trees. The pioneer families would bring their eats and spread out white tablecloths on the ground under the trees. At these Miller picnics Henry Miller liked to carve the meat and pass it out to his friends. He liked to say hello to his neighbors. While he was talking he would cut some meat on a plate and hand it to you. Miller would stand there and carve. He would be the one to put the meat on the plate. Then he would give you cheese. There was nothing stingy about the cheese. He would give you a good-sized piece and make you take some bread.

The coffee was behind him and dished out to the various family groups. People brought their own coffee pots to carry the coffee over to their family group.

Single men were served their plate by Henry Miller and given a cup of coffee. Then they would stand around and talk with Henry Miller while they ate. I think I was about 14 when I first stepped up and took my plate from Henry Miller. I remember that I was too old to be romping around with the other children and that I stayed close to my mother. "Cee his meat used to taste awfully good." The people use to bring their own eats but they would get meat and coffee from Mr. Miller.

The railroad had not yet been built when I went to my first picnic. At all these picnics there were egg races and all kinds of children's races. Also there was a greased pig. I use to think it was so funny to watch the boys trying to catch the greased pig.

Interview August 6, 1944 RLM

Hans Raven, Miller & Lux foreman

I was born near Waterloo, Iowa. I came to California when I was eleven and settled at Watsonville. In 1890 I came to the San Joaquin Valley and I have lived here ever since, 45 years. I lived first four months at Santanella where my father was working and then I went to work for Antone Lopes.

Antone Lopes had charge of driving cattle from Santa Rita to Bloomfield. Sometimes Lopes picked out the fat cattle himself but usually the band was rounded up for him and he simply drove them over the mountains. It was simply that fat cattle were rounded up and got together and then driven over to Bloomfield. Santanella was the headquarters for the cattle driving crew. Santanella was the receiving place and the headquarters for the drive. The drives always started from Santanella; a bunch of saddle horses were always running there and it was there that the drivers changed horses. This bunch of horses were kept there for that purpose.

Antone Lopes and his men were on the road all the time. That was Lopes' job. It took four days to get from Los Banos to Gilroy with fat cattle. Stops were made at San Luis, Mt. House and Bell Station. It took the drive one day to get back from Gilroy to Santanella a forty-mile ride. The outfit had no wagon with them. Each rider simply carried a small roll of blankets, rolled up on the back of his saddle. They slept in the barren or wherever they could. When the outfit arrived at Bloomfield they would turn over the cattle to the people there and they started back for the San Joaquin. Driving cattle went on the year round.

The sheep were driven by an old Frenchman by the name of Joe Bertis. He drove them all alone over the mountains. He had several good dogs. Good sheep dogs are better than men as helpers. He took bands as they came constantly over the mountains the year around except during the winter. At horseback to the various ranches. Sometimes he would even send a man on horseback to Gilroy with a message and the fellow would have to ride some to get there the fifty miles with the message.

Miller & Lux beefed up the tractor by adding dual oil tanks so it could pull larger harvester. 1908

Bloomfield there was another crew that would take the stock the rest of the way to Butchertown.

During the strike about 1893 I helped drive cattle plumb through to Butchertown. We drove the cattle right through the towns on the way. In driving stock from Santa Rita to Bloomfield there wasn't much of anything along the way to bother the driving of stock. There were regular fields at Bloomfield to receive the cattle that came from the San Joaquin.

Henry Miller was one of the great letter writers that ever was. His writing looks like a crow scratching but after you got use to it you could read it easily enough. He would commence writing at five o'clock in the morning. He would have a messenger on hand and have the letter delivered on

Henry Miller used to give Chinamen fifty cents or a dollar and his boots had to be all shined and standing beside his bed when he got up in the morning. He always wore boots. Henry Miller never wore gloves in his life. Seems like he didn't like em. His hands would be all red with cold or sunburn but no gloves.

Miller & Lux ranches were always kept perfectly neat and tidy. Whenever Miller would find an empty sack or a tin can lying around in the yard, he would pick it up and take it to the superintendent himself and show it to him. "You put that away," he would tell them. He would never tell a workman to take care of them but he would take them right up to the superintendent himself.

Miller had a way of going around the barns and

digging down into the mangers to see if they were full of foxtail. The barn men were supposed to keep the mangers cleaned out and not allow them to get filled with foxtail. Miller would go through the barns and throw the hay off the top of the mangers and feel around with a stick to see if there was any foxtail underneath. If he found any he would call the superintendent and order, "Now you clean that out." Miller didn't want the horses to get any foxtail in their mouths nor sores either.

At the ranches they used to have swill barrels on the side. The swill was hauled out to the hogs. I have actually seen Miller take off his coat and roll up his sleeve and run his arm down in the swill barrel to see if there were any bones in there with meat left on them. "See that," he would tell the superintendent if he found any meat had been left on the bones. Miller would always tell the superintendent and then show them the meat left on the bones but would never say anything to us workers. He dealt with the superintendents.

Miller would show you so you could see he was right. He never joked with a man, he never joshed.

It was always Miller luck to see the things that the superintendents wanted to keep hid from him. If there any sick stock or anything that they didn't want Miller to see it would be sure to happen that when Miller was driving over the ranches it would come walking right across the road in front of him. It was always Miller's luck to see such things. He would say, "It is strange how all the things come to me when I come to see them." Miller caught me once. It was the custom on the ranches to kill the sick cattle and cook them up and feed them to the hogs. The hides were saved. The best of the TB cattle were butchered and the meat used on the ranches. Miller believed that by cooking it made TB meat all right. There were some nice cattle and I run them in a bunch down by the San Joaquin River out of sight and had them killed. Miller was riding along with the superintendent and smelled the dead cattle as he went by. He also noticed the buzzards flying over the brush. "Hold on," he told the superintendent. "There must be something in there." They went in and found the dead cattle that I thought I had hidden where Miller would never find them.

Miller would never allow one man to tattle on another. "I don't want to hear it," he would tell them. Nobody works as hard now as they used to on Miller & Lux ranches. Miller always had breakfast on all his ranches at five o'clock each day winter and summer, and everyone worked until six o'clock in the evening.

In time of high water there would often be a few cattle get stranded on some knoll. They were afraid to try to swim out to the mainland and would starve to death unless rescued. They always kept three or four boats on the bigger ranches and when there was a flood they would put a rowboat on a wagon and drive down where there were stranded cattle. They would put the boat in the water and row over to the knoll and get out and drive the cattle into the water and make them swim ashore. If they didn't swim the cattle out they would starve.

Miller lost lots of cattle with anthrax. Probably three or four thousand head a year on the San Joaquin. They didn't know anything about vaccination in the early days. Now they vaccinate and they lose very few cattle.

In the years when other cattlemen were going broke Miller always made money. He always had feed and a place to put cattle and could buy them up cheap. Miller was "hell to putting up hay." He would put up stacks and stacks of wild grass hay. Bunch rakers were used. He would have camps scattered all over the San Joaquin putting up hay. Miller always also stacked the straw. In this way he always had great quantities of feed stored up for any year.

This was Miller's fattening country. They use to fatten them also down at Buttonwillow. The way he used to fatten them was to cut the first crop of alfalfa hay in the spring. There was always lots of foxtail in it. Later on when they were using this hay to fatten cattle they would pile it in little piles and pour water over it. This would make the pile steam and ferment. After about four days the foxtail would be all soft and would not hurt the cattle's mouth. The piles would be forked over and ground barley mixed with it. The cattle liked it and would eat it when it was still quite hot. It general took 90 to 100 days to fatten them. They would get hog fat.

Cattle from the Oregon ranches would fatten here also. Three-year-old picked steers would be shipped down from Oregon and put in the feed yards here. Also nice three-year-old heifers. They would raise lots of them up in the Oregon ranches. They would ship them down here and make them weight 1,200 and 1,300 and make them into 1,400-pound fat cattle here. They never had anything under three years old in the feed yards. From that on up to four and five year old stock. There was no baby beef in those days. They never tried to fatten anything under three years old. From that up to and five year old stuff was all they fattened.

Miller had his own hay and his own barley. Miller used to fatten a lot. Miller used to send fat cattle to Portland and even sent a load to the east. He got his own price for fat cattle when other firms in San Francisco were short and had to buy from him. He sold to the other meat companies as well as supplying his own trade.

Miller bought cattle by the trainload by thousands in Arizona and Mexico. Albert Long was his cattle buyer. In dry years Miller had hay and used to buy cattle cheap. That's why he made money so fast. When Miller would go to buy a band of cattle they would have them all rounded up. Miller would drive through them in his buggy and then tell the owner what he would give for them. "I gives you so much." There was no arguing about the price. If the owner didn't want to take it Miller would tell him I have no time to lose and would drive away. Usually Miller got the cattle at his own price.

Breaking horses went on all the year around. Santa Rita was the main place to break horses. Fernando Polatos was the main horse breaker. Horses were broke to work or to ride and then turned over to the ranches. Polatos was breaking horses at the Santa Rita forty years ago. Miller & Lux used to raise herds of horses there. Studs would be turned out in fields with hundreds of mares. The story of me breaking horses with my arm in a sling is all wrong, wrong. I never had a broken arm in my life.

There is an old board fence about four miles long on the Hermanhouse ranch. The fence is on the left hand side of the road going to Fresno and begins at White's Bridge. It was an old fence when I came here.

Miller used to let people have a cow to milk if they would raise the calf, take good care of the calf for him. Miller would say bring the calf back to me when it is six months old in good shape.

When Miller would be in Los Banos or Firebaugh he would get a lot of coin two and four bit pieces to give to the old stiffs that he knew would strike him for a little money. "Go get something to eat. Don't drink it up, come back to the ranch and do some work," he would tell them. There used to be one of the biggest walnut trees near the headquarter house at Santa Rita. It was a great big immense tree about four feet thick. Nickel had it cut down, people begged him not to. There is an old oak tree at the Riverside ranch that grew from a seed planted by Henry Bader of Dos Palos.

Miller had a great habit of talking to his superintendents and telling them things they ought to do. For instance if there was a board on a fence coming loose he would tell them that they ought to go right then and fix it. Or if there were a fence needed fixing they should fix it before they went any further. If the superintendent says that a man would do it Miller would stop and fix it himself. Miller had a set expression he would always call them young man. "Young man, where there is a will there is a way." That was his great expression and one often heard by his superintendents.

The bull corrals were near Pasqual's ranch. They may have been the old corrals from the very early days. The extreme north part of the grant was McPike. The next field south where the old San Luis Camp adobe is was called the San Luis Field.

There was a bunch of Mexican vaqueros kept at Bloomfield. All were old Mexicans. Each had great big long mustaches. They would come over and receive a bunch of fat cattle. They would drive them over to Bloomfield. They would stop at the Mountain House, then Bell Station then on to the Bloomfield. All they did was drive cattle.

Interview with Hans Raven 3/10/1935 RLM

Will Wiley, Miller & Lux employee

Will Wiley went to work for Miller as a young man for Miller & Lux sometime in 1905-1906. He remained in their employ until the 1920's. Exact dates of his employ are uncertain. He rose from simple hand to superintendent of the canals and worked closely with Miller in his last years. This interview was undertaken when Wiley was in his late 70's.

Henry Miller was wont to upbraid his employees whenever he saw anything untidy about his premises. He hated to see even a tiny bit of tules growing in his canals: "My property looks like it was in the hands of a receiver," he would yell.

Simon Newman wrote to Henry Miller that he had eight hundred head of fat steers at the Linora Stock Farm for sale and wanted Miller to come and look at them. Miller came down and he and Simon Newman drove out in a buggy to see them. When they got to the corrals both got out and proceeded to walk through the corrals.

"Well, what you want for them?" asked Miller. "Make me a price," hedged Newman. "You're selling," replied Miller. "What do you want?" "Eight cents a pound," announce Newman. "Eight cents is too much," declared Miller. "Then how much do you offer?" asked Newman

"Seven cents." "That's too low," replied Newman.

So Newman and Miller could not agree and both walked back to the buggy. "I'll make you a final offer," said Miller. "You can take it or you can leave it. You let me pick out three hundred head and I'll give you eight cents for them. The balance go at seven cents."

"You can have them," said Newman. "Right away Miller pulled out a $500 greenback and handed it to Newman. "I binds the bargain," said Miller. "Oh, you don't need to do that," said Newman. "We both understand the bargain thoroughly." "That's all right," insisted Miller. "I binds the bargain."

So Newman took the $500 dollar bill. Right away Miller sent word to Wade White to come and take delivery of the cattle. "You pick out three hundred head of the smallest, scrubbiest steers there are in the corral," Miller told White. You weight them up and give Newman a check at eight cents a pound. Then you weigh up the balance and pay him seven cents a pound. It was in that way the Miller bought up all of Newman's fat cattle for seven cents a pound.

Newman was furious. But what could he do? "To think that Miller would do anything like that to me," lamented Newman. Here I use to let Miller sleep in my bed in the back of the store and now he treats me like this!"

This was just another cattle deal to Miller. He was often heard to say, "Our business is the cattle business. Everything else is one side."

Julius Kane was a storekeeper in Newman. He had a pretty good store about a block from Simon Newman's store. I (Will Wiley) knew Julius well and always thought he was a pretty nice fellow.

When I was the superintendent of the canals

for Henry Miller I got a letter from Julius asking me to come up to Newman and have dinner with him at the Russ House. I thought that was very nice of Kane and answered back I would be up on the train on the appointed day. In those days people didn't go places in automobiles. Instead if it was any distance they went on the train. When I got down to the depot at Los Banos I found Will White and Johnny Boyle waiting for the train. Will White was the manager of Miller and Lux's department store in Los Banos and Johnny Boyle was the head of the dry goods department.

I asked where they were going. "We are going up to Newman. Julius Kane is going to give a dinner party at the Russ House and has invited us to come." "Why, that's where I'm going, too, I am also invited to the dinner," I said.

So we three Miller and Lux men boarded the train together and soon were in Newman. Julius Kane was right on hand at the depot to meet us. We could just as well walked across the street to the Russ House. But Kane led us across right in front of the Simon Newman store. He was very leisurely about everything. He didn't seem to be at all in a hurry to get his guests over to the Russ House. The four men stood on the sidewalk in front of the Newman store and chatted for a time. "You know I have a store, too, a block down the street," remarked Julius with pride. We three Miller and Lux men knew of course that he had a store.

Soon Julius took us three men over to the Russ House and gave us a wonderful dinner. After dinner Julius insisted that we should all go down to his store while we waited for the train back to Los Banos. "I want you to see what a fine store I have," said Kane proudly. The bunch walked past the Simon Newman store and Wagenheimer himself happened to come out at just that time and saw us. He knew all of us.

We spent a couple of hours or so at Kane's store while we waited for the train. Kane insisted on showing us all through the store. He showed us his hardware department, and the dry goods and the groceries. Finally it came train time and Kane got us up to the depot just in time for the train. We couldn't thank Kane sufficiently for the fine time and splendid dinner we had had.

In a day or two the news spread like wildfire all though the town of Newman that Miller and Lux were going to buy out Kane's store and go in business in Newman. Wagenheim was wild. The last thing on earth that Simon Newman would want would be for Miller & Lux to have a big store in Newman. Wagenheim was almost on his knees begging Kane not to sell out to Miller & Lux, but to let Simon Newman buy him out. A week later it was announced that Kane had sold out to Simon Newman.

Mr. Wiley says that Henry Miller never did anything for spite. He says that Gustine was not a spite town as many claim. It seems that there was a little co-operative creamery about three or four miles from the present Gustine. It was off of the railroad and it was necessary to haul the cream to Newman and ship it. Rudolph Bambauer was the president of this little inland creamery. He wrote to Henry Miller asking if they could buy an acre of ground at the Cottonwood Switch as the Gustine Switch was called and let them move their creamery there so that they could ship their cream without having to haul it to the railroad. It seems that a person could ship stuff from the Cottonwood Switch by leaving it there and the train would stop and pick it up. Cottonwood Switch was so called because it was on the Cottonwood Ranch of Miller & Lux. The real name of the switch was Gustine and had been so named by the railroad company when they built the railroad and had put in a switch there.

Miller wrote to Wiley to go and survey out forty acres of land and pick out an acre of ground for the creamery. Later Miller wrote to Wiley to meet him at Newman. Together the two rode over to the Switch and Miller and Wiley looked over the land

and the best location for the creamery where there would be a ditch for drainage for the creamery. Wiley told Miller that a settlement would naturally grow up there and that more land should be surveyed off. Miller said nothing and when Miller left the next day he didn't give Wiley the least inkling as to whether he intended to follow Wiley's advice. But a few days later Wiley got a letter from Miller telling him to survey off 160 acres of land and to give the creamery a deed to one block. Mr. Wiley says that Miller did all this not for spite but simply to accommodate the New Era Creamery. He says that that was the only reason why Miller started the town of Gustine.

When I asked Mr. Wiley whether the coming of the New Canal had any effect on the country in the way of seepage he said that now I was bringing up a ticklish subject. It seems that when Wiley turned the water in the New Canal it was only a short time until trouble was breaking out everywhere. Several miles below the canal on the Rube Lathrop place a little lake began forming. A few days later a teamster hauling a wagon and trailer loaded with grain along the county road noticed that something had happened to one of the hind wheels of his trailer. He got off to look and there the wheel had broken through the road and gone down to the hub.

Word came that the Simon Newman warehouse at Ingomar was sinking and that a lake was forming all around the house. This was some fifteen miles below the canal. About this time Ben Jeffers down below the Old Canal came out to his pump one morning and there the well was full of water and the water was running out the top of the well.

Everyone was up in arms over the damage being done by the water seeping out from the new canal. But Wiley who was the superintendent of the canal urged people to be patient and see if the canal wouldn't soon silt itself up and the seepage from the canal would be stopped. The next spring the San Luis Creek came down carrying lots of mud. Wiley turned the whole creek into the canal and soon the seepage stopped.

Mr. Wiley says that he has seen it 123 degrees in the shade at Central Point. He says that he went himself and looked at the thermometer on the shady side of the building and it registered 123 degrees.

Mr. Wiley says that one time some of the men from the Miller & Lux city office were down here. They were complaining that Miller never took them into his confidence. He never gave them any advance information. He never told them anything. These office men said that if Miller had just let them know that he was going to build the New Canal down past Cottonwood they could have bought up some of the cheap land and made a lot of money. "That's just why Miller never tells you anything," declared Mr. Wiley.

One time at Christmas Henry Miller happened to be in the store at Firebaugh. The schoolteacher, a young girl, happened to be in the store at the same time. "How many boys and girls you got in your school?" asked Miller. The young woman said that there were about thirty-four. Turning to the clerk Miller ordered: "Wrap up some candy for these children and take it up to the school house."

A few days after Christmas the school marm met Henry Miller in the store and tried to thank him for his thoughtfulness and kindness in remembering all her pupils at Christmas with his gifts of candy. "That's all right," grunted Miller. "They all belongs to me." Miller always seemed to feel that all the people connected with the firm of Miller & Lux belonged in a way to him. Like he was the overlord of all those in his employ and that he was responsible for their welfare.

For years Miller & Lux had the San Luis Gonzaga Ranch rented. In the lease it was specified that Miller & Lux were to keep the buildings and fences in good repair. A short time before the lease expired the barn at the Mountain House burned down. Dr. Fatjo promptly notified Mr. Miller that he must rebuild the barn.

Miller and Fatjo argued the matter vehemently. Fatjo pointed out that the lease said that Miller was to maintain the buildings in good order. "But," countered Miller, "How can I keep a building in repair that doesn't exist?"

Fatjo pointed out that the barn was there when Miller rented the ranch and the barn had to be there when the lease expired. Henry Miller had a way of seeming almost to be crying when he was arguing very earnestly. Fatjo was very firm. He was unmoved by Henry Miller's apparent crying. "I can cry just as good as you can," he told Miller. Fatjo made Miller & Lux rebuild the barn.

When Rouse got let out the Enterprise came out with an article saying that Rouse, after working hard for Miller & Lux for years was about to take a long vacation. Miller in commenting asked, "What does Rouse need of a vacation? He has had one for the last two years. I consider him an embezzler—his salary."

One time someone wrote to Miller asking about a man who claimed that he had been working for Miller & Lux and had given Miller as a reference. Miller wrote back that the fellow as a worthless scoundrel. Eventually the fellow sued Miller for $25,000 for slander.

Some time after Torchiana was no longer in the employ of Miller & Lux, Spreckles wrote asking Miller about Torchiana. It seemed Torchiana had applied to Spreckles for employment. Mr. Wiley overheard Miller dictating his reply. He said, "I won't write you in a letter what I think of Torchiana. I will tell you when I see you."

Miller & Lux employees were very loyal to Henry Miller. They realized that the man at the head of the firm was an extraordinary person and far superior to themselves in ability.

Miller took all the responsibility for everything. All a person had to do was follow his orders. If an employee did just as he was ordered and things went wrong Miller would always take the blame on himself and never try to put it off on someone else.

Wire gates came to be known as Tanner Gates. Wire gates with a stick for drawing them tight were known as Tanner Gates with Schmitz's Improvement.

In going across the country one time with his surveyor crew Wiley saw a grain fire starting up in the distance. It was over near the Jameson land. Wiley rushed over and got the harvester crew organized and fighting the fire. Word was immediately sent to Canal Farm and a team and flat rack with four or five barrels of water soon arrived on the scene.

People began arriving from everywhere and soon there were fully three hundred people on hand. Will Wiley had a team and gangplow soon going along the east side of the fire. His plan was to keep the fire up along the foothills until it should reach a deep dry streambed that he knew would stop it. D. W. Wallis arrived on the scene and rode along on horseback. Wiley explained his plan and Wallis promptly saw that it was practical, so the plow kept on going and the fire fighters kept whipping out the blaze along the east side. The fire ran for some twelve miles until it struck the deep gully where it was stopped. If the fire hadn't been held to the foothills and had got down on the plains it would have burned clear to Tulare Lake.

Fire burns slowly in a grain field but when fire strikes grass land it advances faster than a person can walk. Wiley was so successful in putting out grain fires that Henry Miller gave him the title of fire warden in addition to his other duties. Miller wrote to all his superintendents that in case of fire they were to furnish Wiley anything he called for: men, teams, barrels of water, sacks, and plows.

Eugene Avy was a wealthy businessman in San Francisco who had several thousand acres of grain on the plains along the Little Panoche Creek. Mr. Avy died before the crop was ready to harvest.

Mrs. Avy, who knew nothing of the business,

went to Henry Miller and asked him to buy the crop, as she was utterly unprepared to look after such a harvest. She offered to sell the crop for sixty-five cents a sack. Henry Miller sent Will Wiley to check up on the grain fields and report. Wiley spent two whole days going over the fields estimating the crop. He even shelled out measured scale to verify his seed estimate.

Hotel Los Banos, 1900

Joe Place, Christian and others had grain in the same area. Everything looked like a big crop and people were boasting of a yield of forty sacks to the acre. Wiley, however, reported back to Henry Miller that the yield would be only seven sacks to the acre. Everyone was wild at Wiley's report. But when the crop was harvested he missed the total number of sacks by only 65.

Albert Long liked to tell the story about the time they were driving a bunch of Miller's steers to market. They were camped overnight on the road near San Jose. In the middle of the night Albert was awaken by the screaming of a woman. It seems some people were driving along in a buggy and in the darkness the buggy wheel ran right up on the back of a steer lying beside the road. The buggy was nearly tipping over and the woman screaming with all her might. Fortunately the buggy was gotten off the steer's back before the steer had time to get up.

One time Henry Miller was at Canal Farm and in attempting to walk across an irrigation box the two by four broke and Henry Miller fell heavily against another cross-timber. Several of Miller's ribs were broken.

Miller was taken to the Canal Farm and doctors sent for. But immediately Henry Miller insisted on telephoning Albert Long to come to him at once. Long was in Salinas at the time and rode horseback nearly all night to rush to his uncle's aid. Henry Miller had a peculiar way of humiliating his superintendents. Torchiana was the superintendent at Canal Farm at the time. Miller had become very dissatisfied with him. When Albert Long arrived at Canal farm in the middle of the night he was immediately admitted to Henry Miller's room. "Uncle Henry," asked Long timidly, "Are you hurt?" "No. That isn't what I sent for you for. There is a steer in the Ward Sheds that hasn't had any water for four days. I want YOU! to get that steer some water." The chagrinned Torchiana could only look on and say nothing at Henry Miller's rebuke.

Will Wiley was thirty-two years old. Henry Miller was seventy-six. Will is satisfied now that Henry Miller was grooming him for one of the big superintendent jobs with the firm. He (Miller) had kept piling more and more work on Mr. Wiley. He was constantly giving him more and more authority. He kept sending him out on jobs that required unusual judgment.

One time he sent Mr. Wiley to report on a ranch that Henry Miller had a mortgage on for ten thousand dollars. The owner wanted Miller to take the land and call everything square. Miller sent Wiley to investigate. Wiley found that there were five public roads crossing the ranch. Wiley reported to Miller that all these roads would have to be fenced if Miller was to put any cattle on the place. Even if the firm put sheep on the ranch the feed was poor. So instead of taking the ranch Miller persuaded the owner to borrow the money somewhere else and keep the ranch. Miller got his money instead of taking over the ranch.

Wiley in addition to his constantly increasing load he was the main superintendent of canals. Wiley is satisfied now that Miller was grooming him to be his successor and the head man in the firm when Miller was no longer able to keep up. "But, I was young," states Wiley, "and I had become dissatisfied. I felt that I was being imposed upon. I couldn't at the time see the plans that Miller may have had for me."

So one day Wiley notified Miller that he was going to quit. That was in January. Miller told him that he was foolish. But Wiley was determined to quit. Miller's reply was, "I never ask a man to work for me that doesn't want to." It was not until October that Miller finally got a man to take Wiley's place. He was no doubt giving Wiley a chance to change his mind and remain with the firm. Wiley considers leaving Miller & Lux one of the great mistakes of his life.

Mr. Wiley stated he appreciated the time he spent in the employ of Henry Miller. He still marvels at the load the Henry Miller carried and the intricate knowledge that he held in every detail of his business.

There use to be a fellow by the name of George Phillips who lived on the east side of Center Avenue just south of the canal. Charley Smith lived just south of Phillips. He had a sow with a littler of nice pigs. Charley was in town talking with Wiley and some other men when Phillips came along. "Watch now, fellow," said Smith. "I'm going to have some fun with Philips."

"It's a funny thing," went on Smith when Phillips joined the group, "but do you know my old sow with a litter of pigs got out and the sow came home alone. A day or so later the litter of pigs came home and everyone of them had Phillip's mark in their ears." Phillips feigned surprise. "Why! Did those little rascals get out and go home!"

This same George Phillips noticed one time in the Miller & Lux pasture on Plow Camp that there was a nice unbranded heifer in the bunch of Miller & Lux cattle. It sometimes happened that at branding time a calf was overlooked and didn't get branded. Such a calf was called an "Orowana," meaning an unbranded calf. At the next roundup it would usually be found and branded.

This heifer in the Miller & Lux field may have escaped branding a couple of times for it was now a good-sized heifer. One foggy morning Phillips thought it would be a good time for him to drive it out of Miller & Lux's field and over to his own corral. He was in the midst of Miller & Lux's cattle and was trying to drive the heifer out. Just at that time he ran right in the midst of fog into a bunch of Millers vaqueros.

"My heifer got away and I was just trying to get her home." It happened that the vaqueros didn't know that there was an unbranded animal in Miller's field. It is almost impossible for a person alone to sort out a single animal from a herd and drive it away from that herd. The buckaroos were sympathetic towards this fellow who they thought was trying to get his heifer out of Miller's herd and very obligingly helped Phillips drive the unbranded heifer out the gate.

Antone Lopez was a fence rider and buckaroo for Miller & Lux for years. He never went by the name of Tony. It was always Antone Lopez. He and his bunch of buckaroos would often pass by the Wiley place when they were out rodeoing or branding. Mr. Wiley used to have a big garden patch of maybe twenty acres. The Wileys always planted one part to watermelons especially for Antone Lopez and his vaqueros. When they would come by in the hot summertime they would pick as many watermelons as they wanted and set under some trees and ate them. The Wileys enjoyed hearing these Mexicans enjoying themselves in the shade of the trees. This particular patch of melons were planted every year by Mr. Wiley himself.

There is a hill up near the Basalt Fire Station that has long been known as the Antone Lopez Hill. There used to be a cabin and spring lower down on this hill. The Hill is where Mr. Wiley proposes to have his siphon across the San Joaquin Valley come to an end on the west side of the valley.

Mr. Wiley says that about the last dirty deal he ever did for Miller & Lux was to plug up the James Canal. A man by the name of James was trying to take water out of the San Joaquin River. After James got his canal all dug and his head gate in Miller & Lux attempted to stop him from diverting any water out of the river. A dirt dam was thrown across the canal preventing a drop of water from going down the canal.

A man living near there was paid twenty dollars a month to keep watch and see that James didn't remove the dam. This same fellow was also getting twenty dollars a month from James to help him get water. It was no trouble at all for this fellow to let a trickle of water start flowing across the dirt dam. In no time at all the dirt dam was swept out and the river water went sailing down James Canal. James soon had several thousand acres of grain irrigated.

Wiley states he was sent to investigate. He had two plow teams and plows and fourteen scraper teams and scrapers on hand one night after dark and before morning they had a dirt dam six hundred feet wide closing up the canal completely. So completely was the canal closed that now even the banks of the canal on the James property have almost totally disappeared.

There were several hunters who use to go up on the Haley Ranch and hunt deer. Among them were Miller's men Will Wiley, Jim Huston, Charley Smith and some others. Ranchers in the mountains don't like to have hunters on their ranches on account of the danger of fire. There was an old adobe house on the back part of the Haley Ranch. Each year Haley used to have the place cleaned up and then used to let this bunch of hunters go up there and camp and hunt for a few days. They were required not to have any fire except in the fireplace in the adobe house. Year after year the fellows use to go there for a few days and hunt.

This old adobe house was roofless. These hunters used to fry lots of hot cakes. They became so expert that they could toss up a partly baked hotcake and catch it again in their frying pan. This was their method of turning it. The door of the adobe was right beside the fireplace. In time these hunters became so expert that they could flip their pancake over the adobe wall, rush out the door and catch it in their frying pan before it fell on the ground.

They use to have lots of fun with new recruits that they brought up from time to time to hunt with them. They would get the newcomer to think that he, too, could flip a pancake over the wall and rush out and catch it like the rest. But when the newcomer, looking up in the air, tried to rush out the door, the others would trip him and send the newcomer sprawling in the dust. One time a fellow by the name of Gus Smith came up to camp with them. Gus was sick at the time and so did not go out to hunt with the rest. He simply remained around camp.

One day when the hunters got back to camp they were surprised to see a large deer all nicely dressed hanging in the tree beside the adobe. "How come?" they demanded. Gus replied, "I got tired lying around camp and so about eleven o'clock I went for a little walk. A deer jumped up. There it is." And he pointed to the deer hanging in the tree. "If you fellows will just go up a short ways with a horse you will find another deer laying there. When I shot this fellow another jumped up and I got it, too."

The rest of the hunters had been out all day and had got nothing. They could hardly believe that Gus had shot two deer so close to the adobe. But when they went up and looked there was the second deer, just as Gus had said. For several years the hunters continued to hunt the Haley Ranch. They were very careful and everything went well.

One year when they were packing in Wiley had brought a Miller & Lux horse along. They had an old blanket on its back. To keep the blanket from falling off they tied an old riata around the horse. One of the fellows got tired on the way pacing in. He was carrying a gunnysack in which were matches and a glass jar of coal oil. He decided to tie the sack on the horse and let the horse carry it. He did so, tying the sack by the end of the riata. The weight of the sack soon pulled the riata around and the sack was dangling under the horse's belly. The horse's hind foot, in scrambling up the mountain, happened to strike the matches and set them off. The fire burning underneath the horse set the horse wild and in its efforts to get free its hoof happened to break the glass jar of coal oil. The blazing fire underneath the horse's belly made the horse simply wild. It ran for all it was worth. A string of fire nearly half a mile long was soon burning fiercely. Wiley hollered to shoot the horse.

All the woods were soon on fire. Everybody began fighting fire. Just when at last a day or two later they thought they had the fire all out an odd lot in the night blazed up and set the woods all on fire again. Instead of spending their vacation hunting they spent most of their time fighting the fire.

The horse was so badly burned that they had to shoot it, for which Wiley had to pay Miller $75.00. Miller hated fire.

That season practically ended the hunting in that region for these hunters. They felt they had abused Haley's hospitality so grossly that none of them felt like continuing to hunt on Haley's Ranch.

One day Miller decided to come to Los Banos and settle up old debts. Times were hard and many had gone under. Billy Stockton, who owed Miller some nine thousand dollars, was called in. Miller asked him to settle up. Stockton told him that he couldn't pay. Then Miller asked him if he could pay half. "No," Stockton said, he could not raise even that amount. "Can you pay fifteen hundred dollars?" Miller asked. "Yes, I think I can pay that much." Stockton raised the money and Miller wrote off the rest.

Next Chappell was sent for. "You owe me $5,000, can you pay me?" Miller asked. Like Stockton, Chappell had been hit by the hard times and could not pay. "Can you pay me half?" "No, I can't, Mr. Miller." "Can you raise $500.¿" "Yes, I can do that, Mr. Miller. And so it went and Henry Miller called everything square.

Each pioneer who was head over heels in debt to Miller & Lux was called in. Each paid what he could and Miller wrote off the balances. That day Miller wrote off three hundred thousand dollars that friends and neighbors had owed the firm for years.

The first scrapers that people had here in the country were what were called Chicago scrapers. They were simply a flat surface about a yard square made out of boards. A strip of iron a few inches wide was fastened along the front as a blade. At the back were two handles to hold the contraption sloping toward the front. Two horses were hitched to this scraper to pull it. There were no sides and often by the time a person got to the levy they were building all the dirt they started with had rolled off the sides. Later all metal scrapers with sides were developed and called the Fresno Scraper.

Mr. Wiley's father took a pattern to Billy Stockton at his blacksmith shop for two planks fastened together something like the letter V. This proved very practical. By running along in a furrow the dirt could be pushed up into a pretty good levee. Billy made about 28 of these for Miller's early day irrigators after Mr. Wiley's pattern.

Regent Hotel 562 Sutter Street, SF
Dear Ralph:

Thanks very much for the Los Banos paper. Some water. The storms somewhat equal, as I remember them, were in 1885 and 1906. In 1885 the water came into the old Charleston school house, which was off the ground about 16 inches, two feet deep, stopped school about three weeks and when we cleaned the mud off the floor which was about two 136-15inches deep, we found pencils, pens and books that had fallen off the desks into the mud. Alice Garrison, Billy Stockton's first wife, was our teacher, and a good one.

In 1906 the water in the San Joaquin River ran two feet over the top of the weir at the head works and I had to build a sack dam around both ends of it to keep it from washing out. In 1906 it rained at the old Wright Ranch up in the area near where Manuel Gonzales lives now, something over four inches in one afternoon and evening, and the water ran over the top of the new weir in the Outside Canal at Los Banos Creek about two or three inches, before we succeeded in pulling out enough boards to let it all go through.

At the time we were irrigating and the boards had to be all in, and the flood came at twelve o'clock in the night. Also washed out a piece of the canal bank in the San Luis Creek and drowned some pigs in a pen below, and a few chickens.

But it did give me a chance to fill the canal with muddy water and stop the seepage that was ruining alfalfa fields to Newman.

Thanks and happy New Year.
Yours W.M. Wiley

From interviews with Will Wiley in Los Banos at the residence of Paula Fatjo and Manuel Gonzales at home March 1,2,3 1954 RLM

Ralph Merritt, Miller & Lux employee

There was a big lawsuit going on in Nevada. Henry Miller, Mr. Merritt and Mr. Merritt Sr. were at the trial. Mr. Miller was put on the stand and swore that he owned ten million acres of land. Later in the day when we were alone my father spoke to Mr. Miller confidentially. "Mr. Miller, don't you know that you own only one million acres of land¿" "Yes," replied Miller. "But didn't yous hear that other rancher testify that he owned a million acres of land¿ If I had testified that I owned a million acres of land the judge would have thought that other fellow was as big a man as I."

Henry Miller had a great sense of humor. I drove for Henry Miller for years. In coming through the Pacheco Pass there was a tree beside the road where they would usually stop in the shade of the tree and let the horses have a breathing spell. Mr. Miller used to like to tell me of his early life. He told me many times how he first came over into the San Joaquin Valley. It seems that when Henry Miller was a boy in Germany he often had to drive the calves home that his father had bought for his meat market. One time it was very hot and Henry lay down under a tree and went to sleep. He dreamed that he saw a vast green valley with blue mountains in the distance covered with white snow. In the green valley were lots of cattle. They were red cattle with white faces. What stuck in the young boy's mind was that these cattle had white faces. The German cattle were black cattle. Such a thing as cattle having white faces was unknown in Germany and all these cattle he saw had HH branded on them, and all these cattle belonged to him. When Henry got home he told his parents and his sisters about his dream. They all laughed uproariously. Who every heard of cattle with white faces!

Years later when Henry Miller went to the slaughterhouse to bring a quarter of beef on his shoulder to his butcher shop he came across a fresh beef hide that had the brand HH on it. He instantly recalled the dream of his boyhood in Germany. He inquired around where the steer came from. He was told that there was a big ranch over in the San Joaquin Valley where all the cattle in it were branded HH.

As soon as possible he saddled up a horse and rode through the Pacheco Pass to see this ranch and these cattle. He found everything to be true that he had heard. And the man, Thomas Hildreth, who owned the ranch, wanted to sell. This man wanted $10,000 for everything, the land and cattle, the brand and all. Henry Miller didn't have the money to buy it but he succeeded in getting an option on it. Riding back to San Francisco he stopped under a tree on the Francisquito creek where Stanford University is now. While he was resting who should come riding along but a wealthy butcher by the name of Charles Lux. Mr. Miller knew Lux well. He had often delivered meat to Mr. Lux at his home in Belmont, just south of San Francisco.

"Mr. Lux," said the butcher boy, Henry Miller. Everyone looked on Henry Miller as the German butcher boy. "Mr. Lux, I want you to go in partnership with me." He showed Lux the option that he had on the two south leagues of the Santa Rita Grant. "I want you, Mr. Lux," continued the butcher boy, "to go with me to Mr. Ralston, of the Bank of California and sign a note with me for $10,000 to pay for it." Mr. Lux rode on about his business. When Henry Miller got to the city he managed to get in to see Mr. Ralston. He left the option with Ralston and said that Mr. Lux would sign with him. A few days later when Mr. Lux was in the bank Mr. Ralston said, "What about that option the butcher boy brought in? It looks like a good proposition." So Mr. Lux signed the note with the butcher boy and out of this transaction grew the firm of Miller & Lux.

Of course there are lots of discrepancies in the story. Henry Miller and Charles Lux bought out Hildreth and Dumphy in 1863. The firm of Miller & Lux according to the records was established in 1858. It is perhaps the failing of historians that they take such delight in spoiling good stories. But Henry Miller's story how he began to acquire land in the San Joaquin Valley is an interesting tale and may be as near the truth as any that have grown up about Santa Rita.

Another time, this time in the shade of the porch on the Poso Ranch, Henry Miller told me about his famous Peacock Dinner. As Henry Miller used to tell it, one time he had read a book about a man named Alexander and Miller thought, "I own more land than he ever conquered." In this book there was a picture of a feast Alexander gave to all the great people of the land over which he ruled. In order to show his greatness Alexander gave a banquet of peacocks. In the picture all the people were seated around a large table and husky black slaves were shown carrying the peacocks on their heads on platters. A whole peacock lay on each tray.

It seems that while Miller always felt he was as good as any man, the rich and influential men of San Francisco still saw him as the "butcher boy" even after he had become a multimillionaire. So he determined to correct their impression. There was always a twinkle in his eye when Henry Miller told this story. "I invited Stanford and Crocker and Ralston and all the rest of them to a peacock dinner at the Bloomfield." Miller had peacocks all over his ranch at Santa Rita. The foreman and the buckaroos were sent to capture the birds and it was hard work. The poor cowboys in their high-heeled boots had a terrible time. Finally they just took clubs and knocked their heads off. At the Bloomfield all preparations for the great feast went apace. Miller remembered the picture of the black men, peacocks aloft on their heads and hired waiters from San Francisco to fill the role.

The guests arrived. The waiters marched in, each carrying aloft a peacock. They marched in triumph twice around the table and then deposited the largest, fattest and plumpest peacock in front of Henry Miller to carve. He attempted to drive the fork into the breast of the bird. To his consternation the peacock slid off the platter and on to the tablecloth. Miller got it back on the platter and tried again. Again it slipped off the platter. He tried again with no more success than before. He finally had to give up. He did not know that peacocks are tough, no one can eat them. Quickly recovering Miller said, "These cooks of mine, they know nothing, they can't even cook a peacock! They made them so tough one can't eat them." Thus ended Henry Miller's peacock dinner.

Miller would laugh heartily each time he told this story on himself.

It was Tom Hildreth from whom Miller bought the first 8,000 acres of the Santa Rita and not Henry Hildreth. Henry Hildreth was made up in order to secure the brand HH brand.

Henry Miller is responsible for many of the stories about himself. He dearly loved to tell stories about himself. He use to say, "That fellow Alexander was a great man, I also am a great man. I own more land than he conquered and I give my friends a peacock dinner." The story of his dream is another story he often told on himself, but that great romancer, Billy Stockton, in his speech before the Lions Club embroidered the idea of Henry Miller being tapped to tend St. Peter's flocks in Heaven.

There was a tremendous water fight going on in court at Bakersfield between Miller & Lux and Haggin and Tevis. Miller & Lux had land lying along the Kings River. They claimed all the water in the Kings River as theirs by virtue of riparian rights. Haggin and Tevis, whose land lay away from the Kings River, by virtue of prescriptive rights by prior filings. Their case was fought most violently. Finally the judge gave down his decision. He gave all the water in the Kings River to Miller & Lux.

After the decision was announced Mr. Miller went over to the table where Haggin and Tevis were sitting, stunned by the judgment. With their land without water they were practically broke. "Mr. Tevis," said Henry Miller, gleefully, "I beat you, didn't I?" "Yes," admitted Mr. Tevis, "and your land with no water is practically worthless." "Now listen, Mr. Tevis. You builds me a high levee all around Buena Vista Lake, big enough to hold all the flood waters of Kings River the years that Kings Rivers comes down in floods and I will give you half of all the water in the Kings River the rest of the year."

Tevis was utterly astonished. No man in all the history of the work had every given anyone half of all the water in a river. Haggin and Tevis built the levee entirely around Buena Vista Lake which lasted as storage reservoir and Henry Miller kept his word and gave them water.

When I used to drive for Henry Miller we would have a big beefsteak breakfast and drive perhaps all day, changing horses at the various ranches, without anything more to eat until nightfall. I was driving Mr. Miller one day about 1906; there was no Gustine then. When we drove into Newman Mr. Miller said "Stop. I want to see Simon Newman." I tied the horses and Miller went into to Newman's office.

"Mr. Newman," said Miller "I have more grain this year than I can possibly get in my warehouses. You have a warehouse that is empty. I want to rent it to put grain in." "No sir," declared Simon Newman positively. "I won't rent you anything. Nothing at all." This made Henry Miller mad. He walked out and got in his buggy. We drove down the railroad track about four or five miles. "Mr. Merritt, you builds me a warehouse right here. Have Mr. Wiley draw a picture (plan) for you." I built that warehouse the summer of 1906. That was the beginning of Gustine. I don't know where Miller got the name of Gustine.

Mr. Merritt's first promotion with Miller & Lux came through Mr. Rouse, the superintendent of the Canal Company. He was up in San Francisco and was entertained at the home of C. Z. Merritt, the secretary of Miller & Lux and all their kindred interests. I was only a little boy at the time. I was born in 1884. Mr. Rouse promised me that if I would come down during summer vacations he would get me a job. I was right on hand when summer came and Mr. Rouse began taking me around with him wherever he drove on the canal. A few days later Henry Miller came by. "Who is that boy you got with you?" Miller asked Rouse. "That's C. Z. Merritt's boy." As soon as Miller knew who I was Miller ordered, "Young man, you gets in the buggy here. You ride with me." That was my first promotion. I eventually became vice president and general manager of Miller & Lux.

I am proud to say I grew up in the company of older men who were outstanding to me. I am proud that I was "raised" by Henry Miller in business and I regard him as one of the greatest men of California. You know many of the stories about Henry Miller are not true or are only partly true. There is the story of Henry Miller riding all over in a boat on a wagon in order to get title to some land. There is no truth in that story. It was told about a man in Bakersfield who was nicknamed the "Commodore." The story somehow later got attached to Henry Miller.

One time I was going to go to Fresno and I asked the Chinese cook if there were any supplies he wanted to have me bring back. The Chinaman gave me a long list. At the bottom was a note for me to get "One more long, tell 'em how hot." This was too much of a puzzle for me. It turned out that the Chinaman had put the ranch thermometer out in the sun and it had burst. What he wanted was a thermometer that would register higher without breaking.

4/1/1962 Interview RLM

Mollie Dailey, housekeeper Canal Farm

I came to California from Illinois when I was 17. My husband and I lived at Exeter for six years. We came the 17th day of March 1890. Fifty-seven years ago. She was Mrs. Sherman. After living at Exeter for six years we traveled north for a couple of years.

In October in 1898 we were going through Los Banos and heard that the superintendent's wife, Mrs. Torchiana at Canal Farm wanted someone to work in the house. We went out there and both were hired on the spot. After Torchiana was superintendent at Canal Farm a man by the name of McBride was superintendent for about six months. His wife was the daughter of Henry Lux, who was a brother of Charles Lux, of Miller & Lux. Her maiden name was Lena Lux. After McBride came D. W. Wallis as superintendent. Torchiana was over five years at Canal Farm. We were there three years with Torchiana.

H. A. van Coenen Torchiana, Miller & Lux foreman, attorney, Netherlands Counsel at San Francisco, author

There were big May Day celebrations at Canal Farm ever since we came to Canal Farm. There was an old Spaniard, Mr. Lopez, who took care of the barbecue. He was good at it, too. Henry Miller was right in his glory at the May Day celebrations. They had been going on for years before we came.

Mrs. Torchiana and a son Henry lived in San Francisco. There were two sisters. The younger had trained as a nurse. When she came to the United States she became a nurse at the old Waldeck Hospital in San Francisco. It was destroyed in the earthquake in 1906 and was not rebuilt. The nurse was Mrs. Torchiana and later became the second Mrs. Torchiana.

Torchiana said he was always going to write up the history of Henry Miller. His "gringo" book is not history; it makes him a hero and is a list of slights he felt under Miller.

D. W. Wallis was originally from Alabama. His father died of a broken heart. Lost all his property in the Civil War. D. W. Wallis' mother died when he was very young and he never remembered her. When still a boy Wallis struck out for himself and came to California. Wallis had a sister who taught at Mission High School for many years. The Wallis family were southern Methodists.

D. W. owned the Texacita Ranch near Gilroy. They lived there six years and Mary went to school in Gilroy. The Texacita ranch house is the long narrow house on the hill near Gilroy. It used to be a show place with the hillside all landscaped. Wallis also worked in King City for Spreckles.

Young Henry Miller used to make a slave of his wife. No one ever paid any attention as to when young Henry Miller came or went. Even his father Henry Miller seemed to pay no attention to him. One time young Henry and his wife were three weeks at Canal Farm. In that time he came to but one meal at the table. Young Henry's wife was a very, very nice person. She would always take his tray to his room for him.

There was always great movement when word got around that Henry Miller was coming. In Torchiana's time the main ranch house at Canal Farm burned. It was a very nice two-story house. There was some uneasiness as to just what Miller would say when he came after the house burned. Henry Miller use to wear boots. He had a bootjack in the house at Canal Farm and it was his custom to loosen his boots with the bootjack and then I would pull them off for him. All Miller asked me was, "Did you save my bootjack?" The first day of the San Francisco earthquake was the day the house burned.

Henry Miller's wife was an invalid for over twenty years. I used to know the housekeeper for the Miller family in San Francisco. She told me that when Mrs. Miller died the body was not taken to the undertaking parlor but was laid out in the Miller home. It was only of late that bodies are taken away from the home to the undertaking parlor. Henry Miller got up and dressed and then came and sat down beside the body of his wife. The housekeeper urged Mr. Miller to come and eat some breakfast. "Silvana, I can't eat this morning." Henry Miller died in 1916 not 1912 as Treadwell's book *The Cattle King* said.

Henry Miller told me many times, I remember well, about the great flood in the San Joaquin Valley. He went in a boat all the way from near this side of Pacheco Pass to the Santa Rita. It was only by boat that a person could get to the Santa Rita. Miller said that they had nothing to eat or drink but Hostettler's Bitters. He never could look at Hostettler's Bitters again. He said he got so tired he could have thrown away his boots.

Henry Miller always carried a picture of his daughter Gussie with him after she was killed. It was a small picture, framed and glassed over. Every night on going to bed he would kiss her goodnight. He would leave the picture on the bureau and many a time I has wiped the smear off the glass. Henry Miller carried that picture wherever he went.

It was a regular nightmare when Mrs. Nickel would come to Canal Farm on May Day. She would take possession of the big house. She would bring six or eight ladies from San Francisco with her. She would also bring her secretary and her maid and another maid to wait on the maid and her friends. They would take possession of the big house and I and the superintendent's family would have to go over and stay in the cottage. The clerks would have to get rooms in town wherever they could. Everybody would more or less tiptoe while Mrs. Nickel and guests were here.

Mrs. D. W. Wallis was the cutest, prettiest, little woman imaginable. Mrs. Nickel disliked Mrs. Wallis. When Mrs. Nickel was there the meals for her and her guests had to be served according to her orders. One time she came to me and said she would like to have dinner a little early. When the dinner was about to be served I went to Mrs. Wallis and told her that Mrs. Nickel had ordered dinner served early. When the dinner was served Mrs. Wallis was late in getting to the table. No one took the least notice of her. When Mrs. Wallis had eaten her soup she got up and without excusing herself left the table. No one took the least notice that she had left.

I was with Tapatula when he died. Tapatula was sick and had a room to himself in the bunkhouse. Torchiana had picked on Henry Reutter who was blacksmith at Canal farm to stay right with Tapatula and wait on him. I used to run over sometimes and take things to Tapatula. One time just before supper I went over to the bunkhouse to see him. Henry asked is she was going to be there for a few minutes. I said I would stay and Henry stepped out. Tapatula whispered, "Water" and I got a glass of water. Lifting up his head I gave him a sip and he lay back on his pillow. He gave one great big gasp and was dead. Torchiana gave him a wonderful funeral. He called in all the buckaroos from the ranches and there was a wonderful parade on horseback following Tapatula to the cemetery.

Henry Miller toward the last used to suffer from gout. His feet would swell so he couldn't put his weight on them. Miller would get so mad when they would tell him that he had the gout. "There isn't a foreman I got but what's lives better than I do."

One time just before the May Day picnic he had a terrible spell. He had started out to go to Dos Palos and down to Bakersfield but the next day he was back. When he drove in the yard I took his grip and helped him into the house. He kept getting worse. Wallis wanted to have his doctor from San Francisco, Dr. Reginald Smith. But Miller would declare: "I'm all right, I'm getting good care."

I was out to the park for a little while at May Day. The celebrations were being held in the Park.

Finally Miller kind of gave in. Dr. Smith came. Dr. Wade had been doctoring Mr. Miller. I had been acting as nurse to Henry Miller, putting on bandages as Dr. Wade had ordered. When Dr. Smith arrived on the afternoon train from San Francisco Mr. Miller was sitting up. Wallis, Clyne and Dr. Smith remained with Mr. Miller and I left the room. Presently Mr. Wallis called me back. I was afraid I had not wrapped the bandages, as I should. Dr. Smith asked, "Are you a nurse?" Then he asked, "Where did you learn to put on bandages?" Then he told me, "They are just as good as anybody could have done."

That old man Henry Miller was just as stubborn as he could be. The doctor wanted him taken at once to San Francisco. A special train was telephoned for from Fresno and it was ordered to stop at the Canal Farm crossing and pick up Mr. Miller there. It was planned to carry Miller on a cot and put him on board the train when it stopped. Miller got to thinking, "I'm not going to be put on the train at the crossing." He was determined he was going to take the train at the depot. He declared that if people saw Henry Miller being put on the train at Canal Farm everyone would say that Miller was failing. So to the depot they went. It took four men to lift Mr. Miller from his armchair into the carriage to take him to Los Banos. When the train was coming I hurried out to the crossing. As the train slowed down I motioned to the engineer that Mr. Miller was at the depot and so the train went on to Los Banos. At the Miller home in San Francisco they had to put in an elevator to handle Mr. Miller in his last days.

The May Day celebrations at Canal farm used to end up with a dance in town. Mrs. Nickel and her guests would be coming home in their carriages in the middle of the night. Jesse Potter was a jolly, well met old man and as they would drive in the yard at Canal Farm he would tell everyone to be quiet "else Uncle Henry will be getting up and starting to write letters." I remember Albert Long. He had red hair and was a very pleasant man.

Henry Miller was always very nice around the house. He wanted things quiet and every effort was made to have things quiet for him. When he would get to the house he would put on a house jacket and then stretch out on a couch in his office. After resting a little while he would get up and go to writing letters. He was an enormous letter writer.

Henry Miller at Canal Farm used to dictate his letters to Archie Smith and he would fix them up for him in proper shape. I think it was the same time that Henry Miller was so sick about May Day. He couldn't write letters himself he was so sick and so he would have Archie come out from the office and he would dictate to him. There was a copy of a letter that Miller wanted to refer to very much. It had been written a year or two before and was very important to him. Miller wanted Smith to send up to the city office for a copy. Smith told him he felt sure the copy was in the Los Banos office. "Well, brings it out," was Miller's orders. When Archie would typewrite Miller's letters he would fix them up so they read proper. Smith brought out the copy of the letter and read it to Miller. The old man grinned the way he used to do when pleased—his mouth stretching from ear to ear. "Did I write that letter?" he asked in surprise. "That's a good letter!"

Henry Miller's family was a lot of grief to him. He was embarrassed by the way his son, young Henry, turned out. Young Henry was rather nice looking. He had black hair and black mustache. But he was so dissipated every joint in his hands were swollen. J. Leroy Nickel was always an irritation to him. But Miller thought so much of his daughter Nellie. She was so much like him and was a businesswoman. Henry Miller would hardly even look at George Nickel. At Canal Farm when he would see George around he would hardly speak to him, he would just give him a grunt.

Henry Miller would say such droll things. Mrs. Brown was the housekeeper at Buttonwillow. She was there for years and finally decided to quit. Miller just hated to have her leave. I use to help Mr. Miller when he was an old man acting as valet and helping him get to bed. One night as I was helping him to get to bed I asked Mr. Miller how he liked the new housekeeper at Buttonwillow. Miller was disgusted. "She doesn't know," he said, "whether to serve the cheese with the soup or with the dessert."

Everybody had to be up early at Canal Farm. The main house at Canal Farm burned when Torchiana was superintendent. Then the people from the big house moved into the cottage. It was a seven-roomed house at that time. One morning before daylight I was up starting a fire in the heating stove. It was still pitch dark outside when to my surprise Henry Miller breezed in. "Where is Torchiana?" was Miller's first question. I knew perfectly well that Torchiana was sound asleep in his room. But I suavely replied, "I guess he's out around the barn." At the first opportunity I slipped up to the door of Torchiana's room and spoke quietly, "Mr. Miller is down stairs."

Miller had evidently arrived in Los Banos on the train and stayed all night in town and then got someone to bring him out to Canal Farm early in the morning. Torchiana bounded out of bed, dressed and then slipped out to the barn. Later he came to the cottage and greeted Miller. Miller always expected everyone to be up and going.

Miller & Lux water cooled tractor, 1916

One time I was very sick and was away from Canal Farm for months. I was in the hospital in San Francisco. To my surprise one day in came Mrs. Nickel to see me. She said that her father when he heard that I was sick in the hospital he asked Mrs. Nickel, "Have you been to the hospital to see Mrs.

Dailey?" Mrs. Nickel made many visits to the hospital to see me. She always brought something for me. She said her father wished it so. Sometimes she would bring so much fruit and flowers I could not begin to use it all and so I gave it to the nurses to take up to the children's wards. Henry Miller always remembered his friends. It was wonderful to me that Mr. Miller with all his vast business affairs should remember his housekeeper at one of his ranches.

Torchiana and Miller used to have round after round of arguments. They didn't agree very well. But Torchiana was superintendent for over five years. When Wallis came he didn't run things that much better but he had a way of getting round Mr. Miller.

McBride wasn't any account at all. He would just lay around drunk. He would go down to the office but he wouldn't bother to get out to the ranches. The reason McBride was there at all was because the Lux heirs claimed they weren't being given any hand in running the Miller & Lux business. McBride was the husband of Charles Lux's niece and was gotten in by the Lux heirs. He only lasted about six months and Wallis came in to take his place.

Mrs. Wallis was born in Battle Mountain, Nevada. Her people ran a railroad hotel called the Hancock House. Wallis and his family were at Santa Rita for a short time when Mary was five years old. Mrs. Wallis never made her home at Canal Farm. I always understood that when Wallis came to Canal Farm that Miller had it understood that Mrs. Wallis was not to live at Canal Farm. Mrs. Wallis and Mary lived in San Francisco. Mrs. Wallis would come down to Canal Farm and visit from time to time. But Mary would spend all her holidays at Canal Farm. She loved country life and horseback riding. After Mrs. Wallis died Mary had finished her schooling and she came to Canal Farm to live. She married J. F. Clyne, the canal superintendent. They were married in the living room of the big house at Canal Farm. Mrs. Clyne's brother was a Catholic priest and he came down and performed the wedding ceremony.

Mr. Clyne had come to Los Banos as bookkeeper for the Canal Company but it was not very long before he was at the top. I was housekeeper at Canal Farm for over fourteen years. I came in October. I married Mr. Dailey in January 1912 and left Canal Farm in the spring of 1912.

The great event of the year at Canal Farm was the May Day celebrations. These picnics were held in the grove. There was always an awful crowd of people present. The May Day picnics were meant as a favor to the superintendents and their families and the employees of the ranches. It was the one great event of the year. Miller wanted to show his employees a good time once a year. The public was welcomed to attend. Miller would mingle with his employees and shake hand with them all. Even the least Mexican laborer was treated courteously by Miller who would shake hand with every old blanket stiff that came near. Miller would greet everyone bluntly as was his way.

There was a large dining room at Canal Farm for the ranch laborers. There were usually about 85 laborers on the ranch. In the men's dining room there were immense large dining tables. On May Day all these tables were moved out to the grove. Sawhorses and planks were used for seats. No one sat on the ground to eat. Everyone sat at tables. The laborers of the ranch were suppose to eat out there too. There were tables for everyone.

Miller would furnish meat, bread, cheese, coffee and cream and sugar for all. The Chinese cooks at Canal farm would be up late at night baking the great quantities of bread needed for the celebration. They baked such great big loaves of bread. There was always a cheese maker at Canal Farm and they made cheese there. It was this home made cheese that used to served at the May Day celebrations. At night after the celebration in the grove there was always a big dinner for Miller guests inside. Then came the big ball at night in town. All would tog up and go. But I never attended. I would be so tired by night that I could hardly crawl. At the Miller table at the picnic Henry Miller was always at the head of the table and did the serving for his table. Meat would be brought direct from the pit for his table. The rest of the people went to the barbeque pit to get their own meat.

The cottage burned after I had left Canal Farm. The main house burned the day after Thanksgiving. Torchiana's son Henry used to play with another child at Canal Farm. Both boys hollered fire and I told them not to do that because people would think there really was a fire. "But there is," they shouted. Sure enough the big house was on fire. The fire broke out in Hans's room. It was thought that the fireplace a big old fashion double one had lost some mortar from a repair and the wall caught fire.

The second Mrs. Torchiana used to have pictures of the house. After leaving Canal Farm they moved to Watsonville. I was there the morning the first Mrs. Torchiana was leaving for Holland. She cared for no one. Even her own child Hans. She thought nothing of but herself. After she went to Holland they were divorced.

Mr. Torchiana became a lawyer and was appointed the Consul for the Netherlands in San Francisco. He married his wife's sister, a nurse. It was not a case of one sister cutting the other out. At the Worlds Fair in San Francisco in 1915 Mr. Torchiana was in charge of the Netherlands building. His wife was the hostess of the building. When my husband and two relatives were at the fair Mr. Torchiana invited the four of us to have dinner in the Netherlands Building and everything was served up beautifully. Later in the evening, Hans the Torchiana boy, now grown up, came in. He was in charge of locking up the Netherlands Building. When I was chatting with Hans I wanted to ask about Hans' mother. I inquired, "Do you ever hear anything from Holland?" He then told me his mother was married again and was living in New York. The two sisters were not alike in anyway. After Torchiana's marriage to his second wife they adopted two children who were sisters.

When young Henry Miller's wife died he married Miss Sara Onyon, an old maid schoolteacher. She was the one who used to wait on him hand and foot at the Canal Farm. She later died of cancer. There were three Henry Millers. Henry Miller himself, Young Henry Miller and Henry Nickel III, who froze to death in Nevada.

It sometimes rained May Day and celebrations had to be postponed. Among the crowd of the Miller family that use to come to the May Day celebrations at Canal Farm were Jesse Potter and his wife, there were also the Chamberlains from Merced, her parents. There were always lots of people. The bosses and straw bosses and all their families. One May Day while on the way to their home in Livingston Will White and his family had an accident. It was when automobiles first came out and there was a most terrific wind and the windshield was blown out.

Yes, Henry Miller did have an automobile. One time when automobiles first came out they were talking and Henry Miller expressed himself, "I hope my daughter don't get the idea of getting one of these devil's inventions." But she did and she quickly got one. Later Henry Miller himself got one for himself. He had a chauffeur to drive it. He would chuckle with pleasure that he could get in at Gilroy and be over at Canal Farm in two hours. He thought it was pretty nice. He could cover so much more ground in such a short time.

Pierson was the woodworker at Canal farm for years. After the house burned and Henry Miller's bootjack with it, I said to Pierson, "Do you think you can make a bootjack?" So Pierson made one and when Henry Miller tried it he told me, "Just tell that fellow he better try and make another one. If it is not better than this one he better go off an die."

Tony Lopez was dark and he was very dignified. He seemed to feel that he was one of the important persons around Canal Farm. Tony lived at Canal Farm and had a room in the bunkhouse. He and Tapatula each had rooms of their own. The more important people had rooms of their own in the bunkhouse. The China boys also had rooms there. Also the yard boss.

One time Lindsey the yard boss was awful sick at Canal Farm and it was thought that he was going to die. But he didn't. Later he died in Los Banos. He left $500 to each of Mary Clyne's two children.

I have seen Henry Miller so mad he would pound on the desk with his fist. It seems that something had gone wrong over the telephone. "Before telephones," declared Miller, "business was tended to right, but now nothing is done right."

In Tanner's time Henry Miller and some men where driving around the ranches around Dos Palos. They came to a gate. A man was working on the fence nearby. Miller yelled to him to come open the gate. "Open your own gate," was the prompt reply. "I'm busy working on this fence. You're just driving around and have got time enough to open your own gates."

When the workingman came in at night he asked who had been driving around that day. When he heard it was Henry Miller himself that he had talked to that day he was uneasy. He didn't get fired. Henry Miller left orders, "You makes me that man a foreman." Miller admired people that stuck to their post and went ahead with their work.

When I was at Canal Farm there was an average of 85 laborers, and as high as 19 clerks in the new cottage. There were the superintendents' families. There was way over 100 people living at Canal Farm. There were three Chinamen employed in the kitchen.

Interview April 4 & 5, 1947 RLM

Jim & Marie Knudsen, Miller & Lux employee

There was a storehouse for the ranch dining rooms and it was under the house in a basement. Fruits and meats were kept down there. They would bring a half of beef at a time. Jim the Chinese cook would go down there and cut up the quarter of beef. There were boxes of oranges down there and when Jim would come back he would bring a couple of oranges under his apron and slip them to me to eat. The laundry was a small building to the north of the Clerks Cottage. One time one of the clerks took her picture standing at the door of the laundry room. She still has it today.

The hitching rack was where the teams were tied when the stableman would bring up a team for Mr. Wallis or any others of the family. There was a bell at the stable and they could signal from the house when they wanted a team and buggy. When Mrs. Clyne wanted a team or any of the others wanted a team and buggy they would signal to the stableman and he would drive a team and buggy up to the hitching post and tie them to it so it would be there ready when they were ready for it.

People did not tell Miller the truth. He would ask Mrs. Clyne where her husband was and she would say up and gone when she knew he was eating breakfast in the clerk's dining room at the time. I lost my job at Canal Farm through this jealousy and lying. I had been working there for several years when Mrs. Sherman was housekeeper. I was her helper. I used to tie Mr. Miller's tie for him. He was waited on very solicitously. When Mr. Miller found I was let out of Canal Farm he was very much put out and sent me a letter offering me another position at either Bloomfield or Mt. Madonna. I prize this letter very much. It is dated 1899 and I was offered $25 a month in wages a lot of money then. One Christmas Mrs. Nickel sent me a piece of dress goods to make a dress for a Christmas present from Mrs. Nickel. That same Christmas I got my usual present from the store. Mrs. Clyne found out that I had gotten two presents and was very much put out. She declared, "You shouldn't have two presents." I kept them anyway. I had a splendid lapboard for sewing made by the Canal Farm carpenter as a present to me.

There was a big kitchen range in the kitchen at Canal Farm. The Chinaman use to fry the beefsteaks on the top of the range. Two kinds of meat used to be served at the Henry Miller table. Miller objected one time. "When you serve beef," he ordered, "serve beef. When you serve mutton serve mutton." There was always a big kettle of beans cooking away on the stove. Ellen Olson used to come to visit Mrs. Sherman. Ellen used to eat great helpings of beans. One time the Chinaman cooked carrots. Mrs. Clyne thought they were such sweet carrots. "Didn't she know," the Chinaman grunted to me—"Didn't she know I put sugar in them when I cook them."

When Henry Miller got down on a person he never forgot. He hated Simon Newman. One time Henry Miller bought a large bunch of cattle from Newman. Miller left orders at the Cottonwood ranch for the vaqueros to go and get them. But when they came for them Newman had sold them to someone else. Its seems another buyer came down and offered Newman a cent a pound more and so Newman didn't keep his bargain with Miller, but sold the cattle to this buyer who had offered him a cent more. Miller had business ethics and Simon Newman did not. In the dry years when the settlers would get discouraged and want to leave the country Henry Miller would always see that a family had a good team of horses and a wagon and some money to leave on. Simon Newman would strip the luckless farmer of everything. Miller was a fine man.

Interview December 22, 1954 RLM

John Dugain, cattleman

When they were pretty near through rodeoing Miller came and told me that he had rented the Fatjo place. Miller had agreed to buy all the stock on the place. The ranch included all the San Luis Gonzaga and the San Felipe Grant as well. At that time the San Felipe extended all the way down in the valley almost to Hollister.

"You are to gather the cattle and count them," Miller told me. "Fatjo's men will be there," continued Miller. "Don't take any Mexicans with you. Take your outfit and four or five white men." I did so and pitched camp at the fork of the road below Bell's Station about where the Don Pacheco service station is now. Miller came along. "I've bought everything on the ranch," said Miller. "Look out for my interest. Don't see too much. Send two men with Fatjo's men and take two of Fatjo's men with you." I told my men, "You don't want to see too much."

I made the proposition to Fatjo's foreman that he should let two of his men ride with Miller's outfit and two of Miller's men should ride with Fatjo's outfit. Fatjo's foreman thought it was a good idea. Each outfit went by itself scouring the country for cattle. They began at the west end and worked east.

It took us six weeks to make the round up. Each night they would all report to me the number of cattle that had been rounded up during the day and turned back behind them. One man in each outfit would count all steers, another the young cows, another the old cows. While they were scouring the hills for cattle they would see lots of little bunches of sheep. There were little bunches all over. I asked the Fatjo men how all these stray sheep got here. "Oh, the sheepherders would get drunk and wander off to town. The coyotes would get in the bands and scatter them." Every night I would mark down the number of cattle counted.

There were two or three fenced fields in the middle of the Fatjo ranch at the time. The two outfits got through with the counting of the cattle about noon one day. Both Miller and Fatjo were present and both seemed satisfied. Then came the counting of the sheep, hogs and horses. Miller told me to hand over the sheep to the head sheep man by the name of Vanderlip. The hogs were to be turned over to Tanner. There was supposed to be two bands of sheep. They counted two bands but none of the stray sheep. There was supposed to be 100 head of hogs in the field and they counted these and the number was correct.

After the settlement was all made between Miller and Fatjo, Tanner sent the whole Canal Farm outfit over to count up the sheep and hogs. They gathered about 4,000 stray sheep and about six hundred stray hogs. Soon after I got a letter from Miller telling me to gather all the cattle with the Fatjo brand and make a strict account. I began gathering the cattle. I had 15 or 16 men to help me. The first count was 1,700 cattle. We scoured the whole county. I know I didn't get them all but we gathered 3,000 head in addition to the original 1,700 head. I also rounded up at least 00 veal calves; some were a year old and still sucking. Miller came along to the big field where I had all the cattle corralled.

"You done well. You done well," said Miller when he saw what an enormous lot of cattle we had gathered. I said I thought I ought to have a raise in wages and I told Miller so. Miller wanted me to take charge of the Fatjo ranch for him. He offered to build me a nice house and put me in full charge. I refused. I had not proved up on my land yet. My family was living on the land and I was a little wary about moving off until I had proved up. So I refused Miller's offer. I told Miller that he ought to give me $75 a month. "Well go ahead," said Miller. I was tired out with the weeks of riding in rounding up the Fatjo cattle, I went home to my ranch to rest up. This was in 1872.

Miller used to have teams on the road over the Pass all the time. He had five or six ox teams as well as several teams of horses. He had the most beautiful horses.

Interview 1933 RLM

Mrs. Jeffers, pioneer family member

I remember when I was a child in foggy weather in the wintertime when there was lots of snow up on the tops of the hills the deer often wandered down on the plains. In the fog I have seen deer within 100 feet of the house. They would put up a fight if a person tried to drive them away. They were much larger than the deer we see today. They would follow the fence lines and you could see the prints of their feet as then went along the fence lines.

Germino was the harness maker for Miller & Lux. But he didn't work very long for Miller & Lux before he set himself up and opened up a shop of his own in Los Banos. He used to repair shoes for people for miles around and often when I was there with shoes to be fixed the Miller & Lux men would bring in whole wagonloads of broken harness from the various ranches for Germino to fix. Germino's shop was about where Louie Germino now has his shop at 5th and I streets.

One time when I was a little girl my friend and I were playing along side the road. She had taken off her shoes and hid them in the bushes. We saw a team and buggy coming along the road. I said to my friend it looks like "the big boss," just as Miller and his driver went by. They had only gone a short ways when they turned around and came back. My friend was barefooted but I was wearing my shoes. Mr. Miller gave her two dollars and told her to go to his store and to buy a pair of shoes for herself. Then he drove on. If I had known he was going to give out money for new shoes I would have had taken mine off just to get a new pair.

There was a well on the Reeves place. They had only a hand pump and the children used to take turns pumping water to irrigate their garden. They always raised a large garden and always had two or three rows of corn. Henry Miller always stopped when going by to get a drink from that well. He used to tell us children that their pumping so much water was good for the garden and good for the well. They had a nice garden and it kept the well sweet to pump it so much.

My father was David Reeves. He was born in Napa and when about three months old te family started driving and ended up in the San Joaquin Valley where the land was cheap. They landed around the Hill's Ferry area and settled and he grew up there.

One time when I was a little girl I was out along the road when a sheepherder went by. Two lambs had given out and he was just going to leave them by the road to die. I asked him if I could have them. The sheep man said certainly. I took them home and watered and fed them. The next morning they had revived. I kept them and raised them and the next year one had twin lambs. I sold the two lambs and the two old sheep for five dollars. I was staggeringly rich. I bought material for two dresses at fifty cents apiece. Two dollars in those days would buy a good pair of leather shoes.

When I was about five years old I went to a Henry Miller May Day celebration. This is the first one I went to. I remember there were a lot of hard-boiled eggs. Easter was very late that year and it was a sort of Easter and May Day celebration combined. There were big platters of green tossed salad, with rings and rings of hard-boiled eggs around them. Nobody took anything to eat. Mr. Miller furnished everything. When people left to go home Mr. Miller gave them large quantities of meat to take with them. That was the only time I ever saw hard-boiled eggs at a Millr celebration.

Miller expected all his employees there and all to be clean, shaved and have their hair cut when they came. A good boiled shirt. In those days that meant a white shirt. Miller & Lux teams would haul the men in from the ranches a day or two before the picnic to get shaved and haircuts so they would be spick and span for the picnic.

Interview 7/27/1959

Louie Arburua, rancher

Sentiment played no little part in the recent sale of the Miller & Lux cattle interests to Swift & Co. It was to get possession of the old "mother herd" of Miller & Lux cattle that prompted them to purchase the last of the Miller & Lux cattle. This herd represented the results of seventy-four years of cattle breeding by Henry Miller and his successors. Taking the original wild cattle of the plains and crossing them with Devon and shorthorns Henry Miller developed an ideal beef type, cattle that produced a small boned yet weighty carcass. From the Devon came smoothness and quality, while the shorthorn gave size and weight. In later years the adding of Hereford blood to the strain gave thriftiness and quick maturity.

"If you ever get the opportunity to buy Miller & Lux out do so without fail," the elder Swift advised his son years ago. The opportunity came recently and Swift & Co. became the owners of the "mother herd," the last of the Miller & Lux cattle. There were some nine thousand head in the deal. Swift & Co. will continue the cattle business on part of the old Miller & Lux range, but under a new system of management. Instead of raising steers until they are three and four years old they will simply raise the calves until they are big veals, when they will be shipped to their baby beef feed lots at Collinsville, near Rio Vista, on the Sacramento River. There they will be fattened as quickly as possible and marketed at once as baby beef.

Swift & Co. was never able to buck Henry Miller. In years gone by when the Swifts sought to dominate the butcher business on the Pacific Coast they never were able to compete with Miller. If Swift sold beef cheap Miller could always sell it still cheaper and if Swifts tried to put the price up too high Miller could always under sell them.

In the heyday of Miller & Lux prosperity the firm owned as many as 50,000 head of hogs and a 100,000 sheep and nearly 200,000 head of cattle.

The old Los Banos Division of Miller & Lux territory provided for 28,000 head of cattle; the Santa Rita Division, 30,000; the Poso Division 35,000 head and the Southern Ranches around Bakersfield 30,000 more. This was in addition to cattle kept on their Oregon and Nevada holdings. Three years ago when Miller & Lux sold out their hogs they disposed of some 22,000 at one time. Swift & Co. got some 9,000 head of stock; Miller & Lux still have about 1,000 head.

Albert Long, a nephew of Henry Miller, was a marvel in his understanding and care of cattle. He was to have been Henry Miller's successor. J. LeRoy Nickel, on gaining total control, proceeded to cut his salary from $150 month to $75 a month. Another cattle company quickly grabbed him up. His wage went to $300 a month and the Miller & Lux business under Nickel went to ruin.

Grant Pyle of the firm of Allen & Pyle was the man whom Henry Miller wanted to be his successor. One time Henry Miller was sick and not expected to live. He sent for Grant Pyle. When he came Miller told him that he didn't know whether he would ever get well or not, and told him that he wanted him to take charge of Miller & Lux and run the business. "Why, Mr. Miller," Pyle told him, " I can't hold a candle to a man you have already got." Miller asked him who it was. Pyle told him it was his nephew Albert Long. "All right," said Miller, "you can go."

Undated RLM

James Huston, Miller & Lux employee

Tanner was superintendent for Miller & Lux for years and years. He was a good man and was of such a tough physique that he was called "Rawhide." He was nearly seventy-five years old when finally he quit and went back East to retire. Torchiana was his successor and shortly before he quit he took Torchiana around with him to show him the various ranches and what was to be done. Coming to the ditch that used to supply Los Banos with water he found about ten Italians trying to mend a break in the ditch between the present bridge on the Pacheco Pass Highway and the paved way from town. The Italians were not making any headway in stopping the flow of water through the bank. Tanner jumped out of the buggy, threw himself in the mud and water across the break and ordered the Italians to shovel in dirt until the break was stopped. This the Italians did with a will and soon the break was filled. Tanner got up out of the mud and water and went on the rounds with Torchiana. Torchiana did not last long as superintendent. He went to San Francisco and later became a Consul for the Netherlands.

No deal was ever too big for Tanner to handle. He would give a sharp decisive answer no matter if the deal amounted to a $100,000. Tanner was boss of everything in those days, lumberyard, store and all. When Tanner retired it is said he had at least $200,000. When people used to tell Miller that Tanner was stealing from him, Miller would reply, "That's all right. He make me lots of money." Miller used to be very abusive to everyone in his employ and finally one day Tanner told Miller to his face as they were riding in a buggy: "Every time you call me a name it costs you $500!"

It was the custom in the days of horses and buggies to keep good driving teams at Santa Rita, at the River, at the Cottonwoods and other ranches so that Tanner could drive up, have a fresh team hitched up and drive on without delay in getting over the country. Tanner would sometimes drive all night, going along sound asleep, letting the horses keep to the road. Tanner was tall and slender, much like Bert Acker. He was as tough as a pine knot and had a great physical endurance. He talked quick and decisive and used to repeat himself much the same as Ogle who was superintendent at the Poso.

Torchiana lasted only four or five months and then came Wallis. Wallis had been headman for Spreckles on his estate near Soquel over at the coast.

Huff was the carpenter who built many of the big warehouses for Miller & Lux. His wife was Mrs. Tanner's sister. He now lives in Stockton. He built the Dilse house and used to live there. Since Wallis' time there have been many so called superintendents at Canal Farm, but they are for the most part simply fence riders.

The main fields of the Miller & Lux lands around Los Banos are the San Luis field that contains 40,000 acres under fence and extends along the river from the San Luis house to the McPike Ranch near Newman. There are two parts to the field, called the upper San Luis and the lower San Luis, although no fence separates them, there is a country road running from the Kragenhagen Gate north to the river.

All on the west side of the fence is the Upper San Luis field. That on the east is the Middlefield and surrounds the Arroyo Ranch. The Pick Anderson field is that field along the river east of the San Luis field and north of Middlefield. The Lone Willow field is that around the Lone Willow Ranch and lies west of the upper San Luis field. The little Mexican field is that field north of the racetrack.

One time Henry Miller wanted that field near the present Sweeney Gilt Edge ranch plowed and put into barley. Tanner didn't want to put it into grain, claiming that the land was no good and that the seed wouldn't even come up. But Miller had his way, and Tanner had to put it in. When they were nearly through seeding Tanner was going by and stopped to see how things were going. They used drills in those days and there were several of them working in the field. Tanner went over to one of the drills and looked into the seed box. There wasn't a thing in it! The driver was simply driving around the field with an empty seeder! "Good man," grunted Tanner. He knew the seed would not come up anyway.

1/5/1926 Interview RLM

Tom Hallahan, Miller & Lux employee

As a 9-year-old orphaned boy from San Francisco I started working as a vaquero for Miller & Lux at Firebaugh in 1892. I was paid the traditional $30 a month and board. Shortly after that I was joined by my brother.

What kind of man was Miller you ask? "Many people who did not know Miller from Adam's off ox but who have heard stories about him being a hard man and land grabber have the wrong idea about him." To his thousands of employees he was a good man, a tough man but a fair man. Sure he demanded loyalty and results, but what businessman doesn't? I remember when the Texas fever broke out on the West Side in the 1890's. There were more dead cattle than live ones scattered about the plains. Miller brought in an expert from Fresno. Miller in his heavy accent asked the man, "What do you think?" The doctor started into a explanation of the causes and results of Texas fever only to be cut short by Miller. "Doctor I know the symptoms and results. Have you got a cure?" The doctor started to say it would take time but Miller again cut him off. "You haven't got a cure! Then go back to Fresno. You have wasted my day."

Miller may have cut a few sharp corners in the matters of ownership of land and cattle, but the corners were so wide in those days that it was pretty hard to tell sometimes which were square and which were round.

With eight other vaqueros, all Californians and Sonorans, I was commission by Miller to drive 1,800 cattle from Santa Barbara to his New Columbia Rancho west of Madera. The Newhall Land and Cattle Company owns it today. From the Rancho Espagnole at Santa Barbara across the Carissa plains to Templor, down the West Side plains to the Devils Den, up the Kettleman plains to the Fowler Ranch for water, then paralleling the Kings River to Straford, to Whites Bridge, to Mendota to Los Banos, arriving there early on the Fourth of July for the traditional Henry Miller Los Banos Homecoming Day picnic.

We drove from 2 to 10 am stopping because of the heat, and resuming the drive at 4 o'clock in the afternoon and go until 10 at night. It took us 12 days to make the trip. On rating types of cowboys Hallahan stated, "I was talking about that one day with Will Rodgers" and he said, "The Texans and Oklahomans are wonderful riders and ropers, the native Californians know cattle better, but there is no one who can handle cattle like a Sonoran."

One time I needed 900 steers for the San Francisco market that I had to take from a herd of 3,000. I would send a Sonora man among them with his horse and he would get the 900 out without even getting the rest of the herd on their feed. Those Sonorans not only knew how to ride and rope they also knew and understood cattle. Many people do not know that Miller & Lux also had hogs and sheep. They sheared 120,000 head of sheep a year and shipped the fleeces from Firebaugh to San Francisco by boat.

Most people think that Mercey Hot Springs in the foothills southwest of Los Banos were so named because they are in Merced County. This is not so. They were named for John and Henry Mercey who ranched there in the early days and as far as I know the ranch is still in the Mercey family.

Fresno Bee 2/24/1951 clipping RLM

Frank Arburua, sheep and cattle rancher

When the Arburuas were married eleven years ago there was still about seventeen antelope in the hill southeast of the house out towards the Plains. We saw them that summer but they disappeared and were never seen again. Probably slaughtered by some band of hunters.

One time Wallis was in town early one morning and saw Prosper Idiart going along the street with his hands in his pockets. "Take your hands out of your pockets," Wallis shouted to Prosper, "and put them in someone else's pockets and maybe you will prosper." Turner is an old timer and related to the Turners that Turner's Island is named after and could probably give you lots of facts about early day river days and river steamers. Arnell is an old timer and always has a little dog "Prince" with him. He lives back of Mrs. Rodriquez's house. Mrs. Lassart who was Mrs. Ward can tell some very interesting stories of Henry Miller and the way he used to insist that his employees should always stand a broom up with the handle down.

One time a teamster driving a big team hauling for the store happened to be met by Tanner and Miller driving together. It just happened that one of the bellybands of one of the horses was unfastened and dangling. Miller and Tanner happened to notice this as they drove past and the promptly came back. "What is that strap for?" asked Miller. "Don't know," replied the teamster, thinking the best way to get out of the difficulty was to plead innocence. "Then cut it off," replied Miller as he drove on.

There was a breaking barn at Canal Farm and it was run the year round. Fernando Perlattos was the colt breaker for years and years. He was a wonderful roper and Charles Darr says that in his early days Perlattos used to be able to lasso even a jackrabbit. Perlattos used to fall off the water wagon at times and Wallis would fire him and declare that he would never have Perlattos on the place again. This was all for effect as Wallis thought the world and all of Perlattos and would never think of letting Perlattos get away. So when Perlattos would sober up and come back and humbly ask to be taken back Wallis would always give in and finally take him back provided of course that Perlattos would promise never to get drunk again.

Where Los Banos stands today was once the saddle horse field of Miller & Lux. It was a great life for the cowboys when they started out in the spring to round up and brand the calves. It took all summer and progressed from one ranch of Miller & Lux to the next. A cook wagon and about fourteen or fifteen riders with about eighty or a hundred head of saddle horses comprised the outfit. About 4 o'clock in the morning the saddle horses would be driven up to camp, the riders would have their breakfast and then pick out the horses they wanted to ride for the day. If some horse was wild or unbroken, a free Wild West show would start immediately. If any of the Arburuas came over to the camp wagon and wanted to stay a day or so the boss would always provide them with a horse and let them out with the riders as long as they liked.

Wallis was a great horseman and always drove splendid teams. The Buttonwillow country is some thirty miles wide and sixty miles long. Rafael Quinn used to be the boss of the wagon and buckaroos. When Miller & Lux rounded up the elk and shipped them out I wanted to be in on it, so Quinn promised to send me a telegram to come. The date for the round up was kept secret for there were so many people who wanted to come out and see it. A stockade of seven-foot wire had been built with wings on each side of the opening. I got the telegram to come. I put my saddle in a gunnysack and packed half dozen or so bottles of whiskey in my grip and started for Bakersfield.

The drive started in the evening. Just after supper about fifteen riders including myself started our for the hills. We got in back of the bank of elk and kept following them towards the enclosure. The riders were strung out in a long line with one of the bosses on one flank and the other on the opposite end. Every once in a while the bosses would light up a sage brush on each end of the line and the riders would get in line and get their bearings. We succeeded in getting a hundred and fifty or so in the stockade that first night. That was the best drive we made. Elk are vicious animals and have no mercy on a crippled elk, but will pounce up and down on the hapless animal and kill it. The elk were shipped all over the country. Many of the elk corralled were injured and were killed by the other elk or by the cowboys to put them out of their misery. Several drives were made but none so successful as the first night. Finally the elk became wary and the remnant gradually worked off toward the Lost Hills country.

Berenda is Spanish for antelope. Fred Strubble is a buyer for Johnson since 1895. In his early days he used to drive the lambs from the Avery place in Panoche to San Francisco. In the early days there wasn't enough sheep to supply the market in San Francisco for mutton and they used to drive sheep all the way from Los Angeles. Miller & Lux got in the sheep business. In wet years they would come by the San Joaquin Valley and cross over the Pacheco Pass on their way to San Francisco. The drive from Los Angeles occupied several months and the sheep were fattened as they traveled though the valley. In dry years Miller would have them driven up the coast instead of through the San Joaquin. Chris Padia first came up into this country with a band of Miller sheep from Los Angeles.

Interview 4/24/1929 RLM

Mrs. Howell Safstrom, wife of M&L Foreman

Howell Safstrom was the long-time superintendent at the Santa Rita Ranch. His wife contacted me in 1969 wanting to donate several of the photos he had taken over the years while at the Santa Rita. I asked her at the time to write down any stories or memories she thought was important about her days at the Santa Rita Ranch. These photos and comments are from her. Notes RLM 1969

My First Visit to the Santa Rita

It was during Christmas week of 1905 that my sister Alice Gamboni and her husband Joe came up to Marin County to spend the holidays with their families. As a young couple not married too long they were employed at Santa Rita, Alice as the housekeeper and Joe as the head blacksmith.

Miller & Lux Store in Los Banos after the earthquake, 1906

On their return to the ranch I was permitted to accompany them, to remain there for a couple of weeks. They had told us so much about the ranch and the people there and I was so very excited for I never before had been far away from home. They talked so much about the superintendent (Mr. Safstrom) and had his permission to have me go back with them.

My sister Maude had come home to be with my mother during my absence. A stage (two horses hitched to a spring wagon) carried mail and passenger to Coloma—two miles over a long hill, meeting trains. After kissing them goodbye and boarding the stage I said, "I'll be back with the superintendent." In reality there was no such thought in my mind.

In South Dos Palos we were met by Joaquin the stable man with two horses hitched to a buggy. He led a saddled horse, which he rode back to the ranch. It had rained the week they was away. The road were full of chuckholes and it was a pretty bumpy ride over all the deep ruts. The wheels were caked with adobe.

Arriving at the ranch and after unpacking Alice and I proceeded to develop some films of pictures she had taken while at home. The window shades were drawn and the room was in darkness. It was then about 3:30 in the afternoon. In a little while there was a knock on their bedroom door. It was Mr. Safstrom there to greet them home to ask her where she wanted to place a crate of oranges he brought from the Canal Farm. She introduced us, as I still sat on the bedroom floor with all the developing equipment about us. He said, "Hattie, I feel that I already know you. Alice has talked so much about you." I wasn't in any romantic mood but I thought, "Isn't he nice?" And later he told me his first thought was, "Isn't she cute?"

As the days went on, he at times, would take Alice and I with buggy and team also on horseback and several times Alice couldn't go, so I went alone. It was not until Alice became ill that my mother consented to my stay for a third week in order to be of help to her that Howell and I had really "discovered" each other and became engaged. In another month I would be sixteen years old but I still felt it was pretty young and too young to tell Alice and Joe and my folks at home, so we made no immediate plans. He told me he would come up soon to see my folks. I used to hide the little box with diamond engagement ring behind a picture frame in the "sitting room" and I would get the mail and would hide his letters. I always felt my folks would like him but there would be a question of disparity in our ages and the little time I knew him. But Alice and Joe had helped a lot, in all the nice things they told about him.

About six months later Alice and Joe reluctantly left Santa Rita. Joe's father was in the butcher business in Olema for many years was retiring and wanted Joe to return and take of the meat market, which served several adjacent communities. They made Howell promise to come up to visit them but he didn't tell them he already had plans.

On his first trip he took a train to San Rafael to engage a team at a livery stable to drive the distance of about twenty-five miles to Olema. Alice and Joe told him what a beautiful drive it was through the hill and the Lagunitas canyon of pines and redwoods and past dairy ranches. In those days dairy cows grazed all over the Marin hills. A Mr. Dave Geary operated the livery stable in San Rafael and his father-in-law a fine old and retired dairy farmer of Pt. Reyes, then living in San Rafael, was at the stable about ready to drive up to the country to visit friends and he invited Howell to ride along with him. The North Western Railroad operated trains along that route but Howell was eager to see more of the country and Mr. Butler having been of resident of Marin for the many years could tell him a lot. Mr. Butler's son Edward was Superior Judge of Marin County for many years and the Geary family operated a dairy farm also at Pt. Reyes.

While at the Nelson hotel where Mr. Butler put up his team, Howell asked Walter Nelson if he could tell where Mrs. Gamboni lived, and it was just at that moment that Alice came walking past on her way up to our house and Howell called to her. She came in all excited and said, "I'll bet you could never guess who I just saw—Mr. Safstrom from Santa Rita and he will be up here in a few minutes." I don't know what kind of an expression I had on my face pretending I too, was so surprised but knowing by his letters that he was coming. After being introduced to my folks who I could see readily liked him, he said, "Mrs. Fiori, Alice and Joe and Hattie told me so much about this place and the country side, I had to come and see it."

Within a little while Joe came. My mother said, "Joe, Mr. Safstrom can stay over for only one night, couldn't you hitch up your horse and buggy and drive him around a little before we have supper—he said he would like to see more of this country" Though he wanted to see more of me he had to go. After their ride through Bear Valley, a beautiful scenic one, we all had a nice dinner and in the evening visited Joe's folks. So we were never alone. About two months later he came up again by train to "ask for my hand," my mother joking asked him he wanted to see more of the country. His third visit was when he came for me, thirteen months later in February of 1907, after I had just passed my seventeenth birthday. We were married in the minister's home.

My mother was only sixteen when she married and she and my father, though both born in Switzerland didn't meet until after they came to America. They had a family of nine children—eight girls and one boy. Their first and last babies, girls, died in infancy and I was the youngest of the living. They wanted me to wait at least until I had attained my seventeenth birthday. Now seeing girls of that age I married at, what we knew. We were taught all the phases of homemaking even to baking bread and sewing our own clothes. It was a part of our duties a part we expected to do. My mother would have thought it a bad reflection on her if her daughters weren't prepared at marriage to care for a home and the man she married, a girl who would make him a good home.

On that first visit to the ranch in 1905 Howell's father, William Safstrom, was a foreman for Howell. He slept in a spare bedroom of the old house. He was a fine man, born in Sweden, a graduate of the Stockholm Agriculture College and immigrated to America a very young man. He settled in Georgia where he met Howell's mother at church. He used to say that most Swedes settled in Minnesota, that in fact he saw so few Swedes in the south that he forgot a lot of his own language. Howell was their first child and two years old when they came to California and settled in Gilroy where his father farmed Miller & Lux fields around the Bloomfield Farm on shares.

Going back to his time as foreman at Santa Rita one day while helping my sister, I was sweeping the hall when he came to his room for something. He chatted a while and asked me if I liked housework and I told him "very much so." Howell told me later that his dad spoke to him about me, that he said I would make a good wife for either a rich man or a poor man.

I always regretted that my mother didn't live to see our son—she contracted Bright's disease and died two months before Arthur was born in 1909. Howell being a farmer was an outdoor man. And I was born on a diary farm near Nicasio and though I had many young friends I always enjoyed the quiet of country life and I was happy when with older folks. So the disparity of ours ages (31 to 17) was never noticed.

As I write this today (March 28) had Howell lived, we would have been together for sixty-two years and I'm sure we would have had many happy moments reminiscing of our days and happenings while at Santa Rita. It is thirty years today that he left us. God seems to a way in molding our lives or shall we just call it "fate."

Shooting of George Nickel

There was a day of great suspense at Santa Rita the day a small Italian by the name of Lapadula (I think his first name was Vincenzio) waylaid and tried to kill George Nickel in a field near the Delta Farm.

It began when Howell phoned me from the Delta Farm, to tell me George had been shot and was in a serious condition and at that moment were waiting for a doctor to arrive from Los Banos. Ernestine (his wife) was soon to have a baby and had come down from the city to supervise the plans of a large new house, being built at the Delta for them. George was then superintendent of the ranch and Pasquale Bisignani his foreman. That day George had driven Ernestine to Los Banos in a Ford to take the train back to San Francisco, and he was returning to the Delta when he saw a man at the gate not far from the ranch. Lapadula had worked for Miller & Lux off and on but George didn't know him. Thinking it was a man waiting to ride on to the farm he (George) waited for the man to open the gate, but instead the fellow leaped up on the running board and shot George in the chest. George later said he didn't hear or feel the shot and not until he saw the pistol in Lapadula's hand did he realize what had happened. George leaped out at him and they struggled for the possession of the gun, which was knocked to the ground.

Mr. Perkins, foremen of the Midway Ranch, was on horseback riding over to the Delta and seeing two men fighting he spurred his horse on, soon to see one was George. He thought they were just fighting and seeing the gun on the ground reached for it just as Lapadula did. His chief concern was getting George back into the car and to the ranch, and lucky that he did. His father J. Leroy Nickel was immediately notified at his office in San Francisco and he ordered a special train made up at Fresno to take George on to a hospital in the city. It was said the train made up of an engine, baggage car and passenger car made record time, carrying George, a doctor and two nurses from Los Banos. He had asked that his wife not be notified—he didn't want her to suffer a shock and he didn't want his boots removed when they laid him on a bed at the Delta. He said if he had to die, he wanted it to be with his boots on (George most always wore special tan leather riding boots). He was met at the Oakland pier by an ambulance and his wife was there too. I can't remember the hospital to where he was rushed, but there was deep concern for many days.

Going back to Lapadula, while Perkins was giving aid to George, he vanished from near the gate, but men from the ranch were dispatched at once—the sheriff and deputies soon on their way from Merced, Los Banos and Dos Palos. On arrival they spread out in all directions. By that time they knew Lapadula had a good head start. And up to this time no one had any idea who would be the assassin. A cornfield near by had been harvested but the tall stalks were still standing and cattle had been turned into the field. Pasquale and another man with him saw the cattle standing—all with their heads in a single direction and as if frightened by something. They did not go far before finding the fellow hiding in a ditch and he didn't resist his captors. In very little time and manacled he was in the sheriff's car. The sheriff wanted to phone to Merced so they drove into Santa Rita and quite a crowd had gathered about the car. Some one had called out, "Leave him here with us, we'll take care of him with a rope," but the deputies kept close watch. Howell's niece (a school teacher) from San Jose was visiting us at the time and she and I went out to look at him. We stood some distance from the car. He glared at us and with a sneer said, "Take a good look." Howell was in an extremely angry mood and said, "You little_____, why did you shoot George for?" Pasquale was standing right besides the car calling him everything he could think of in Italian. It made but little effect on both Howell and Pasquale when Lapadula said, "Wait until I get out, I'll get the two of you." But I surely was fearful that he might escape somehow and return to carry out his threat. I felt so relieved when I knew he was taken to San Quentin and killed by a guard on the wall. A single rifle shot felled him as he was raising a piece of pipe or something ready to strike a guard in the prison yard. The guard's back was turned—those in the yards carry no weapons. From all accounts he had been a troublesome inmate all the while he was there. Some of his countrymen who knew him and worked with him on the ranches said he was always quarrelsome and created a lot of trouble.

He was seeking vengeance when he shot George—vengeance for losing the sight of an eye years before when as a carpenter's helper and hammering a nail in a fence or something the nail flew up and stuck him in the eye. His idea and purpose that day was to kill the first member of the Miller family who came along. He had worked out at the Delta at one time so I guess he pretty well had George in mind. He was so small in stature; one would wonder how he could hurt anyone as big as George. He had made George helpless having grabbed George in the groin. If it hadn't been for Mr.

Perkins who was a big man too, George would have died.

Hobos under San Joaquin River Bridge

One summer's night, at about 2:00 o'clock in the morning (the date escapes me) I heard footsteps on the garden walk, beside our bedroom windows. It was a warm night and we had all three windows open close to our bed. The footsteps continued on to the front porch and soon the doorbell rang. I immediately thought it to be the night watchman. Howell was sound asleep and before he awakened to know what was happening I called from within the living room, "Who is it?" It alarmed me for I knew Charley Riley (the night watchman) would walk up close to the windows whenever he wanted Howell for something. On this particular night, Riley was at some other spot and didn't see anyone until later. There were two large spotlights on light poles in the yard and all the gates entering the yard were locked each night.

When I asked, "Who is it?" a very young voice answered, "We want to telephone the sheriff. Some fellows down at the river bridge tried to hold us up." Just at that moment I saw Riley walking towards the office. Any direction the boy went out to meet him and after very little conversation both entered the office. Immediately I heard the phone ring. The house and office phones were connected. I didn't listen in as I remember I was too frightened, but I knew Riley was phoning for the sheriff. Howell said, "If it is anything of importance Riley will call me." So I went back to bed. After we were up (that same morning) Riley said while he was phoning the sheriff an older man appeared at the office door. Thinking it was a planned scheme to hold him up, he drew his gun and told the man to stay where he was. He told Riley he had tied his horse (hitched to a buckboard) outside the gate when he found the gate locked that he came in to see why the boy hadn't returned. And he told Riley some men under the bridge tried to hold them up on the east side of the San Joaquin River Bridge there was a regular hobo camp down under the bridge on the bank—a mile or more from the ranch. Riley advised them to drive back to the bridge to meet the sheriff there. Instead they drove on towards Los Banos, trying to make a getaway.

When the officers arrived at the "hobo camp" they found several men there—two had been shot to death—one was wounded and they crawled through a barbed wire fence in the dark and hid in some tules. What happened was the other men were there when the man and boy drove down under the bridge, unhitched the horse, preparing to spend the night there. They had a jug of wine they bought in Merced. During the evening there was an argument and the man hitched up his "rig," drove up on the bridge from the railing on the bridge, shot down on the heads of the men, killing two and wounding one. It seems some of the men already there had some wine while sitting around a fire, and they told them they couldn't see too far in the darkness. Driving up to the road and seeing the lights at the ranch prompted the man and boy to drive in and pretend they were seeking help. Riley said the boy seemed almost scared to death. The man he thought was about 45 years old. He was a "wagon-stiff" as they called them. The boy was a runaway from Stockton and was picked up by the man and 15 years old.

They hadn't gotten too far up the road when apprehended and taken to jail in Merced. Their rig was brought back to Santa Rita, the poor old horse taken care of and turned out in a field, the wagon (a buckboard) with their belongings wheeled into a shed until after the trial.

The trial wasn't too long after their apprehension. The case was tried in Superior Judge Rector's court in Merced. The constable John Campbell of Dos Palos came to the ranch with a subpoena for me to appear as a witness for the defense. They were assigned an attorney to defend them and he wanted to make it appear they were seeking help and that I had talked to the boy, even if it was through a closed door. Needless to say I just dreaded the thought of appearing as a witness for a murderer and his caliber. Howell went to see Judge Rector. I was so relieved when he dismissed me with without having to appear. The boy was placed in juvenile custody—the man sentenced to life in San Quentin. The "raggedy" wagon and all blanket rolls and other gear burned and long before now the poor old horse must have gone to happier pastures.

Lux Family

The only members of the Lux family I knew were the Chamberlain family in Merced and Sheldon Potter. I had been to the home of the Chamberlains in Merced—they only had one child, a son, Fenner, and I think he is still in Merced. Their properties, the Chamberlain Ranch, was leased by Miller & Lux for grain farming and joined on to the Ehlers ranch. The Chamberlains and Potters were very close friends of the Ehlers family who seemed almost a part of the Miller family. The Chamberlains had one of the fine homes of its day. They were about the first to have a car. While cranking the car one day, he had left the car in gear and it ran over him—his death resulting from the injuries he sustained.

More of Mr. Miller's Generosity

At Christmas time all superintendents and foremen received Christmas gifts—some got checks and some increase of salary. Howell and I each got generous Christmas checks with letter asking that they be not discussed. And I must add for many years I was gifted with a very lovely vacation check each summer. When I first went to Santa Rita one Christmas Mrs. Nickel (shopping for Mr. Miller) had 12 yards of beautiful black silk taffeta sent to me from the "White House" in San Francisco. In those days it was "the store" and one could be so proud to say, "It came from the White House." I had a lovely suit made from the material by a lady Mrs. Cunningham of Dos Palos and enough was left over for a separate skirt to wear with pretty shirtwaists. And one Christmas a gift from Mr. Miller was a set of lovely "Beauty" pins—gold bars with a genuine pearl at each end and a matrix stone in the center. They were used in the days shirtwaists were worn. Instead of buttons, a large pin was to fasten the collar and two smaller ones (of same design) fastened the shirtwaist down the front. I still have the two smaller ones. Years back when women were wearing wide brimmed hats with veils fastened tightly from them to around one's neck. My veil caught in the pin. I was in the ferry building in San Francisco waiting for a ferryboat for Marin (on my way home to visit my folks). On my arrival my mother noticed one of the pearls gone. Remembering the catching of my veil and pulling it loose I must have pulled the pearl out of the pivot. So on my way to the ranch I stopped at Shreve's where the set came from to have a new pearl set in. When they told me it would be $30.00 I decided to wait till later. I wrapped the pin in tissue paper and placed it in my purse. Howell met Arthur and me in the city where we remained at the hotel for a couple of days. Before leaving I cleaned out the shopping tags from my purse into the waste basket and I fear that is where my pin went for on reaching the ranch I did not have it.

Every time I wear the smaller ones I am reminded of the occasion and I've always felt so badly about it. Not only do I not know what happened to it I knew they were very good pins because they came from Mr. Miller.

When Mrs. Nickel went abroad with her daughter Beatrice, who graduated from a girl's school in San Francisco, she brought back gifts to most all the women on the ranches. Mine was a very lovely sewing basket from Germany. I forgot to add that foremen who had children were asked to send in names and ages of their children at Christmas time and appropriate gifts were sent to them from the store in Los Banos. Each time Howell received an increase in salary, Mr. Brown (office manager in the city office) would write saying Mr. Miller did not want it discussed.

Crops

There was a large acreage of Miller & Lux lands rented by a large melon company from the Imperial Valley. They raised acres and acres of luscious watermelons, cantaloupes, casabas, Persian melons and citrons. Almost everyday a good-sized paneled truck would bring melons to the ranch and unload them alongside the fence on the slough close to the men's dining room. The men could help themselves to any amount they wanted. Some would place the melons in barley sacks and stake them in the slough near the bunkhouses. The water from off the snows was pretty cold and it made for better eating melons.

Another firm from the Imperial Valley experimented with growing peppermint on several acres of land close to the river bridge. From the highway the smell of the mint was very prevalent and pleasant. In appearance it looked like a field of alfalfa. The peppermint people constructed a press or may I say "peppermint still" to extract the oil. Somehow the experiment ended with the first planting but the oil was wonderful—so pure. It was bottled in small vials, being expensive—a drop or two going a very long way. I made delicious peppermint candy. The picture of the still, I'm sure it's with the group of shots I am sending you. Howell was standing on the raised platform.

We were sent quart bottles of the freshest and purest olive oil extracted from the olives at the Holland Farm. No matter what fine brands were sold in stores, there was no comparison to this very refined clear oil. It makes me hungry for a nice French dressing on a salad right now—it had the real taste of olives.

Henry Miller had a lot of peacocks on the Santa Rita Ranch. They were trapped and shipped to various parks in other states as well as to Golden Gate Park in San Francisco. Also, some were sent to friends of the firm for their private estates.

Hogs

They also use to trap wild hogs along the river, which were then hauled to Santa Rita where they were altered, their long tusks were sawed off and they were fattened for slaughter. I can't remember if the tusks were also knocked out or not, I think some were. I have sent you two pictures of Pachuco on horseback with his dogs. We called them "coyote dogs"; they also ran down coyotes, which were plentiful. The mournful (coyote's) howl in the night used to frighten me. They roamed the fields adjoining the San Joaquin River. Charley Ehlers used to trap many of them. The county paid a bonus for the pelts.

Pachuco's dogs were exactly like the "greyhound dogs." Since I am reminded of Pachuco and his dogs, I can't remember of him being in a camp but he may have been. I do remember one day driving to the Holland Farm to visit Mrs. Leonard. I was on the what we call the back road to Dos Palos, when Pachuco's dogs sprang up from a deep ditch that parallel the road, scared my horse into a regular run away. I was driving a gentle palomino mare and by the time I got her stopped I was scared stiff and that poor mare just trembled. I think the dogs were asleep in the dry ditch and the sound of the horse's hooves scared them. It was some distance from the ranch and Pachuco was nowhere about. But those darned dogs always came back. I can't remember what became of Pachuco or his dogs. He was an elderly man even then. I can't remember the boys who Pachuco hired to hunt the wild hogs. But I do remember the vaqueros when at slack times they would go to trap the hogs. To them it was a sport though it was a dangerous one. I seem to recall a horse having its belly split open by the tusks of a

Flooding at the Belmont Ranch, Dos Palos, early 1900's

Flooding at the Santa Rita, 1907

boar. They use to frighten me even though I was standing on the fence of the corrals near the slaughterhouse. They were truly vicious. If a man stepped from the fence into the corral he soon scurried back to his high perch.

Mr. Miller on Horseback

I had never seen Mr. Miller on horseback or ride a horse at all the whole while I was at the Santa Rita. Howell had seen him ride and many times he went amongst the cattle and in corrals. In later years he used a buggy and driver.

Floods

I have written you about the floods and the June rise when we had to use boats getting to and from Santa Rita. Howell at one time said he was nearly drowned when the horse he was riding (a good swimming horse) floundered and Howell was forced to leave the saddle and swim to safety. Miller & Lux opened up all the floodgates in the ditches and canals, taking all the floodwaters over their land in order to save the farms of the colonists in Dos Palos is true.

Mr. Bates and the $1000.00

Miller & Lux got the $1000 back, found in the overcoat pocket. It was established before several of the witnesses who were there when he collapsed, that it was their firm's money and especially when Mr. Whelan (office manger of the Dos Palos office) told Howell receipts turned in to the officer were $1000 short of usual.

As I said I wanted to get these things to you and these are the thoughts that came to mind. I don't know if this material will be of much use to you if I can afford you a better scope on the subjects let me know.

PS: Paqual Bisiganani of the Delta Farm was up to Fairfax last Friday night and he phoned me that he had come up with some friends from Los Banos to attend the funeral of their uncle Mr. Guaglianoni and wasn't able to come and see me. I told him I had written you about Lapadula and George Nickel. He said to tell you if you need some other facts to call on him. He may be able to make corrections on what I've written, as he was the one who found Lapadula in the field.

PPS: You asked if I believed in tipping. "Yes." I am on Mrs. Milliken's side.

Post Note: Ernestine couldn't see why Perkins, (when he had the gun) didn't shoot Lapadula, rather than let him get away. The three men were out in that field alone. Had Perkins shot Lapadula and George had died who was there to prove that Perkins having the gun in his possession didn't shoot both men? She still thought Perkins was stupid. He instead deserved a good reward in our estimation and I would bet she didn't even thank him. Anyway as I recall the bullet was never removed from George's chest, fearful it would move and be fatal. I may be wrong. If you talked to Pasqual he could give you a better version of this affair, also the dates.

Frank Marks, early pioneer, Dos Palos

In April 1892, Bernhard T. Marks and his elder son Howard B. Marks secured a contract from Miller & Lux to purchase six sections of land for the purpose of colonization. The price $40.00 per acre to be resold at $60.00

Henry Miller had never sold land—his life in California had been for the development of the cattle and sheep business. Year after year the profits of the business had been put back into more land to raise more stock to make more money to buy more land to increase the business. Marks' approach happened to be at a time when Miller was being harassed by the heirs of Charles Lux for a settlement of the estate. All they wanted was cash! If he could go into court and say he was trying to sell the land he could gain time. This is undoubtedly what led him to make the contact with Marks.

I learned to know Henry Miller very well and

found him a man of integrity, and I do not believe he would have put Marks on such worthless land if he thought the land would really be sold. Bernhard Marks had three years previously tried to colonize land in Kern County owned by Haggin & Carr. A few families had been settled there only to learn the next summer that they could not get enough water to irrigate their crops owing to the fact that Miller & Lux controlled the right to the water. It was this experience that induced Marks to look for land that had secure water rights (in his next venture) and the Miller & Lux lands in Fresno and Merced County met the test.

This was one of the main reasons why the original location south of the Southern Pacific Railroad was selected, as this land was some of the first to be irrigated when the canal system was completed by the Scotch Syndicate's engineer—J. Ham Hall. Another reason for the selection was the fact that the land was perfectly level and the distribution ditches were already built.

A dozen families were brought out from the Middlewest during the summer and fall of 1891, and most of them put up houses on the land immediately. While Marks was in the middle west, Iowa, Kansas and Nebraska, interesting more people in the idea of moving to California—he received the news from his local agent, Wm. Ogelvie, that the settlers were dissatisfied—no well water—no alfalfa or tree soil with the land on which they had been located and were demanding his return. He came back to learn that he was held responsible for the "swindle" as they termed it. A meeting had been held and Marks had been "burned in effigy" and the demand was made for redress.

Marks went to San Francisco the next day and met Miller. After a verbal battle lasting from seven in the morning to noon secured the promise of an exchange in land and the moving of houses and barns at the expense of Miller & Lux. At first Miller said he would settle with the people for cash and they could go to some other part of California. Marks reminded him that this was what the Crocker Huffman Company had done a few years before with the people from Holland, who had been sold worthless land, and that it had given the county of Merced a bad reputation from which they were still suffering. Neither Miller nor Marks could afford to lose their reputations by such a deal and the agreement just mentioned was reached.

The new location just north was all pasture land, never having been plowed, and of course no ditches. The colonists inspected the soil and were pleased, especially with the fact that it was virgin land. Miller & Lux agreed to build a new main canal to furnish water for irrigation and to build distributing ditches for which Marks was to pay. The buildings were moved in 1892 and a lot of volunteer hay was cut on the land being abandoned thus furnishing feed without cost except for labor for the stock.

Good well water was obtained at twelve to thirty feet and in a few months 95 percent of the people were happy and contented. The move had cost Marks all his profit on the land sold and he was broke. He had to borrow money from Miller & Lux to continue the work. It takes time to organize a school district and there was no time so Marks erected at his own expense a schoolhouse in what is now South Dos Palos. This building was moved with the other buildings by Miller & Lux to a location on the new land now occupied by the cemetery. The school was used one year and then when we organized the first creamery I gave it to the company for a creamery building. When this enterprise failed, after having been given as security for a loan of $500 from Miller & Lux, the building was moved to Colony Farm and used as a bunkhouse.

I took charge May 15, 1893 and during the next four years the colony prospered. The people were happy and this resulted in their inducing their relatives and friends to join them. By this time Henry Miller had become interested in the success of the colonists. To aid them in starting on the new location cows and horses were to be had from the Santa Rita ranch for their keep. The only rental asked was the return of weaned calves and the horses could be returned or purchased on time.

Every child in the schools received a present at Christmas from Henry Miller. In 1891 there was probably not more than 300 acres of alfalfa on both the Poso and Santa Rita ranches and it was irrigated by means of contour levees. No leveling had been done and the water stood two or three feet deep on the lower side of the checks. Henry Miller told me one day when riding around the colony that it was the results obtained by the colonists that induced him to begin leveling Miller & Lux land.

While the colony was a success Bernhard Marks on the expiration of the contract was financially bankrupt and I only saved myself by taking a contract on 100 acres of land and going into the diary business.

Interview 2/15/1954 RLM

Frank Marks

J. W. Schmitz told me that Henry Miller never did any experimenting. The way he would do he would watch farmers all over the country and see how they did and then follow their practices. It seems that it took Schmitz about six years to convince Miller that it would be better to put wooden floors in pig pens. But Miller would not give his consent to letting Schmitz go ahead. Miller claimed that he had watched a farmer over around Gilroy who always fed grain to his hogs on the bare ground and raised the finest kind of fat hogs. Schmitz tried in vain to make Miller see that he was losing grain by feeding grain on the ground instead of on wooden floors. Finally after about six years Miller gave in and let Schmitz go ahead. Schmitz claimed that Miller never did any experimenting himself; he always watched others then formed his own conclusions.

Miller was not as great as people think. I say that the way the Colony Farm and the Colony Division got started was when J. W. Schmitz took back a thousand acres of land that had been set aside for the Dos Palos Colony. It seems that six sections had been originally set aside for the Dos Palos Colony. All at once Miller & Lux moved in and developed a thousand acres that belonged to the tract set aside for the Colony. This became Colony Farm. Eventually the Colony Division was created. Before that the country where Dos Palos is located was under the charge of J. W. Schmitz.

The way D. A. Leonard came to be superintendent was that Mr. Leonard was the livery stable man in Bakersfield. Miller used to go down on the train to Bakersfield and have Leonard drive him around to the ranches. In that way Leonard got acquainted with Miller and when Miller offered Leonard the job he took it. The livery business was not paying very well and Leonard was only too willing to go to work for Miller.

For a long time Leonard never had any use for the Colony and its future.

12/31/1954 Interview RLM

Frank Marks

I often helped unload emigrant cars that arrived at Dos Palos. Many of the settlers from the East came in freight cars. There would be maybe a couple of horses in one end of the car and maybe a cow. Maybe a pig also. And of course there would be the family dog. The people themselves with their household possessions would be in the other end of the car.

When the high school was established in Los Banos the high school pupils from Dos Palos came to Los Banos. They came and boarded here. Only one boy, George Christian, rode his bicycle along the Santa Fe Grade and went to the high school in Los

Banos. That was only one season. The next year his father needed him at the store. What helped to get a high school started in Dos Palos was that fact that Mrs. Marks declared that when their daughter Blossom was ready to go to high school that she wouldn't let her come to Los Banos to board. Wilmer Scott was the principal of the Dos Palos High School at one time. It was through Mr. Marks' influence that he was appointed. Marks and the others started to organize the Dos Palos High School in 1907. In 1908 the school was started in the Rochdale Hall above a store. Prof. Evans was the principal and he had two assistants. The first graduation class consisted of only two members.

The corner stone of the Dos Palos High School was laid in 1911. The building and furnishing cost $35,000. The Southern Pacific and Miller & Lux paid practically all the costs of the building in taxes. At that time practically none of the colonists had obtained a deed to their property. Direct taxes on railroad property in the county was paid directly to the county at that time. A year or so later things were changed and railroads no longer paid their taxes directly to the counties. It was the direct taxes levied on the railroad company and on Miller & Lux lands that paid for the most part of the cost of the building. The Dos Palos school district embraced land in three counties: Merced, Fresno and Madera.

The industry of the Dos Palos Colony was alfalfa, hogs, and dairying. Some of the settlers rented Miller & Lux land to raise grain. It was Dan Leonard, the superintendent for Miller & Lux, who raised the first rice in the Dos Palos country. He planted quite a large field, perhaps a hundred acres or so. It was on the Dutch Boys Ranch, later known as Holland Farm. He planted a big field and raised a splendid crop. It was the first rice ever grown in the Dos Palos Colony. Dan bragged about the tremendous crop he raised for years. Later Floyd Redfern was the first person to raise rice in a commercial way. He was followed by the Japanese.

Henry Miller was easy to approach if you had any business. He was always ready to talk business with anybody. The first meeting Mr. Marks had with Henry Miller was in 1892. Marks was a young man of twenty and was bookkeeper in San Francisco. One of the Dos Palos colonists was having difficulty getting to his land on account of a slough that needed a bridge over it. Mr. Marks went to Miller & Lux's office in San Francisco to see Miller and see if he couldn't get Miller to build a bridge over this slough.

Marks didn't know at that time what a wonderful knowledge of every part of the county Miller had. When Marks started in to try and to tell Miller the need of the bridge he found that Miller had a vastly greater knowledge of the Dos Palos country that he had ever supposed. Miller knew every inch of the country in question. Much more than Marks knew. Marks was trying to build up his request for the bridge. Miller never liked any beating around the bush. "Make your point, young man," he told Marks impatiently. Poso Schmitz had charge of the Dos Palos Colony. Bernhard Marks was the originator of the Dos Palos Colony.

A traveling printer by the name of Woodruff sold Bernhard Marks on the idea of starting a newspaper. Marks bought the outfit from him and started the *Colonian*. Bob McDonald was an artist with a slingshot. A fly on a barn he would knock off four out of five times.

At the time of starting of the Dos Palos Colony Miller & Lux had only two or three hundred acres of alfalfa. These patches were on the Santa Rita and the Poso. It was simply contour checks. No leveling was done. Sometimes the water would be two or three feet deep on one side in order to get water over to the other side of the check. After Miller saw the Dos Palos people leveling their land and planting it to alfalfa he went to leveling land and planting alfalfa. Steamers used to go up the San Joaquin River as far as Sycamore Point, now called Herndon.

When I came down to Dos Palos and took over my father's business I knew that the settlers must have some means of making money. So I got a creamery established. Bernhard Marks had built a schoolhouse for the colony when it was still along the railroad. When the colony was moved the schoolhouse was moved right along with the colonists. School was held for one year in the schoolhouse in its new location. Then I got my father to give the schoolhouse for the use of the creamery and it was moved to a new location. Later when the creamery went out of business Miller & Lux who held a mortgage on the property moved the old building to Holland Farm where it served as a bunkhouse for years until it finally burned up if I remember correctly.

4/8/1954 Interview RLM

Lillian Altnow Hoffkencht, early pioneer

My father John Altnow was a foreman for Miller & Lux at the Juntura Ranch in Juntura, Oregon in Harney County. This was a pretty large ranch, with good grazing ground, lots of shade trees and the Malhuer River running through it. As I recall, there were always from 6 to 12 cowboys hired to round up the cattle in the fall and spring. After being on this ranch for several years, my father, when shoeing a horse, got a piece of steel in his eye, which necessitated his going to Sacramento to have his eye operated on. This was in 1893.

Henry Miller, who was a frequent visitor at our home, helped make arrangements for the operations. Mr. Miller suggested that since we were coming to California that my father run his ranch at Gilroy, or one of his Westside ranches if my father would rather do that. But my father decided to go into ranching for himself and he purchased a ranch near Sacramento. Although my father no longer worked for Miller he never lost contact. Mr. Miller was a frequent visitor to the Juntura and in fact we always kept a room ready for him.

My memories of Mr. Miller are of a very kind and generous man, but a very shrewd and brilliant one.

While we were on the Juntura Ranch, even through I was a small child, I remember the Agency Ranch, which was about 20 miles from us, where my uncle Ranse Drake was a foreman and Tom Arnold was also a foreman. There was also the Indian Reservation Ranch on which Mr. Miller employed many Chinese for various jobs. I recall many trips with my parents to these ranches. My main recollection is that the Indian Reservation Ranch was a beautiful, large alfalfa ranch and of eating delicious meals prepared by the Chinese cooks.

Copy of letter donated to museum by Pat Westfall, granddaughter.

Bascom Newton, early pioneer family member

Henry Miller was the most observing person Mr. Newton ever knew. It was only Henry Miller who could have built up the vast Miller & Lux property. Henry Miller was always looking for water. The story about Miller peddling sausages was true. He heard the story told on a lawyer present at a Masonic banquet in San Jose. The speaker was telling the story on a prominent lawyer present. It seems that it was in a water lawsuit and this lawyer wanted to get Henry Miller mad. So he asked Miller, "Didn't I use to know you when you used to peddle sausages in a basket on your arm?"

Miller's retort was, "Yes, and if I didn't know anymore than you do I would be peddling them yet." This was the opposing lawyer in a Kings River water suit. I found an Indian skull one time out on the plains on the Santa Rita field. I turned it over to a government man who made a report on it.

Henry Miller knew the value of water. I drove for Henry Miller for about three months. I took Buggy John's place. Buggy John's shoulders gave out. It was over here at the Los Banos farm. Buggy John was sent back to Bloomfield where he got well.

Tanner picked up on me to drive for Miller. Tanner knew I would drive any team that ever was hooked up. I had driven for Tanner before that. Tanner gave me my instructions. It was not known whether Buggy John would get well and maybe the job as driver for Miller would be a permanent one for me. Tanner told me never to speak to Miller unless Miller spoke to me. Tanner told me to make my answers to Miller as short as I can. I followed instructions and got along fine.

It was not uncommon for Miller and I to set out at midnight and be at breakfast at Canal Farm in the morning. Miller was always looking to find out where things were going wrong on the ranches. Miller never could stand waste of any kind. His greatest effort was to prevent waste. He would go and look in the swill barrels to see that nothing was being wasted. He would also watch all the mangers to see that no feed was being wasted. I remember particularly one evening at a ranch house. Miller called the landlady and showed her how he wanted bread cut. It seems that the bread was being baked in very long loaves in a big round skillet. The loaves were being cut in big slices which were really more than a person wanted to eat and part of the slice would be thrown away and wasted. Miller took a carving knife and split a loaf. Why do that? The other way makes too large a slice. By cutting the bread in half slices a man would not take more bread than he wanted and so part of the slice would not be wasted.

I have driven Mr. Miller all night long many a time. In hot weather they drove more at nighttime than they did in daytime. It was very seldom that Miller slept as they rode along, but I have known him to do so. I ever saw any man that could beat Miller at a bargain. Getting the best of the bargain with Miller was seldom done. I have seen Miller ride through a bunch of cattle at Santa Rita on horseback. He would ride through the bunch and then say, "I gives you so much a head." Miller would never raise his price. The seller usually took the offer right there. If he didn't Miller would simply say, "All right, we can't deal."

It was a common saying that Miller made Lux a millionaire against Lux's will. The story is told that one time Lux wanted to quit. I think we have gone far enough, he told Miller, and he wanted to divide up. He wanted to divide up the money on hand and the land they owned. Lux happened to say that the firm had a hundred thousand dollars in cash on hand. "You say we got one hundred thousands dollars on hand?" queried Miller carefully. Miller went away. When he came back the firm no longer had a hundred thousand dollars in cash. Miller had spent it all on land.

I have no doubts the story about Miller riding all over in a boat over land that he wanted as swamp land. Miller always told the truth and stayed with it when he could. If he said he had done it, he had done it.

Miller was the most kind-hearted man that ever lived. He asked me a great many questions. When he learned that I wanted an education and wanted to go to college he was greatly interested. The desire for education was the supreme trait he liked to see in a young person. Years after Miller and I were talking. "Well, how far did you go?" I told him I had gone through the high school in San Jose and to the State Normal. "You have done better for yourself than anyone could have done for you," said Miller. "The things that you do for yourself are the things that help you." I later graduated at Santa Rosa Methodist College.

The nicest overcoat that I ever had was given me by Henry Miller. I was driving for Miller and they were riding in a storm. Miller noticed that the wind was blowing down my coat collar. When we came to Fresno we drove up to a big dry goods store and Miller ordered me to stop. Miller handed fifty cents to a man standing there to hold the horses. Miller took me into the store. "I want the biggest, heaviest overcoat you have got," said Miller, "for this young man." Miller bought me the nicest coat I ever had. It had a fur-lined collar and must have cost $45 or $50. I was never cold again and was warm as toast as I continued to drive Miller. I wore that coat for 10 years afterward.

Miller also bought me a pair of beaded buckskin gloves that reached up to my elbows. I was used to buying my own clothes and fully expected that all these gloves and overcoat would be charged to me. At the time I did not even know how much Miller was paying me. When we left the store Miller had the clerk wrap up my old overcoat and gloves and then took them along with him. "You may need them again," he told me.

Finally after about three months Buggy John got well and was ready to come back on the job. "We're going home now," said Miller one day. I knew that Miller meant that we were going to Gilroy. We got to Gilroy about noon. "You have been a good boy," said Miller. He gave me two twenty dollar gold pieces as a present and sent me home on the stage to Los Banos.

Miller had regular stopping places among the ranches. At each stopping he had his own bed. No man, Miller didn't care who, was ever allowed to sleep in Miller's bed. He also kept a nice suit of clothes and underwear at each of his stopping places.

4/7/1939 Interview RLM

Miller & Lux sheep being driven near top of Pacheco Pass, 1906

CHAPTER SIXTEEN

End of an Era

Miller's decline and death came just as America was on the brink of a major shift from an agricultural to an industrial nation. Early in the 1900's Miller realized that he needed to shift gears and got out of the sheep and hog business. He was still a stockman and held on to the cattle business to the end. He also realized that food had to be produced and so went more and more into farming and sold off lands that had in the past been grazing land. The cattle business was moved into feedlots where he could continue to produce.

When his son-in-law Nickel started to take more control of the business, the trouble began—family fights, lawsuits, threats and counter threats. Nickel was not well liked nor did he have the respect that Miller had with friend or foe. Nickel was not a stockman but was more like Charles Lux, a man of business and the face of the company. He was not interested in the nitty gritty of the daily grind. He wanted things done, he wanted it done cheaply, and he didn't care about the overall long range view of the operations.

As Miller became more disabled and unable to manage his affairs, Nickel embarked on replacing men who had been with Miller for years. Many he flatly fired and many left due to differences of opinion on how things should be done. Among the first to go was Albert Long, Miller's nephew and well known as only being surpassed by Miller himself in the judgment of stock. Miller had wanted Long to replace him and take care of the cattle business of Miller & Lux. Nickel had other ideas. This article from the Dos Palos Star tells part of the story.

Dos Palos Star
12/4/1909
TROUBLE IN THE LAND

For the past week the air has been full of rumors of changes in the organization of the Miller forces in the towns and settlements south of Newman, from Gustine to Dos Palos in which the company is the most important business factor. According to Dame Rumor, practically every department head on the West Side has been summarily removed. There is certainly more or less foundation for the reports, though just how many decapitations have been made cannot be ascertained.

Managers White of Los Banos and Barker of Gustine, who have been in charge of the company's general stores at these places, are to give way to new men at once. Mr. White goes to Livingston, where he has secured an interest in a store.

Ranch Superintendents Wallis and Leonard, Superintendent Clyne of the canal company, Manager Simpson of the Los Banos lumber yard, Land Agent S. J. Shannon, Messrs Boyle and Johnson of the Los Banos office and perhaps Cashier Heron of the Los Banos bank are among the number said to be slated for a change of occupation on or about the first of the coming month. Whether the reports are correct or not remains to be seen.

Many theories are advanced to account for the sweeping changes. One of them is that Leroy Nickel has taken total charge of the general office and proposes to have new subordinated in charge of the various departments.

The above was taken from the Newman Index. It seems that the above news if it can be called news originated from a long hungry traveling man working on commission doing reporting on the side.

W. T. White quit the firm of Miller & Lux to go into business for himself and Mr. Johnson of the bank has accepted a position as cashier of a new bank down south. As to Messer Leonard, Shannon and Wallis, Mr. Wallis is still on the job, Shannon is still selling land and Leonard not only retains his position but also had taken on the duties of the late Franz Misch.

We can see the twinkle in Mr. Miller's eye as he reads this piece of news in the Index until he reaches the place where he is said to be superceded by J. Leroy Nickel and then the blue smoke will fill the room.

Nickel had started to become more in control of the firm around 1908, as Miller began having health problems and slowed down. The years between then and 1912 were a seesaw of control between Miller and Nickel. By 1912 Nickel was in total control. Long's departure and Miller's failing health left the main source of Miller & Lux income, the cattle operations, vulnerable. Battles still were being waged with outside meat companies from the east, more and more farmers had come in to farming Miller's lands. This had become a two edged sword. While Miller & Lux sold the land and made a handsome profit the farmer in turn raised crops that began to affect Miller & Lux operations across the board. With the increased need for food and feedstuff caused by World War I Miller & Lux was able to continue on.

By the mid 1920's the cattle had all been sold to the Swift interests. Nickel then began the sale of lands in earnest. The depression followed and all businesses were affected. Family members became unhappy with management, and by the time of Miller's daughter's death in the mid 1940's, the firm Miller & Lux and the various branches of the family were involved in lawsuits that would not be settled until the 1960's.

Merced Express 12/31/1926

Real Estate and financial operations of Miller & Lux Inc. have reached major proportions during the past twelve months, according to announcement by company officials in this city. Land sales have totaled $3,500,000 during the year and negotiations in progress represent an approximate expenditure of $2,000,000 for additional land.

Proposal of the directors of the San Joaquin River Water Storage district to purchase canals and water rights from Miller & Lux involves a consideration of more than $10,000,000, it is stated. Water rights in Kern River are also to be sold to the Buena Vista water district for $1,500,000 it is announced.

Favorable prices for cattle are helping the firm's current position materially and as a matter of fact, operating costs for the current year promise to be better than those for past years. Cotton lands of the company have been yielding this year on an average of one bale per acre. During the past year the firm has reduced its bonds and notes outstanding in the hands of the public by more than $2,000,000.

Farming continued during that time with part of the family taking the land around Los Banos and another the land around Bakersfield. In the years between the land values had so increased that some land was sold for development, which was far more profitable than farming, while other land just as valuable was kept in farming.

February 22, 1965 Los Banos Enterprise

MILLER & LUX Directors Vote End of Firm

LOS BANOS, Merced Co. The directors of Miller & Lux Incorporated have voted to dissolve the company, which at one time owned more than 1 million acres of land. The assets of the company, estimated to total $40 million, will be divided among the heirs of the founder, Henry Miller.

Half of the assets will be divided among Henry Bowles, George Bowles and Mrs. Amy Lawrence, all of the bay area, and the other half will be divided among Mrs. Sally Nickel Mein, Mrs. Mary Nickel James, George W. Nickel Jr. and two children of the late James Beverly Nickel.

Perhaps the best known and most widely seen photo of Henry Miller. Taken at the last May Day celebration he attended in 1904

The first group is children of the late Mrs. Beatrice Bowles Morse; the second are children of the late George W. Nickel, Sr., Mrs. Morse and Nickel were grandchildren of Henry Miller.

The land taken over by the Bowles interest will be farmed under the name of the Bowles Farming Company of Los Banos. The land taken over by the Nickel interests will be farmed under the name of The San Juan Ranching Company of Los Banos. The present Los Banos headquarters of Miller & Lux will be discontinued after completion of all the details of the dissolution.

Today the family has come full circle. The family has formed a corporation to handle its vast business affairs and also had been brought together by the marriage of Miller's great-great great-great grandchildren.

Merced Sun Star
5/10/2001
Miller Descendants
Return to Los Banos

Cannon Michael and Heidi Nickel, great-great-great-great grandchildren of Henry Miller have returned to Los Banos to live. Their marriage has joined the families and brings Miller descendants to live full time in Los Banos and to work on the Miller properties.

Of Miller's children, only his daughter Nellie married and had children. Her four children in turn married and the family circle expanded and soon became involved in a court battle over Miller's estate, which had been initially left to the third generation.

There are many other descendants as well and most are holders of shares in the Bownick Corporation which the family formed to handle the remainder of the Henry Miller's holdings that at one time claimed to be over a million acres of land.

During the 1940's the Nickel and Bowles families where were the main branches of Miller's descendants became involved in a lawsuit that was finally settled in 1962.

Cannon Michael is from the Bowles side of the family, Heidi is from the Nickel side. Heidi was working at her father's San Juan Ranch office when she met her several times removed cousin Cannon. He had come to work at Bowles Farming to learn the family business so that eventually he could manage it when his uncle retired. Though both knew some family members, they had never met.

When they decided to marry, they also decided to invite all the family members on both sides of the family. As a wedding present for Cannon, Heidi had a family tree drawn showing the distant cousins who still participated in the income from the Miller properties but did not necessarily participate in the farming or management of the properties. The poster was a great hit at the wedding as people gathered round to see who was related to whom. Heidi says that family ties are strong since all family members have shares in Bownik Corporation and share in the profits of the farms in Los Banos as well as the profits from other corporation holdings.

Cannon and Heidi love the people of Los Banos; because it is small "We are fast meeting and becoming friends with people of many generations. If we still lived in the city we would only have friends our own age."

Their first child is expected in June of this year, a boy, and because he will rejoin the two main family lines, it is possible he might eventually manage both of the farms in Los Banos. Cannon Nickel Michael was born June 18, 2001.

The legacy of Henry Miller and his empire will continue on. Yet there are still unsolved mysteries in Henry Miller's story. Of his years in Holland and England there is no record. His time spent in New York is only recorded in what he later said he did. There is nothing on record. His first years in San Francisco are shrouded as well. He arrived there, at his own account with six dollars, in ill health and set to work that same day. He claimed to have worked for a year at $2.00 per day and selling offal for extra income. Yet with the exorbitant prices of the time he somehow had the funds needed to rent a lot in the center of the business district in less than a year. At the time lots were selling for $10,000 and renting at $1,000 a month. Miller not only rented the site, he put up a building and had cash to spare, and within the year purchased the site. There is no doubt Miller was a hard worker and was extremely frugal. Butchers at the time were making $10-$15 a week and by selling the offal he could double that.

Was Mrs. Otto his main backer? He always visited her but never acknowledged her as a backer yet just before his death he sent her a rather large sum ($15,000). Did he get early backing from miners, friends, customers? Did he like many others sweep the floors of the gaming dens and saloons picking up the gold dust discarded on the floors and under the floorboards? Perhaps, as he alludes to, the Menches Brothers backed him early on as it was their lot he started out on. No one knows how he financed his start in San Francisco. By the mid 1850's he was able to invest $33,000 in a cattle deal with Charles Lux. Had Lux, who was established, previously backed him earlier as a silent partner? What does matter is that he took what he had, did whatever it took to build his empire and live his boyhood dream, leaving the San Joaquin a better place than he found it.

Map of Henry Miller's heirs' current holdings, approximately 80,000 acres

Trained hunting cows often used
by hunters to surround ducks

Miller & Lux cooks Sing and Chung

"Sasto" one of the Basco's sheepdogs.

Bibliography

Brewer, William - *Up and Down California 1860-1864* Univ. of California Press - 1966

History Company – *Fowler Interview*
Bancroft Library Univ. of California
Berkeley, California

Ingler, David - *Industrial Cowboys*
Univ. of California Press – 2001

Lawrence, David – *Uncompleted Writings*
Property of Ralph Milliken Museum

Miller, Catherine - *Flooding the Courtrooms*
Univ. of Nebraska Press – 1993

Pimentel, Wayne - *Dog Town and Ditches*
Loose Change Publications - 1988

Rose, Gene - *The San Joaquin A River Betrayed*
Word Dancer Press – 2000

Soule Frank – *Annals of San Francisco*
Appelton, New York – 1855

Torchiana H.A. Van Coen
California Gringos
Paul Elder & Company – 1930

Tredwell, Edward - *Cattle King*
Reprint Western Tanger – 1981

NewsPapers:
Alta Californian – Daily Alta
San Francisco Call Bulletin
San Francisco Chronicle
San Francisco Examiner
Fresno Weekly Expositor – Merced Express
Argus and the Star early Merced papers
Los Banos Enterprise
Dos Palos Colonian – Dos Palos Star
Bakersfield Californian

Collections of interest:
Bancroft Library Univ. of California
Berkeley, California
History Company – Fowler Interview
Miller and Nickel family papers, photos and documents.

Gilroy Museum
Gilroy, California
Collection of photos, documents and family papers

Henry Madden Library
California State University Fresno, CA
Letters and miscellaneous Miller & Lux documents

The Huntington Library, Art Collections, and Botanical Gardens
San Marino, California
Collection of Miller & Lux documents

Ralph Milliken Museum
Los Banos, California
Documents, personal belongings and items relating to Los Banos, Dos Palos, and surrounding ranches, towns, development of the Westside of the San Joaquin, Miller & Lux documents, papers, deeds and miscellaneous related items.

Index

A
Acapulco 6, 74
Acker, Bert 286
Ackerson 132, 133, 141
Adams, R. 233
Agard, Henry 211
Agency Ranch 302
Aguila, Antonio 174, 175, 177
Alamo Hotel 189
Allen & Pyle 286
Allen, Mrs. 155
Almaden 120
Alsace 10, 78
Altnow, John 302
Alvarado 134
Alvarado, Juan B. 16
Amabiles 174
American Cattle Trust 34
Anderson, Bill 165
Anderson, Kee 165
Anderson, R. H. 108
Antioch 28, 199
Antone Lopez Hill 270
Arbios, Peter 104
Arburua 289
Arburua, Frank 288
Arburua Grant 177, 185
Arburua, Louie 285
Armour 34, 84
Arnell 288
Arnold, Tom 302
Aromitas y Agua Caliente Rancho 18
Arques, Luis 156
Arroyo 252
Arroyo Ranch 287
Athlone 166
Auzerais House 120, 148
Avery place 289
Avy, Eugene 267
Avy, Mrs. 267

B
Bachle and Gaynor 241
Baden Ranch 91
Baden Station 15
Bader 248
Bader, Henry 104, 263
Bader, Mrs. Henry 104
Badger Flat 54, 150, 172, 173, 217, 240
Baker, B. B. 118
Baker, John 44, 47
Bakersfield 20, 44, 45, 51, 108, 143, 179, 186, 274, 275, 277, 286, 289, 300, 306
Bambauer 195, 259
Bambauer, C. D. 174
Bambauer, Charlie 255, 259
Bambauer, Harry 259
Bambauer, Rudolph 265
Bancroft 87, 88
Bancroft Library 69, 117
Bank of California 92, 94, 273
Bank of Gustine 106
Bank of Los Banos 106, 242, 255
Bantas 199
Bantas Station 146
Barker 305
Barron, Edward 8
Barry, Elizabeth Long 104
Basalt Fire Station 270
Battery Street 74
Battle Mountain 177
Bay of San Francisco 14
Beale, Genl 144
Bear Flag revolt 6
Bear Valley 291
Beaugardest, Dr. 165
Beddell, William 105
Bell, L. F. 119
Bell Station 58, 126, 161, 196, 260, 264, 283
Belmont 273
Belmont Ranch 297
Bensley 38, 218
Bensley Canal 125
Bensley, J. B. 126
Bensley, John 37, 39, 40, 218, 219
Bensley, John B. 81
Bensley, John D. 81
Bentley 141
Berenda 161, 165, 196, 237, 289
Berenda Station 130, 138, 139
Bereton, Mr. 132
Bernhard Marks 302
Berry, Lizzie 103
Bertis, Joe 260
Berwald Brothers 157
Bibby 150

Bibby, Enoch 188
Bibbys 173
Bicken, Joe 74
Big Panoche Plains 200
Bills 162
Bills, J. L. 140
Bird 231, 250
Bisiganani, Paqual 298
Bisignani, Pasquale 292
Black 120, 145
Blackmore, Mrs. 224
Blankenship, Noble 141
Blankenship, Zachariah Taylor 174
Bliss Ranch 165
Bloomfield 96, 98, 162, 191, 212, 213, 217, 223, 246, 254, 260, 261, 264, 274, 282, 303
Bloomfield Farm 78, 98, 99, 102, 122, 123, 124, 125, 126, 128, 131, 136, 137, 140, 141, 142, 146, 147, 153, 156, 224, 292
Bloomfield Ranch 92, 95, 130, 174, 187
Blue, Rupert 35
Bolsa de San Felipe Rancho 17
Bolton, John 144, 145
Bolton, John H. 137, 141
Bolton, Mr. 129, 139
Bonita Farm 233
Booley & Johnson 122
Bowles, Beatrice 106
Bowles family 99, 308
Bowles Farming Company 307
Bowles, George 306
Bowles, George McNear 109
Bowles, Henry 306
Bowles, Henry Miller 109
Bownick Corporation 307, 308
Boyle 305
Boyle, Johnny 265
Brackenheim, Germany 3, 4, 69, 99, 103
Bradford, George L. 108
Breen, Pat 132
Brem, Mrs. Richard 224
Brem, Richard 105
Brereton 142, 145, 166, 218
Brereton, Mr. 143
Brereton, R. M. 139
Brereton, R. R. 40
Briggerman 126
Briggs's store 162
Brown 296
Brown, David 104
Brown, Mrs. 278
Brown, Mrs. David 103
Brown, Sara Long 104
Browning, Bill 182
Brundage 49
Brundage, Benjamin 46
Buchwald 69
Buena Vista 46
Buena Vista Lake 38, 44, 45, 46, 49, 51, 274
Buena Vista Lake oil field 108
Buena Vista Slough 128
Buena Vista water district 306
Buffalo Bill 163, 250
Buggy John 162, 190, 198, 210, 303, 304
Bull Pens 257
Bunker, N. E. 150
Buri Buri 14, 17, 34, 68, 78, 91
Burkett, Mrs. Charles Watson 107
Burnett, W. F. 156
Burrell, A. W. 156
Burrell, Mr. 155
Bush, Joseph 120
Butchers Board of Trade 35
Butchertown 174, 245, 246, 247, 261
Butler, Edward 291
Butler, Mr. 291
Buttonwillow 233, 246, 250, 262, 278, 279, 289

C

California 4, 7, 10, 14, 18, 19, 22, 24, 27, 28, 30, 33, 35, 36, 37, 46, 73, 76, 82, 91, 98
California Home for Feeble Minded Children 95
California Street 8, 92
Callison 141
Calloway 49
Calloway canal 45
Calloway weir 45
Camp 13 192, 255, 258
Campbell, John 294
Canada de San Lorenzo 92
Canal Bridge 150
Canal Company 117, 128, 132, 139, 140, 144, 149, 165, 166, 169, 188, 199, 219, 222, 275, 280
Canal Farm 64, 65, 79, 130, 136, 138, 141, 143, 145, 149, 150, 157, 161, 162, 167, 169, 171, 172, 178, 180-182, 184, 194-197, 211, 212, 216, 217, 238, 241, 242, 250-252, 254, 255-257, 260, 267-269, 276-279, 282-284, 287, 288, 290, 303
Cantua 194
Caran, Fred 142
Carayago 131, 132
Carissa plains 287
Carlisle 155
Carlisle Ranch 155
Caroline (steamer) 127
Carr, W. B. 44, 47, 50
Carrisa ranch 58
Case 130

Castro, Narcisio 105, 175, 211
Castro, Narcissco 217
Cattle King 105, 182, 203, 211, 239, 240, 277
Center Avenue 269
Center, John 9, 75
Center Point 173, 177, 179, 181, 195, 219, 222, 255, 258
Central California Mutual Water District 242
Central California Valley 105
Central Point 172, 266
Central Point store 180
Central Ranch 169
Central Valley Project 242
Cerubi, Antone 173
Chamberlain 140, 149, 281
Chamberlain, A. 149, 237
Chamberlain, Eliza 237
Chamberlain family 172, 295
Chamberlain, Fenner 190, 295
Chamberlain, J. F. 237, 239
Chamberlain Landing 122
Chamberlain, Millisia 127
Chamberlain Ranch 140, 149, 295
Chamberlin 148
Chamberlin, A. 147
Chamberlin Landing 119, 146
Chamberlin, Mellisa 147
Chandler 173
Chandler, Tommy 162
Chandois 139, 169
Chapman 19, 121, 125, 126, 127, 133, 144, 165, 218
Chapman Field 130, 136
Chapman, William S. 42, 81
Chappell 271
Chargres River 74
Charleston 196, 257, 272
Charleston District 177
Charleston School 258
Charleston stage station 258
Cheatham 255
Cheatham, Bill 195
Chester 155, 156
Chester Bridge 156
Chestnut Avenue 91
Chewago Land 120
Chicago 84
Chicago scrapers 272
Chico 126
Chinese Camp 180
Chism, Dutch 245
Chowchilla Canal 42, 192
Chowchilla Ranch 165, 231, 250
Chowchilla Slough 42
Chowchilla Y 182
Christian 195, 268
Christian, George 300
Cirimele, Joe 173, 199, 203
Clara Bell (steamer) 132, 134, 136, 137, 141
Clark, Sam 123
Clark, T. B. 149
Clarke, W. F. 173
Clarks Point 9, 74, 76
Clay Street 74
Cleveland, Grover 185, 220
Clifton, Zena (Place) 241
Clyne, J. F. 197, 278, 280, 305
Clyne, Mary 282
Clyne, Mr. 211
Clyne, Mrs. 211, 251, 282
Coast Range 14, 236, 238
Codington, Loretta 108
Coe, Mr. 233
Coffee, Dave 50
Coffman 165
Cole Levee 49, 51
Coles oil field 108
Collier 125
Collinsville 285
Colm, W. W. 108
Colma 106, 107
Colony Division 300
Colony Farm 299, 300
Colp, Joseph 169
Columbia 231
Columbia Ranch 42, 123, 143
Colusa County 205
Commercial Street 74
Conroy 165
Cornett, S. C. 54
Cornwell, Al 161, 162, 168, 169
Cornwell, George 50
Cothran 189
Cothran, Supervisor 197
Cotton Field Section 162
Cottonwood 137, 150, 161, 187, 195, 196, 199, 221, 241, 266
Cottonwood district 220
Cottonwood Farm 259
Cottonwood Ranch 266, 283
Cottonwood Switch 265
Cox family 173
Cradle of Innocence 31
Crane, George W. 17, 79
Crocker 128, 175, 186, 212, 274
Crocker, E. M. 122
Crocker First National Bank 107
Crocker Huffman Company 299
Crocker, J. C. 44, 117, 119, 143
Crocker, James 44, 47

Cross Creek 129, 138, 174
Crossmore, George 138
Crow, Brad 188
Crow Creek 219
Crowley 187
Crow's Landing 219
Cudahy 34
Cunningham, Mrs. 295
Cupertino 259
Cuyama 58
Cymric oil field 108

D
Dailey 280
Dailey, Mollie 275
Dailey, Mrs. 184
Dalton, Mrs. 99
Dalzell, Bob 251
Dalzell, Mrs. 242
Dalzell, Robert R. 242
Darr, Charles 288
Davidson, Charles 120
Davis 66, 128, 169
Davis, David 66, 156
Davis, F.A. 81
Davis, George 136
Davis, Thompson 66, 156
Deacon, Tomas 118
Deeney, M. V. 105
Delta Farm 292, 298
Delta Ranch 107
Devils Den 287
Dias, Thomas 10, 78, 120, 121, 124, 127, 128, 137, 146, 147, 148, 175
Dickenson 231
Dickenson, Mr. 229, 230
Dickinson, Chas. 154
Dickinson, Chester 164
Dickinson's Ferry 154, 164, 177, 182, 196, 238
Dickinson's Ferry Road 196, 197, 198
Dietz, Louise 224
Dillon 246
Dillon, John 104
Dils 189
Dilse 287
Dipple 246
Dirty Plate Route 207, 208, 211, 252
Dolores Street 75
Don Pacheco Service station 283
Donahue, Newhall & Co. 165
Dos Palos 31, 54, 103, 182, 196, 197, 204, 205, 208, 238, 239, 241, 254, 255, 258, 263, 277, 282, 293, 295, 296, 297, 298, 300, 301, 302, 305
Dos Palos canal 42
Dos Palos Colony 31, 300, 301
Dos Palos High School 301
Dos Palos Ranch 169
Dos Palos Slough 131, 149
Drake, Ranse 302
Drum, J. J. 102
Drummond, J. Q. 41, 43, 165, 173, 180, 195, 196, 219
Dugain, John 198, 199, 283
Dumphy, William 14, 16, 78, 79, 92, 177
Dunbar 102
Dunbar, Harry 177, 182
Dunn, James 79
Dupont Street 8, 74, 75, 76
Dustin 165
Dutch Boys 169, 207
Dutch Boys Ranch 301
Dutch Corners 133, 218

E
East Side Division 232, 233
Eddy Street 84
Edmondson 91
Edmondson, Alfred 15
Ehlers, Charley 296
Ehlers ranch 295
El Camino Real 14, 17
Elis, Mr. 122
Elkhorn Ranch 119
Enau Brothers 239
Enau, Frank 239
Erskins 9, 13, 77
Essex Street 96
Evans and Sontag 245
Evans, Prof. 301
Evergreen 161
Exeter 275

F
Fairfax 298
Fairplay stage line 167
Farmers Camp 131, 132, 133, 169, 256
Fatjo, Dr. 189, 266, 283, 284
Fatjo, Paula 272
Federated Clothing store 258
Fickett, James E. 107
Fifteen Mile House 174
Fifth Street 9
Fillmore Street 246
Fiori, Hattie (Safstrom) 291
Fiori, Mrs. 291
Firebaugh 54, 64, 81, 106, 131, 136, 139, 147, 155, 160, 163, 164, 166, 167, 169, 177, 181, 183, 184, 192, 194, 208, 212, 229, 231, 232, 233, 237, 241, 246, 248, 250, 251, 263, 266, 287, 288

Firebaugh Division 248
Firebaugh Ferry 40, 120, 121, 122, 125, 126, 127, 130, 131, 132, 139, 140, 145, 155
Firebaugh, Jerry 164
First Avenue 245
Fischer, Kate 224
Fisher, Henry 121
Fisher, John 69
Fleming 123
Flint, Senator 204
Flood, James L. 106
Flournoy, Geo. 46
Folsom 167
Folsom, Mr. 141
Ford Farm 233
Fort Tejon 144
Foster, Conrad 169
Fought, Bill 255
Four Creeks 174
Fowler 78, 84, 87, 173
Fowler, Edward 88
Fowler, Mr. 69
Fowler Ranch 287
Fowler, Thomas 203
Francisquito creek 273
Frederica 69
Freeman, Carl 188
Freeman, Francis 188
Freer, John 239
Fremont 163, 164
Fresno 55, 184, 199, 241, 263, 275, 278, 287, 293, 303
Fresno (steamboat) 135
Fresno County 17, 20, 23, 24, 36, 42, 169, 184, 205, 219, 226, 299, 301
Fresno Scraper 272
Fresno Slough 40, 42, 184, 218
Fresno State College 245
Fresno State Library 117
Friedlander, Isaac 19, 40, 81, 165, 218
Front Street 180, 195
Fuller, Mr. 233

G

Galbrath, Geo. 150
Galindo, Charlie 198
Gamboni, Alice 289, 290, 291
Gamboni, Joe 289, 291
Garber, John 46, 48
Gardner, Jake 189, 194
Garrison, Alice 272
Garvey, W. O. M. 149
Gaynor, Ed 235
Geary, Dave 291
Geer, John 126
Geir, Captain 127
Gellar, Mr. 122
Georgia 74
Georgia (ship) 5
German Bros. 155
Germany 10, 40, 71, 72, 78, 96, 171, 175, 189, 212, 224, 273, 295
Germino, Louie 284
Gier, Jack 126
Gilcrest, John 104
Gilroy 14, 15, 17, 39, 40, 63, 64, 66, 78, 79, 80, 82, 83, 87, 92, 95, 96, 97, 99, 102, 103, 105, 117, 120, 124, 125, 128, 129, 130, 134, 136, 138, 141, 143, 144, 145, 147, 148, 149, 150, 156, 162, 164, 182, 183, 185, 187, 189, 196, 211, 223, 229, 236, 240, 259, 260, 276, 281, 292, 300, 302, 304
Glen County 205
Glen Ranch 92
Gody, Alexander 153
Golden Gate Park 296
Gomes, Jose B. 149
Gonzaga Ranch 266
Gonzales, Louis 105
Gonzales, Manuel 272
Goodwin, John 129
Goose Lake Canal 49
Gough Street 94
Government Land Office 22
Gower, Hugh T. 104
Gracey, Robert 258
Graham, Jim 133
Grant Avenue 8, 76
Great Canal 42
Great Earthquake 245
Great Valley 14
Grecian Bend 162
Greenlaw, H. P. 235
Greer, Thomas 108
Gregg 173
Grimstead, J. R. 102
Grove Farm 252, 255, 256, 259
Guadalajara 182
Guadalupe Valley 9
Guadeloupe Valley 77
Guaglianoni, Mr. 298
Gustine 40, 96, 196, 224, 265, 275, 305

H

H & F Farming Company 108
Haggerty, Jim 250
Haggin 46, 51
Haggin and Carr 45, 49, 299
Haggin and Tevis 38, 274
Haggin, J. B. 44, 51

Haggin, Louis 46, 50
Haight Street 246
Hairpin Bend 162
Hale 180
Hale, Mrs. 180
Haley Ranch 270, 271
Hall, J. Ham 299
Hall, James 39
Hallahan, Tom 287, 288
Halstead 73
Haly, Salisbury 16
Ham, I. A. Jr. 145
Ham, I. H. 150
Ham, J. H. 149, 150
Ham, Mr. 170
Hamilton, Captain 134, 138, 155
Hamilton, Mr. 132, 136
Hamilton, William 38
Hammond, Mrs. 190
Hancock House 280
Hansen, L. 139
Hansen, Pete 252
Hard, R. B. 117, 118
Harris, Dan 131, 132
Harrison 143, 144, 157
Harrison Street 84, 94
Harrold 143, 144
Hart, Johnny 190
Hassel 120
Hawthorn, J. J. 119
Hearst, Mrs. 95
Hegera, Pacheco 129
Hells Half Acre 31
Helm Place 164
Heminghouse 199
Henden, Capt 132
Henderson 76
Hermanhouse ranch 263
Herminghouse 250
Herndon 301
Heron 255, 305
Heron, Mr. 252
Hessedormstadt 189
Hick and Folkerson 187
Hiderman 150
Higgins, J. F. 143
Higuera, Jesus 230, 231
Higuerra, Pachuca 105
Hildreth 92, 166
Hildreth and Dumphy 273
Hildreth, Henry 274
Hildreth, Mrs. 177
Hildreth, Thomas 14, 15, 16, 17, 78, 273
Hildreth, Tom 150, 163, 164, 177, 274
Hillman House 74
Hills Ferry 123, 132, 134, 139, 145, 150, 153, 167, 171, 172, 194, 204, 218, 219, 285
History Company 69
Hoffkencht, Lillian Altnow 302
Hoffman 177, 181, 184
Hoffman, Ed. 235
Hoffman, G. L. 125
Hoffman, Mr. 189
Hog Camp 165, 247
Hogarth, Joseph 236
Holland Farm 169, 248, 296, 301, 302
Hollister 156, 164, 218
Hollister, Col. 164
Holm's Saloon 195
Home Guards 163
Home Landing 136, 138, 140
Homestead Act 30
Hopeton 167
Hornitos 250
Horseshoe Bend 162
Hostettler's Bitters 277
Hotan 164
Hotel Los Banos 185, 239, 268
Hoten 169
Houchin, C. Elmer 108, 109
Houghton 51
Houghton, Mr. 131
Houghton, R. E. 49
Houghton, R. R. 46, 48
Howard, Nigger 182
Howard Street 9, 76, 245
Huff 287
Huffman's Ranch 204, 205
Hughes, Owen 187
Hunsaker 141
Hunt, Judge 50, 51
Hunter Building 205
Huston, James 42, 286
Huston, Jim 270
Hyde 19
Hyde Ranch 229

I

Ias, Thomas 147
Idiart 177
Idiart, Mrs. 254
Idiart, Prosper 288
Imperial Valley 296
Independence 175, 176
Indian Reservation Ranch 302
Ingham & Sons 155
Ingomar 250, 266
Isthmus of Panama 74

J
J.R. McDonald (riverboat) 160
Jacks Ranch 198
Jackson 76, 94
Jackson Street 9, 74, 122
James 50, 270
James Canal 270
James, Jeff 136
James, Mary Nickel 306
James, Walter 47, 50
Jameson 257, 267
Jameson, Carrie 96, 260
Jameson, Z. G. 257
Jeffers, Ben 266
Jeffers, Mrs. 284
Joaquin 120, 121, 122, 125, 126, 128, 129, 130, 137, 142, 146, 147, 171
Joaquin (stable man) 290
Johnson 256, 289, 305, 306
Johnson, George 235, 255
Jordan 173
Jordan, Edwin M 104
Juntura Ranch 302
Juristac Rancho 17

K
Kahn 184
Kane, Julius 264, 265
Kearny Street 8, 9, 73
Kelley, Yale 208
Kentucky 188
Kern County 20, 23, 24, 36, 44, 46, 47, 48, 49, 67, 80, 92, 108, 153, 226, 235, 237, 299
Kern Island 51
Kern Island Canal 44
Kern Lake 44, 49, 51
Kern Property Corporation 108
Kern River 44, 45, 46, 47, 50, 51, 306
Kern Valley Canal 44
Kern Valley Water Company 44, 46, 49
Kettleman plains 287
Keyes Ferry 16
Keyes Ford 178
Keyes, John 178
Kiltons 150
Kincaid and Livingston 27
King City 276
Kings County 226
Kings, Homer 91
Kings River 37, 129, 138, 141, 274, 287, 302
Kings River Canal and Irrigation Company 214
Kinyoun, Joseph 35
Kirby 182
Knight, Bill 169
Knight, Orin 255
Knudsen, Jim & Marie 282
Knudsen, M. P. 240
Korn 142, 150
Korn & Hirschfeldt 139, 150, 162
Korn's Store 142
Kragenhagen Gate 287
Kreiser, Christian 3, 69
Kreiser, Christine 3
Kreiser, Elizabeth 3, 99
Kreiser, Frederike 3, 99
Kreiser, Heinrich 3
Kreiser, Henry 4
Kreiser, Karoline 3, 99
Kreyenhagen 126
Kreyenhagen Store 124, 129, 133
Kreyenhagen's 142, 167, 172, 187
Kreyenhagen's Gate 182

L
La Laguna Rancho 17
Lagunitas canyon 291
Lahare 185
Lake Pepin 18, 19
Lake Tulare 82
Langdon family 185
Langdon, John 185
Langdon, Mike 185
Langdon, Tom 185
Lapadula 292, 293, 298
Las Animas Rancho 14, 18, 66
Las Juntas 164, 169, 184
Las Lomas Muertas Rancho 17
Lassart, Mrs. 288
Lastsine, Gus A 104
Lathrop, Rube 266
Latta, Frank 174
Laurel Hill Cemetery 102
Lauterwasser 120
Lavder, Fred 136
Lawler, Joseph 9, 77
Lawrence, Amy 109, 306
Lawrence, David 68
Lees 252
Lees, William 255
Lemoore 239
Leonard 166, 305, 306
Leonard, A. 169
Leonard, D. A. 104, 300
Leonard, D. O. 229
Leonard, Dan 248, 301
Leonard, Mrs. 296
Leonard Ranch 192
Levee oil field 108

Lincoln 82
Linda 4, 5
Lindsay, Alice June 108
Lindsey 282
Linora Stock Farm 264
Lions Club 274
Little Panoche Creek 195, 257, 267
Littleton 155
Livermore 30, 46
Livermore-Chester 45
Livingston 171, 281, 305
Livingston & Kincaid 9
Livingston & Kindman 77
Lomas Muertas 16, 79
Lomas Muertas Ranch 92
Lombard Street Wharf 245
Lombardi, Mary 107, 109
Lone Tree 149
Lone Willow 117, 120, 136, 139, 172, 250
Lone Willow field 287
Lone Willow Ranch 257, 287
Lone Willow Slough 126, 182
Lone Willow Station 118, 124, 164
Long, Albert 33, 99, 103, 104, 199, 263, 268, 278, 286, 305
Long, Henry 99, 103
Long, Mary 103
Long, Raymond 103
Long Valley 176
Lopes, Antone 246, 248, 260
Lopes, Antonio 257
Lopes, Jesus 105, 246
Lopes, Jim 199
Lopez 276
Lopez, Antone 270
Lopez, Jesus 230
Lopez, Tony 282
Los Angeles 77, 175, 183, 207, 289
Los Angeles County 37
Los Animas 120
Los Banos 14, 20, 22, 30, 40, 41, 54, 61, 64, 65, 103, 106, 129, 133, 137, 138, 140, 142, 145, 148, 150, 154, 164, 167, 168, 172, 177, 178, 180, 181, 187, 188, 194, 195, 198, 200, 204, 205, 207, 211, 214, 216, 219, 222, 224, 232, 235, 236, 237, 239, 240, 242, 244, 246, 247, 250, 251, 253, 255, 257, 258, 260, 263, 265, 271, 272, 275, 278, 279, 280, 282, 284, 286, 287, 288, 290, 292, 293, 294, 296, 298, 303, 304, 305, 306, 307, 308
Los Banos Creamery 240
Los Banos Creek 172, 211, 217, 218, 221, 241, 272
Los Banos Division 248, 286
Los Banos Farm 153
Los Banos Homecoming Day 288
Los Banos Hotel 244
Los Banos Opera House 242
Lost Hills 289
Luning 40
Lux & Miller 10
Lux, Charles 9, 11, 14, 15, 28, 34, 53, 75, 77, 78, 80, 82, 84, 86, 87, 91, 92, 93, 94, 95, 97, 102, 106, 117, 118, 120, 122, 123, 125, 126, 127, 128, 129, 130, 131, 132, 133, 134, 136, 137, 138, 139, 140, 141, 142, 143, 145, 146, 149, 150, 175, 177, 236, 238, 273, 275, 280, 298, 305, 308
Lux family 190
Lux, Henry 9, 91, 275
Lux, Lena 275
Lux, Mrs. 77, 95, 127, 237
Lux School of industrial training 95
Lux's ranch 174

M

Maas, Gustav Rembold 104
Maas, Phillip Rembold 104
Madera 161, 176, 192, 196, 198, 248, 257, 287, 301
Madera Colony 192
Madera County 24, 36, 184, 204, 226
Malarin 133, 146
Malarin, I. 124
Malarin, Mr. 124, 142
Malarin, Mrs. 190
Malhuer River 302
Maples & Co 141
Mare Island 164
Marin 295
Marin County 289
Mariposa 163
Mariposa County 203
Mariposa Island 150
Market Street 245, 249
Marks, Bernhard 31, 298, 299, 300, 301
Marks, Blossom 301
Marks, Frank 298, 300
Marks, Howard B. 298
Marks, Mrs. 301
Marquardt, Mrs. Emil 105
Martin, Mrs. Kays 188
Mason, Mr. 80
Mason, N. B. A. 141
Mastick 123, 131
Matthias, Edward 224
Matthias, George 224
Matthias, John 224
Maxwell, James 125
May Day 178, 180, 181, 190, 194, 196, 203, 240, 242, 276, 277, 278, 280, 285, 307
McAllister, Hall, Sr. 46
McBride 237, 275, 280

McBride, Rex 107
McCabe 199
McCarthy, John 187
McClelland, D. T. 99
McClelland, Dr. 67
McCray, Frank 42, 47
McDermott 199
McDermott, James 187
McDonald, Bob 301
McGovern, Vincent J. 108
McIntyre 164, 184
McKittrick oil field 108
McPike, John 122
McPike 172, 264
McPike Field 132, 145
McPike Landing 140
McPike Ranch 287
Means, John R 117
Meany, Sheriff 41, 168
Mears landing 140
Mein, Sally Nickel 109, 306
Menches, Adam 75
Menches Brothers 308
Menches, Fred 75
Mendota 184, 195, 196, 220, 222, 242, 287
Menjoulet, John 173
Merced 41, 109, 126, 127, 129, 130, 132, 133, 137, 138, 140, 141, 142, 143, 145, 147, 148, 161, 164, 165, 166, 168, 177, 178, 180, 181, 182, 184, 196, 197, 205, 241, 250, 281, 293, 294, 295, 301
Merced Band 180
Merced Bridge 177
Merced County 17, 20, 23, 24, 36, 37, 42, 54, 63, 92, 118, 150, 153, 155, 157, 177, 203, 204, 205, 206, 213, 219, 224, 226, 288, 299
Merced Falls 56
Merced River 37, 156
Merced Security Savings Bank 258
Mercey, Henry 288
Mercey Hot Springs 200, 288
Mercey, John 288
Merchants Exchange Building 231, 233
Meredith, George 259
Merritt 194, 272
Merritt, C. Z. 104, 275
Merritt, Ralph 212, 272
Michael, Cannon 307, 308
Michael, Cannon Nickel 308
Middlefield 287
Middlefield road 106
Midway Ranch 252, 257, 293
Miller & Lux 10, 12, 15, 16, 18, 19, 23, 25, 27-30, 32-35, 37, 38, 40, 42-47, 49, 52-54, 62, 63, 67, 78, 88, 89, 92, 96, 102-108, 119, 129, 135, 142, 144, 149, 150, 153-158, 160, 161, 163, 169, 171, 174, 176, 177, 180, 181, 183, 186-188, 190, 192, 194-199, 204, 207, 211, 213, 214, 216-218, 222, 224, 229, 231-233, 235, 236, 238-242, 245, 246, 247-255, 258-267-276, 279, 280, 282, 284-289, 292, 295, 296, 298, 299-302, 304-306
Miller & Lux Divided Lands Inc. 11, 24
Miller & Lux income 306
Miller & Lux Store 290
Miller, Benjamin 104
Miller, C. A. 218
Miller, Gussie 190, 277. *See also* Sarah Alice Miller
Miller, Henry 3-5, 7, 9, 11, 13-16, 18-20, 22-25, 27, 28, 32-35, 37-44, 46, 47, 49, 51, 53-57, 59-61, 63-70, 84, 87, 88, 90-92, 94, 95, 98, 103-109, 117, 118, 120-122, 124, 126, 127-134, 136-145, 148-150, 153-157, 161, 162, 165, 167-172, 174-181, 183-192, 194-200, 203, 204, 207, 209-213, 217-219, 221, 222, 224, 229, 231, 233, 235-242, 246, 247, 250-256, 258-269, 271-283, 285-288, 295, 296, 298-303, 306-309
Miller, Henry, Jr. 67, 84, 96, 102, 106, 114, 127, 155, 187, 191, 216, 239, 276, 278, 281
Miller, Mrs. 162, 188
Miller, Mrs. Henry, Jr. 95
Miller, Nellie 67, 84, 96, 97, 106, 127, 216, 278, 307. *See also* Mrs. J. LeRoy Nickel
Miller, Sara E. 104
Miller, Sara Onyon 95
Miller, Sarah Alice 67, 96, 100, 101, 102, 106, 153. *See also* Gussie Miller
Miller, W. J. 218
Millerton 165
Milliken, Mrs. 298
Milliken, Ralph 68, 161, 199, 207, 210
Mills 140, 148, 162
Mills, James 133
Milpitas 10, 78
Miner, Stacy 257
Misch, Franz 306
Mission area 9
Mission Dolores 175
Mission High School 276
Mission Street 75
Mitchell estate 222
Mitchell, William 104
Modesto 222
Molton 246
Monrozhay 75
Monte 233
Monterey 77, 79
Monterey County 17, 24, 92, 149
Monterey-Fresno Railroad Company 236
Montgomery 92
Montgomery, Captain 6
Montgomery, Judge 156

Moore, Capt. 131
Mora, Santos 245
Mora, Tommy 245
Moran, Mrs. 254
Morgan 188, 189
Morris 34
Morse, Beatrice 109
Morse, Beatrice Bowles 307
Morse, Mora 81
Mosby 164
Moss, Mr. 145
Mount Diablo 14, 189
Mount Madonna 65, 94, 97, 98, 103, 282
Mount Shasta 14
Mountain Brow Lodge 180
Mountain House 183, 260, 264, 266
Mountain View 120, 174, 246
Mt. Hamilton 161
Mt. Madonna State Park 14
Mud Slough 217, 257
Muddy Bend 126
Murphy 253
Murphy, Dan 186
Murphy's Slough 119
Murrieta 184
Muth, J. F. 205
Myer 185
Myer, Ike 182, 185
Myer, Jake 162, 169, 184

N
Nance, L. 33
Napa 170
Napa County 50
Nelson, Walter 291
Nevada 82, 92, 177
New Canal 220, 241, 266
New Columbia 247, 250, 251
New Columbia Rancho 287
New Era Creamery 266
New Idria mines 217
New York 4, 5, 7, 8, 10, 72, 73, 91, 154
Newhall Land and Cattle Company 287
Newhall, Major 163
Newman 157, 178, 195, 203, 204, 218, 219, 220, 237, 238, 240, 241, 258, 264, 265, 272, 275, 287, 305
Newman, S. 136
Newman, Simon 196, 264, 265, 266, 275, 283
Newton 200
Newton, Bascom 302
Neylan, John 109
Nicasio 292
Nickel 63, 211, 246, 251, 252, 254, 257, 263, 305
Nickel, Beatrice 295
Nickel, Beatrice Sara 105
Nickel, Charlotte 107
Nickel, Ernestine 292, 298
Nickel, George 232, 278, 292, 298
Nickel, George W. 98, 105, 106
Nickel, George W. Jr 306
Nickel, George W. Sr. 109, 307
Nickel, Heidi 307
Nickel, Henry III 281
Nickel, J. LeRoy 84, 97, 103, 105, 106, 219, 231, 232, 278, 286, 293, 305, 306
Nickel, J. LeRoy Jr. 108
Nickel, James Beverly 306
Nickel, John Beverly 107, 109
Nickel, Mrs. 99, 277, 278, 279, 280, 282, 295
Nickel, Mrs. J. LeRoy 67, 103, 105. *See also* Nellie Miller
Nickel, Sally 98
Nickels, Gladys 108
Noel, F. C. 108
North Beach 9, 76
North Camp 207
North, Captain 91
North Midway oil field 108
North Western Railroad 291
Nourse, John P. 108

O
Oakland 30, 95, 156, 187, 293
Odd Fellows 180
Oddfellows Lodge 178
Ogden, Bill 195
Ogden, James 104
Ogden, Will 196
Ogelvie, Wm. 299
Ogle 230, 233, 248, 286
Ogle, E. F. 104, 229
Ogle, Jim 237
Old Canal 219, 266
Old Center Point 222
Old Gilroy 14
Old Jimmy 170
Old Los Banos 172, 174, 177, 180, 254
Oldham, Maude 96, 102
Olema 291
Olsen, A. R. 108
Olson, Ellen 283
Onyon, Sara 281
Opera House 185, 195, 244
Ora Loma 258
Oregon 80, 82, 92
Orestimba Rancho 18
Orognen, Ben 188
Orognen, Mrs. 189
Orognen's 195

Ostrander, F. M. 157
Otto, Dorothy 105, 223
Otto, Irving 224
Otto, John 224
Otto, Mrs. 308
Overfeldt 92
Overfelt Company 29
Overhauser, Mr. 125
Overland Stages 163, 164
Oviatt, T. M. 102
Owens River 175, 176
Oxalis 166, 169, 246, 248
Oxalis Ranch 247

P
Pacheco 198
Pacheco, Francisco 14
Pacheco Highway 187, 190
Pacheco Pass 10, 13, 14, 17, 39, 58, 61, 66, 77, 87, 118, 156, 161, 163, 164, 167, 172, 187, 189, 210, 213, 218, 236, 254, 272, 273, 277, 289, 304
Pacheco Pass Highway 286
Pacheco Pass road 162
Pacheco Turnpike 119
Pachuco 246, 247, 248, 296
Pacific Livestock Company 80
Padia, Chris 289
Page, Tim 130
Palmer, F. L. 142
Palo Alto 13, 106
Palo Solo 171
Paloma oil field 108
Panama 5, 74, 75
Panama City 5
Panamint 167
Panoche 130, 145, 195, 198, 289
Parker, Paul 209
Parkin 139
Parrott, John 28
Pasqual 293, 298
Pasqual's ranch 264
Patman, Mr. 170
Payne 33
Peach Tree Ranch 92
Peachtree 121
Peacock Dinner 211, 212, 273
Pebley, Robert 117
Peck, Jas. 235
Peck, Jim 257
Peklam, Judge 120
Pemberton, W. A. 235
Perkins 294, 298
Perkins, Mr. 293
Perlattos, Fernando 288
Pfam, Mr. 76
Pfeiffer 155
Pfeiffer, Albert Jr 104
Pfeiffer, Alfred 99
Pfeiffer, Henry 104
Pfeiffer, Henry John 104
Pfitzer, Antone 172
Phillips 191, 270
Phillips, George 269
Pick Anderson field 287
Pierce-Arrow 211
Pierson 184, 281
Place, Joe 104, 177, 192, 241, 268
Plainsburg 166
Plainsburg Switch 166, 168
Plow Camp 171, 181, 184, 241, 255, 269
Polatos, Fernando 250, 263
Pollard 122
Porter, Harvard 207
Portola 27
Portsmith Square 7
Portsmouth Square 75
Poso 286, 300, 301
Poso Division 286
Poso Ranch 63, 188, 212, 248, 273
Poso Warehouse 177
Potter 10, 120, 121
Potter, E. K. 108
Potter, J. S. 239
Potter, Jesse 91, 95, 96, 121, 122, 137, 140, 141, 142, 147, 175, 190, 278, 281
Potter, Jessie 10
Potter, Mrs. Jesse 237
Potter, Sheldon 295
Presidio 6
Pritchard 164
Pritchell 167
Probasco, Nancy 104
Pt. Reyes 291
Pyle, Grant 286

Q
Quinn, Rafael 246, 250, 289

R
Rainly, Bob 209
Raisina School District 177
Ralph Milliken Museum 117
Ralston 40, 81, 92, 165, 212, 273, 274
Ralston, W. C. 38
Rancho de los Californios 184
Rancho Espagnole 287
Rancho San Emigdio 184
Rancho Sanjon de Santa Rita 14

Raven 232
Raven, H. J. 103
Raven, Hans 247, 248, 251, 260, 264
Rector, Judge 294, 295
Reddy, Charles 104
Redfern, Floyd 301
Redington 46
Redwood City 174
Reeves, David 285
Refuge 31
Rembold, Gustav 104
Rembold, Martha 104
Renson 19
Republic (ship) 6, 74
Republican 82, 185, 242
Reutter, C. 235
Reutter, Henry 277
Reynolds 121
Rhode Island 10
Riley, Charley 294
Rio Vista 285
Ripperdan, Thos 154
Ritchie, William 66, 156
River Ranch 248
Riverside ranch 263
Roberts Island 199
Roberts, Mace 257, 258
Roberts, W. 133
Roberts, W. G. 132
Robinson, Anna Sibley 104
Robinson, C. Ray 107, 109
Robinson, Charles 104
Robinson, Mrs. Perry 103
Robison 177
Robison Ferry 177
Robison, Mr. 177
Rochdale Hall 301
Rodgers, Will 288
Rodriquez, Mrs. 288
Roehme, Edward 132
Romero Creek 182
Rouse 267, 275
Rudolph, Edwin G. 104
Russ House 265

S

Sabranes, Panifilo 149
Sacramento 27, 32, 97, 157, 224, 302
Sacramento River 285
Sacramento Street 74, 92
Sacramento Valley 38, 106
Safstrom 229, 251, 290
Safstrom, Arthur 292, 295
Safstrom, H. C. 104
Safstrom, Hal 209
Safstrom, Henry 187
Safstrom, Howell 289, 292, 294, 295, 296, 298
Safstrom, Mrs. Howell 289
Safstrom, Red 248
Safstrom, William 292
Salandy 75
Salau, Adolph 259
Salinas 217, 268
Salinas City 149
Salinas Valley 15
Salivpuedos Rancho 17
Salt Lake 13
Salt Slough 62, 64, 132, 134, 138, 141, 172, 183, 196
Salt Slough Warehouse 19, 64
San Antonio Rancho 18
San Benito County 17, 24, 92
San Bruno 78
San Diego 9, 77
San Emigdio 184
San Felipe 196
San Felipe Grant 283
San Felipe Valley 161
San Francisco 5-10, 13, 15, 17, 34, 35, 38-41, 44, 49, 50, 54, 66, 67, 74, 75, 77, 84, 91, 97, 99, 102, 103, 105, 106, 121, 134, 141, 144, 150, 153, 163, 164, 175, 181, 189, 197, 198, 207, 211-213, 217, 219, 222, 223, 229, 231, 233, 237, 240-242, 245, 246, 255, 267, 273-279, 281, 286-289, 292, 293, 295, 296, 299, 301, 308
San Francisco Bay 6
San Francisco County 14
San Francisco Home for Incurables 103
San Francisquito creek 13
San Joaquin 47, 56, 57, 58, 61, 77, 80, 97, 106, 131, 136, 149, 150, 207, 261, 308
San Joaquin and Kings River Canal 218
San Joaquin and Kings River Canal and Irrigation Company 38
San Joaquin and Kings River Irrigation Company 242
San Joaquin Bridge 182
San Joaquin Canal 81
San Joaquin Canal Company 242
San Joaquin County 24
San Joaquin River 14, 17, 37, 38, 40, 42, 44, 64, 92, 106, 117, 136, 153, 154, 155, 156, 160, 165, 166, 172, 177, 178, 183, 184, 197, 198, 199, 203, 204, 205, 211, 213, 214, 217, 218, 222, 238, 242, 247, 254, 262, 270, 272, 296, 301
San Joaquin River Bridge 294
San Joaquin River Water Storage 306
San Joaquin Valley 13, 14, 15, 17, 20, 22, 23, 27, 28, 29, 30, 32, 36, 37, 38, 44, 52, 53, 63, 77, 164, 165, 177, 184, 197, 203, 211, 214, 218, 224, 235, 242, 258, 260, 270, 273, 277, 285, 289

San Joaquin Valley Bank 143
San Jose 13, 14, 74, 78, 120, 148, 157, 163, 167, 174, 181, 187, 217, 236, 246, 268, 302, 303
San Juan 131
San Juan Ranching Company 307
San Justo Rancho 17
San Lorenzo Rancho 18
San Louis Obispo County 79
San Luis 196, 260
San Luis Camp 54, 121, 122, 124, 131, 136, 137, 146, 148, 169, 170, 171, 174, 210, 264
San Luis Creek 266
San Luis Field 134, 137, 264, 287
San Luis Gonzaga 119, 120, 123, 146, 190, 283
San Luis Gonzaga Ranch 14, 142, 266
San Luis Hotel 189
San Luis Island 136, 146
San Luis Obispo County 226, 237
San Luis Ranch 122, 123, 133, 162, 181, 190
San Luis Slough 124, 132, 136
San Luis Slough Landing 140
San Mateo 78, 174
San Mateo County 24, 92, 106
San Quentin 293, 295
San Rafael 107, 291
San Ysidro 14
Sanchez, Isidro 91
Sandborn, Mr. 121, 122
Sandwich Islands 9
Sandy Mush 178
Sanger 169, 239
Santa Barbara 28, 131, 132, 287
Santa Clara 14, 24, 153
Santa Clara County 103, 105, 156, 236
Santa Clara Valley 15, 177
Santa Cruz 167, 242
Santa Fe Grade 183
Santa Nella 170. *See also* Santanella
Santa Rita 17, 22, 23, 34, 54, 59, 80, 164-167, 169, 171, 175-179, 181, 182, 187, 190, 198, 212, 213, 216, 230, 232, 246, 248, 250-252, 254, 257, 260, 261, 263, 273, 274, 277, 280, 286, 292-298, 301, 303
Santa Rita Bridge 247
Santa Rita County 203, 204, 235
Santa Rita Division 286
Santa Rita field 302
Santa Rita Grant 273
Santa Rita Ranch 16, 17, 21, 28, 29, 42, 68, 78, 92, 119-121, 124-128, 130, 131, 133, 134, 136, 139, 141, 143, 147, 149, 153, 197, 208, 241, 289, 300
Santa Rita Slough 166, 172
Santa Rita Store 187
Santa Rosa Methodist College 303
Santanella 133, 137, 260. *See also* Santa Nella
Santanella Hills 139
Santanella Ranch 132
Saratoga 164
Saratoga Springs 154
Sargent Station 66, 156
Sawyer 195, 255
Scanlon 177
Schafer 128, 130, 133, 134, 137
Schaffer, Henry 196
Scheurlante 74
Schmiedte 233
Schmitz 137, 138, 141, 145, 170, 195, 250
Schmitz, J. 130
Schmitz, J. W. 300
Schmitz, Poso 191, 192, 301
Schornhorst 177
Schottky, T. 235
Scott, Wilmer 301
Shackleford, Dan 199
Shafer 148
Shafer, Dan 194
Shafer, Henry 126
Shannon 64, 181, 253, 306
Shannon, S. J. 305
Shannon, Sidney 55, 181
Shaw 133
Sheldon 217
Sheldon, Minnie Pfeiffer 104
Sheldon, Mrs. Steve 99
Sheldon, Nancy 96, 106, 107
Sheldon, Nancy (Miller) 91
Sheldon, Nancy Wilmont 77
Sheldon, S. W. 154
Sheldon, Sara Wilmarth 106
Sheldon, Sarah 96
Sheldon, Sarah Wilmot 84
Sheldon, Sarah Wilmot (Miller) 67
Sheldon, Stephen W. 103
Sheldon, Steve 188
Sheldon, Tommy 162
Sheppard, Charles 123
Sherburn, Mrs. 254
Sherman, Mrs. 275, 282, 283
Shippee Ranch 238, 250
Shirley, J. C. 150
Shreck, Captain 121
Sierra Nevada Mountains 14
Simmons 38
Simmons, Captain 81
Simmons, Orrin 37
Simon & Harrell 240
Simpson 305
Smith 131, 165, 214
Smith, Archie 252, 278

Smith, C. W. 169, 174, 240
Smith, Charles W. 54
Smith, Charley 269, 270
Smith Ferry 177
Smith, Gus 180, 195, 271
Smith, Johnny 251
Smith, Lillian 165, 182
Smith, Lon V. 108
Smith, Martin Miles 240
Smith, Morris 76
Smith, Mrs. 171, 174
Smith, Mrs. C. W. 169, 177
Smith, Old Man 182
Smith, Reginald 277
Smith, S. A. 235, 239
Smith, Sam 221, 241
Smith, Wallace 53, 245, 246, 247, 252
Smith's Ferry 182
Snelling 167, 217
Snelling Bridge 156
Soap Lake 14, 137
Soap Ranch 122, 124, 145
Sobranes, Francisco 16
Society Jim 251
Soper 195
Soper, Frank 257
Soper, George 235
Soquel 286
South Dos Palos 290, 299
South Fork Valley 48
South Park 94
South San Francisco 91, 94
South San Francisco Land & Improvement Company 34
South San Francisco Land and Investment Company 35
Southern Pacific 143, 183, 299, 301
Spreckles 212, 267, 276, 286
Spreckles, Charles 182
Stafford, Charles 255, 257
Stanford 212, 274
Stanford, Mrs. 95
Stanford University 273
Stanislaus 203
Stanislaus County 40, 42, 44, 204, 205, 219
Staples, I. M. 120
State Normal 303
Sterns, Abel 28, 80
Stevens 48
Stewart, Chas. A. 232
Stillman 166
Stinson Slough 119
Stockton 22, 28, 32, 80, 97, 123, 126, 132, 136, 138, 143, 160, 167, 183, 196, 239, 254, 287, 294
Stockton, Billy 20, 54, 102, 172, 174, 188, 193, 195, 196, 222, 271, 272, 274
Stockton, Mr. 220, 222
Stone, Mr. 69
Straford 287
Strait of Carquinez 14
Stringtown 164, 169, 258
Strubble, Fred 289
Sturgeon 195
Stuttgart 71
Sullivan, J. J. 235
Sullivan, Jack 257
Sullivan, Jim 195
Sutherland, Jack 174
Sutherland, James 119
Sutherland, John 119
Swamp and Overflow Land Act 20
Sweeney Gilt Edge ranch 287
Swift 34
Swift & Co. 285, 286
Swift, Cyrus B 104
Sycamore Point 301

T

T.C. Walker (ship) 123
Talbot 258, 259
Tanner 152, 157, 170, 177, 180, 184, 186, 188, 194, 195, 255, 256, 282, 284, 286, 288, 303
Tanner Gates 267
Tanner, H. G. 83, 150, 235
Tanner, Mrs. 155, 162, 170, 195, 287
Tapatula 216, 277, 282
Tassahara Valley 189
Taylor 258
Taylor, Dr. 251
Taylor, Mary 105
Tehachapi 29
Tehachapi Pass 14
Temple 145
Temple Field 130, 167
Temple Ranch 146, 175, 248
Temple Slough 42
Temple Slough Weir 251
Temple Warehouse 19, 169
Temple's Landing 140
Templor 287
Tequesquite 79, 92
Tequesquite Rancho 18
Tevis, Lloyd 44
Texacita Ranch 276
Texas Company 107
Texas fever 287
Theircof, Ella 238
Theircof, Joseph A. 238
Theircof, Mr. 196
Thomas, M. M. 105

Thoms, O. A. 119
Thornton 120
Thornton, Harry 167
Tofalo, J. Rock 105
Torchiana 257, 259, 267, 269, 277, 279, 280, 286
Torchiana, H. A. van Coenen 276
Torchiana, H. A. W. 242
Torchiana, Henry 276, 281
Torchiana, Mrs. 239, 275, 276, 281
Torchiana, Superintendent 237
Toscano, J. V. 217
Townsend Depot 246
Tracy 197, 199, 238
Tranquility 31
Treadwell, Edward F. 102, 103
Truckee 167
Tucker, John 235
Tucker, Tom 235
Tulare 129, 138, 237
Tulare (steamboat) 134, 135
Tulare County 92, 226
Tulare Lake 38, 44, 45, 46, 56, 242, 267
Tulare Lake Canal 218
Tule Plains 80
Tule River 138, 140, 141
Tully, John 255
Turner 184
Turner, Ed 248
Turner, Minnie Elisabeth Pfeiffer 104
Turner, P. H. 229, 231, 232, 233
Turner Ranch Bridge 247
Turner's Island 288
Twelve Mile House 174
Twenty-one Mile House 174

U
Upper San Luis field 287

V
Van Dusen field 251
Vanderlip 284
Vandiveer 229
Vandiveer, James 103
Vandor, Paul 24
Vasquez 165, 170, 217
Vasquez gang 184
Villa Fairview 242
Visalia 22, 27, 118, 128, 138, 140, 141, 142, 174
Visalia stage 121, 124, 146
Voight 218
Volta 172, 195, 241, 255
Volta Station 172

W
Wade, Dr. 170, 180, 278
Wagenheimer 265
Waggoner, Chance 259
Waldeck Hospital 276
Waldenberg, Sol. 155
Wallis 211, 252, 257, 278, 280, 282, 286, 287, 288, 289, 305, 306
Wallis, D. W. 104, 267, 275, 276
Wallis, Mary 280
Wallis, Mrs. 280
Wallis, Mrs. D. W. 277
Wanzer 42, 195
Ward 128, 141
Ward, Al 166, 167, 199
Ward, Alvin 139, 145
Ward, Art 125, 128, 146
Ward Field 146, 186
Ward, Mrs. 288
Ward, Thomas E. 241
Ward, Tom 186
Warfield, Charles 34, 62, 125, 136, 141, 142, 150, 154, 155, 167, 182, 188, 195, 198, 199
Warfield, Sheriff 240
Washington Market 4, 73
Washington Street 75
Waterloo, Iowa 260
Watkins 177, 192
Watkins, Marion 188, 191
Watson's Ferry 130, 145, 167
Watsonville 239, 260, 281
Watts Valley 169
Webb, Joe 187
Weir Camp 247, 257
Welch 254
Wells Fargo 165
Wells Fargo Express 121
West Side 218, 222, 236, 238, 240, 287, 305
West Side canal 237
West Side Union High school 240
Western Meat Company 34, 35, 54, 222
Westfall, Pat 302
Westir 9, 77
Westmooreland, Z. 216
Whelan, Mr. 298
White 120, 145, 187, 195, 305
White, Billy 188
White, Christiana 171
White, Henry Wade 171
White, Mrs. 181
White, W. T. 177, 306
White, Wade 171, 177, 181, 264
White, Ward 120, 121
White, Will 171, 216, 265, 281

White, William T. 54
Whitehurst Funeral Chapel 107
Whitehurst, J. A. 241
White's Bridge 237, 238, 263, 287
Whiting, E. 118
Whiting ranch 192
Whittington, Mrs. Grant 103
Widman Place 174
Wieble, S. W. 44, 46, 50
Wiener, J. 142
Wilcox, F. 235
Wiley, Will 258, 264, 266, 267, 268, 269, 270, 272
Wilkins, F. 235
Williams, Mr. 120, 123
Wilmath, Sarah 103
Wilson 34
Wilson, Bert 181
Wilson, Dick 173
Wilson, John 76
Wilson, M. C. 105
Wilson, W. 123
Wollery, J. E. 108
Wood 172, 183
Wood, David 22, 139
Wood, Davie 154, 183, 185, 186
Wood, Jim 194
Wood, Uriah 172, 186
Wood, Will 30, 167, 183, 185, 186, 187
Wood, William 76
Woodland Cemetery 67
Woodlawn Cemetery 106, 107
Woodman's Hall 188
Woodruff 301
Woods, Chas 122
Woodward & Bennett 103
Worthington 173
Wright 254
Wright ditch 149
Wright Ranch 272
Wrightmeyer 165
Wurtenburg 224

Y

Yerba Buena 6
Young 161

Z

Zimmerman 9, 76

RALPH LEROY MILLIKEN

Oct. 26, 1880 – Oct. 26, 1970

Born in Lampeer County, Michigan, Ralph Leroy Milliken was the first of two children born to James R. Milliken and the former Ella Thomas Lyman. A sister, Lottie, arrived two years later.

Before his fourth birthday, young Milliken moved with his parents to Alpena, Dakota Territory. He grew up in his father's general store and adjoining bank. Home schooled he upon moving to Sioux City, Iowa entered high school. On graduation, he received a scholarship which he used to attend Leland Stanford Junior College now Stanford University. Following his first year of studies, he retuned to Sioux City to enter Morningside College. He spent his summers working so he could return to his studies in the fall.

After graduation, Milliken came to Los Banos to work as a bookkeeper for Miller & Lux, California's largest land and Cattle Company. He married Lucybeth Wilson a former teacher, in the Stanford Chapel at Palo Alto.

In Los Banos Milliken invested in property with his business partner, Dr. James Mc Clelland. In addition to a walnut orchard the partners raised chickens, turkeys and ducks in the Sunset Road area.

Milliken is best remembered for his work on local history in the Los Banos area. Very early in his career he realized the importance of recording history by interviewing pioneers. His earliest interviews were carried out while working on a rural mail route. All through his young manhood, Milliken kept diaries and wrote many manuscripts on topics such as his boyhood in Dakota Territory; work at Yellowstone National Park, travels abroad and many other topics. A prolific writer, he wrote numerous short stores as well.

During Milliken's later years he spoke to many public gatherings and to many radio audiences. He continued writing and published articles in newspapers as well as writing the book California Dons, the story of the Larios family and early California.

The Los Banos Historical Museum, which had been founded in 1954, was renamed in honor of Ralph Milliken in 1968. The numerous papers and books Milliken had collected, his many manuscripts and object of historical value, were bequeathed to the museum, which now bears his name.

In early 1970 Milliken lost his wife. He survived her by less than six months.

THE MILLIKEN MUSEUM SOCIETY

Built in 1954, the museum was originally called the Los Banos Historical Museum. After a slow start with a few WW1 posters, Indian grindstones and no display cabinets the museum, under the direction of Robert Kuhn Sr., its second curator, expanded to its present facility.

Donations by the Nickels family (Henry Miller's heirs) of his saddle and memorabilia from Miller & Lux, donations from various early pioneer families became the nucleus for the current collection. Milliken had collected a set of Bancroft's 1880 history of California as well as many other early editions of books about California and its pioneers. Personal papers of many early families as well as numerous of Mr. Miller's papers are also to be found. Editions of the Los Banos Enterprise dating from 1891 are on file, in addition to collections of materials on the schools, farms, organizations and May Day memorabilia dating back to the 1880s. On Ralph Milliken's death all of his papers became the property of the Museum and the name was changed to the Milliken Museum Society.

The City of Los Banos turned over the building and collection to the County in 1979. In the past 15 years the museum has been considerably updated. The current curator and volunteers have installed new exhibits. With help from the Turlock Centennial Foundation and the National Historical Publications & Records Commission, the Ralph Milliken papers have been sorted and organized for use by researchers.

Now nearly 30 years after Milliken's passing the museum continues to delight the researcher as well as the visitor.

Milliken Museum - (209) 826-5505
Tuesday thru Sunday 1:00pm-4:00pm
Closed Mondays
Located in Los Banos Park
Hwy 152 at 7th street